Cost Planning & Estimating for Facilities Maintenance

WILEY

John Wiley & Sons, Inc.

RSMeans

Cost Planning & Estimating for Facilities Maintenance

- *Evaluating & Budgeting Operations*
- *Maintaining & Repairing Major Building Components*
- *Applying Means Facilities Maintenance & Repair Cost Data*
- *Special Requirements by Building Type*

Specific coverage of:
- *Offices*
- *Hospitals/Health Care*
- *Educational*
- *Manufacturing*
- *Retail*
- *Apartments/Hotels*
- *Historic Buildings*
- *Museums/Libraries*

The editors for this book were Phillip Waier, Stephen Plotner, and Suzanne Morris. The managing editor was Mary Greene. The production coordinator was Marion Schofield. Composition was supervised by Karen O'Brien. The book and cover were designed by Norman R. Forgit.

For general information on our other products and services, or technical support, please contact our Customer Care Department within the United States at 800-762-2974, outside the United States at 317-572-3993 or fax 317-572-4002.

Wiley also publishes its books in a variety of electronic formats. Some content that appears in print may not be available in electronic books.

For more information about Wiley products, visit our web site at www.wiley.com.

Library of Congress Cataloging-in-Publication Data:

ISBN: 978-0-87629-419-2

20 19 18 17 16 15 14 13 12 11

Table of Contents

Foreword

Good maintenance management happens when competent facilities professionals join their customer's team, and do the right thing—together. In today's business of maintaining buildings and grounds, doing the right thing requires expertise far greater than practiced in the past. As facilities managers, we must keep pace with the professionalism, competence, and almost unbelievable astuteness of our dynamic customers who live and work in buildings entrusted to our care.

Our customers *are* professionals! They too have met the economic storms thrust upon their businesses. They have accepted and indeed mastered the inevitability of change. They have ventured into areas completely unrelated to their work to find new ways of getting things done better and at less cost. Our customers have emerged from this tough world of change as much more than just survivors. They are leaders in a new way of life—doing more with less and doing it well—every day. Facilities managers, as key players on their customers' teams, must emulate their record of accomplishment. We must continually learn to work smarter and get more done with less, or else we lose credibility. And credibility, once lost, is difficult to regain in today's facilities business. Most important, credibility is the catalyst that inspires customers to give us the maintenance funding we need for our buildings.

Our customers' remarkable success in meeting the economic challenges of the past few years creates justifiable expectations in their minds of how well others should conduct their business, especially facilities managers who ask for large sums of money to maintain and improve buildings. Facilities managers, and maintenance managers in particular, are expected to plan better, a lot better, than ever before. Our customers rightfully demand that we develop comprehensive strategic maintenance plans as well as prioritized budget-year plans. They insist that we justify (and defend) our budget requests with credibility and clarity. We are expected to underpin these plans and budgets with automated facilities data that accurately defines needs and costs. Our customers want to base maintenance alternative decisions on investment considerations, rather than emotional or historic arguments. Today, customer decisions increasingly compare the life-cycle cost of capital improvements to continued annual maintenance and repair of building components. These decisions include the consequences to the backlog of deferred maintenance and the size of maintenance reserves. Facilities management has changed! The days of simply adding or subtracting two, four, or six percent to last year's maintenance budget ended when our customers reduced their annual budgets by 20 percent or more.

Yet, as always, we must still fix or replace the leaking roof before offices get wet, and make sure indoor temperatures remain constant at 72 to 74 degrees. That much has not changed. How can we catch this fast-moving, customer-driven train, while at the same time learn new applications of old maintenance principles, and still keep the locomotive fueled, oiled, and repaired, and the passengers comfortable — everyday?

Cost Planning and Estimating for Facilities Maintenance tells us what we need to know, and how to do it in today's challenging world of customer-oriented maintenance management. Twenty seasoned facilities professionals have produced a maintenance manager's encyclopedia of knowledge and practical applications. Facilities professionals at all levels will find comprehensive insights into the four "must do" management functions: We must continually enhance our competence in applying state-of-the-art *maintenance management concepts*; We must acquire *detailed knowledge* about the condition of our facilities, use it wisely and share more of it than ever before with our customers. We must become *"Master Planners" of maintenance*. And we must improve our *problem solving skills* and produce maintenance alternatives backed by cost estimates for our customers' decisions. *Cost Planning and Estimating for Facilities Maintenance* presents each of these "do the right thing" functions in easy-to-understand language. Needs and costs form the backbone of each chapter.

The book starts with intuitive discussions of today's management concepts. "The Building Proforma" documents the costs of acquiring, owning, and operating a building. Facilities professionals stand to gain much from this review of total building costs as they build their maintenance budgets. The reader is then enriched with current theories and practices regarding Capital Planning, Outsourcing, Benchmarking, Computerized Maintenance Management Systems, and Code Compliance. These are the management concepts of today. And it is the application of these concepts upon which our customers judge our worth and the size of the maintenance budget. A thorough review of the major Building Components provides the knowledge and perspective needed to accurately assess the operating requirements and costs of maintaining buildings. The importance of the building audit is emphasized as the basis for developing needs-based budgets for each building component.

Maintenance managers will be reinvigorated by this book's approach to traditional programs involving Maintenance and Repair, Cost Estimating, Preventive and Predictive Maintenance, Value Engineering, and Capital Improvements. These programs are expertly brought together as a process, each interdependent on the other as well as on the constraints of our customer's budget. Excellent discussions about Deferred Maintenance and Maintenance Reserves bring the serious professional up to par with leaders in the facilities world, and convey with constructive reality the most significant aspect of our funding problems. The unique approach taken by the authors becomes apparent as each topic focuses on creating prioritized, needs-based budgets for decision-making by customers. This is where the facilities professional learns to reshape traditional methods of maintenance planning.

Perhaps the most unique contribution of this book is found in the section on different building types. Here we are presented with need-to-know information about the corporate perceptions of customers and facilities managers alike who manage different building types and functions. Our family of facility professionals can take pride in recognizing the great diversity of building types and uses in our career field. The authors know their building types well, and share the special maintenance considerations required by each. The wide spectrum of building types includes: Apartments, Educational Facilities, Health Care Facilities, Historic Buildings, Hotels and Convention Centers, Libraries and Museums, Manufacturing Facilities, Office Buildings, and Retail Facilities. Many readers will have responsibilities in just one of these building areas; others may have a need to manage within two or more of these building functions. Still others may need to quickly learn the characteristics of a building soon to be constructed, purchased or renovated by their customers. Reaching beyond our immediate information needs, a review of different building types will enhance our overall perspective and ability to better plan for maintenance in our own facilities.

Cost Planning and Estimating for Facilities Maintenance concludes with a comprehensive chapter that ties it all together. Management concepts, knowledge, maintenance planning, and problem solving are all expertly combined to help us develop a better maintenance budget.

This is the book today's facilities and maintenance managers need to make good maintenance management happen on the customer's team.

Charlie Buddenhagen, PE,
Executive Director, Planning & Project Management,
Georgetown University

Preface &
Acknowledgments

In 1986, R.S. Means published the first edition of *Facilities Cost Data*, to provide a construction cost resource to budget for capital improvements and renovations. The book was well received and stirred an interest in the development of additional cost data that would support the *maintenance and repair* estimating and budgeting process. After several years of research and development, and work with outside contributors, *Facilities Maintenance and Repair Cost Data* was introduced in 1994. Through subsequent seminars and daily contact with facilities professionals, we became aware of a need for a how-to reference that would not only explain proven estimating techniques, but also put maintenance into context with the overall operating budget.

To cover these topics and add current strategies for ensuring the health of the physical plant while meeting today's demands for new efficiencies, we turned to the experts — an impressive group of practicing professionals interested in supporting their colleagues and their industry by sharing their expertise. Each of our authors brings a somewhat different perspective, based on their respective practice in the fields of facilities management and related work in planning, design and construction. The approach and format of the chapters vary, sometimes considerably, reflecting necessary and natural differences in management style suitable to the authors' particular discipline or application. The book is truly a collaborative effort. The authors are listed below, in order of the book's chapters, with brief descriptions of their professional activities.

William H. Rowe III, AIA, PE, is the author of several chapters in this book, and also provided valuable assistance in developing its original concept. Mr. Rowe is a principal of William H. Rowe & Associates, an architectural and engineering firm in Boston specializing in building system design and property and facility management. He is a proponent of a design approach that emphasizes coordination of architectural and engineering systems with quality enhancement strategies. Mr. Rowe is a frequent speaker at professional conferences and has lectured at several leading colleges, including the Rhode Island School of Design, Cooper Union, The Boston Architectural Center, and Franklin Institute. He is a graduate of Columbia University, New York University, and Cooper Union.

Jack F. Probasco, Lisa H. Macklin, and David C. Marsh coauthored Chapter 2, "Capital Planning." Prior to forming Comprehensive Facilities Planning, Inc., Mr. Probasco was Senior Facilities Planner with Ohio State

University. He has been involved in the budget preparation and planning for nearly $1.5 billion of capital projects and currently provides facilities planning consulting services to a variety of clients. Mr. Probasco has written extensively for industry publications on topics such as space planning and capital planning/budgeting, and has made many presentations at conferences including the Society for College and University Planning, the Council of Educational Facility Planners International, and the Association of Higher Education Facilities Officers.

Lisa Macklin has been a Facilities Planner at Ohio State University since 1988. Prior to OSU, she was Senior Interior Planner with Banc One Corporation. Currently, Ms. Macklin is responsible for developing long-range goals, total resource planning, space utilization and allocation, and preparing and coordinating facility plans for colleges, departments, administrative and academic support units. Her current projects involve over $150 million in new facilities and facility rehabilitation.

David C. Marsh is a principal in Comprehensive Facilities Planning, Inc. Previously, he was Assistant Vice President for Facility Planning at Ohio State, where he was responsible for land use planning, capital plan development, capital project planning and space management and inventory for the Columbus and regional campuses.

The authors of Chapter 3, "Maintenance Outsourcing," are Christopher Park and Mark Peacock, both with the Deloitte & Touche Consulting Group in Detroit. Mr. Park is a Manager in the National Construction Industry and Capital Programs Practice. He earned an M.Arch and an MBA at the University of Michigan, and is an associate member of the American Institute of Architects and a member of the International Facility Management Association. Prior to joining Deloitte & Touche, Mr. Park was a facility programmer for The Environments Group in Chicago. Mark Peacock is a Senior Manager with Deloitte & Touche. His responsibilities include assisting clients in outsourcing facility management, telecommunications and computer operations.

"Benchmarking Facility Maintenance" was written by Mary Greene, a coeditor of this book and Manager of Means Reference Books, with valuable input from IFMA, APPA, NACUBO, and AFE (formerly AIPE).

Dan Hounsell and Chris Keller coauthored "Computerized Maintenance Management Systems." Mr. Hounsell is Managing Editor of *Maintenance Solutions* magazine, a publication for commercial and institutional maintenance management professionals, by Trade Press Publishing in Milwaukee, Wisconsin. Chris Keller founded Integrated Data Solutions, Inc. in 1993, a company that provides CAD, CAFM and multimedia services to the design, engineering and construction markets. The company's primary focus is computer-aided facility management. IDS assists Fortune 500 corporations, design professionals, facility managers, and architecture firms in selecting, customizing, installing and training personnel in the use and maintenance of CAD and CAFM systems.

Stephen C. Plotner addressed Foundations, Electrical Systems, Preventive Maintenance, and Retail Facilities in Chapters 7, 12, 16, and 29, and contributed to Chapter 15. Mr. Plotner is an Engineer/Editor at R.S. Means. His previous experience includes managing the implementation of capital improvements, maintenance and repairs, general and preventive maintenance, and annual facility audits, as well as the development of annual capital budgets as the Northeast Regional Facility Manager of Payless Cashways, the fourth largest home center/lumberyard retail chain in the U.S.

Ken Brzozowski, MS, Ph.D., is the author of Chapter 8, "Roofing." He is currently director of Research and Development for Hickman Systems. Mr. Brzozowski also spent over 12 years with Tremco, where he directed the development of roof coatings, single-ply membranes, modified bitumen products, kettle-modified asphalts, and nondestructive testing systems. Mr. Brzozowski is a member of the Roof Consultants Institute, ASTM, American Chemical Society, ASHRAE, Roof Coatings Manufacturers Association, and The Association for Facilities Engineering (AFE).

Interior Finishes is covered in Chapter 9 by Robert Mewis, Engineer/Editor at R.S. Means and Mary Greene. Mr. Mewis is the senior editor of Means' annual publication, *Residential Cost Data* and contributing editor to *Means ADA Compliance Pricing Guide*. He regularly presents seminars on estimating and other construction project management topics.

John Corcoran wrote Chapter 10 on Conveying Systems, with a focus on elevators and escalators. Mr. Corcoran is Vice President of Service for the Schindler Elevator Corporation in Morristown, New Jersey.

William H. Rowe III, AIA, PE, John Moylan, and J. Walter Coon, PE, combined their expertise in HVAC, plumbing and fire protection to produce Chapter 10, "Mechanical." Mr. Rowe's credentials were listed earlier in the Preface. John Moylan, formerly the senior editor of *Means Plumbing Cost Data*, is currently a cost consultant to Means. Before his tenure at Means, Mr. Moylan was a partner in a plumbing contracting company for more than 15 years. J. Walter Coon, PE, has over 40 years experience working as a fire protection design engineer. He has also taught fire science technology to sprinkler apprentices, and has written many articles on fire protection for national trade publications. He is presently serving as a staff consultant to Black & Veatch, Engineers/Architects in Kansas City, Missouri.

John L. Maas, ASLA, and Al R. Young coauthored "Landscaping." Mr. Maas is a landscape architect with 50 years of experience in landscape architecture, construction and maintenance programming. Trained as a horticulturist in the European tradition as a master gardener, Mr. Maas also holds a Master of Fine Arts. In the course of numerous projects ranging from large-scale recreational developments to public utilities and corporate and municipal facilities, he has developed a design approach that emphasizes attention to maintenance programming early in the design phase. Al Young is a software engineer with more than 15 years experience in corporate information management, communications, and software engineering.

Phillip R. Waier, P.E. collaborated with Stephen C. Plotner on the estimating chapters in Part 3. Mr. Waier is Means' Principal Engineer and has technical and editorial responsibility for five cost data books, including *Facilities Maintenance & Repair Cost Data* and *Building Construction Cost Data*. He is also active in international cost and database consulting projects, and in the presentation of professional development seminars on facilities and construction topics. Prior to his work at Means, Mr. Waier was President and Chief Engineer for a mechanical contractor specializing in process piping, petroleum piping, and oil storage tank repairs, and was Project Manager with the John Mutch Company, overseeing industrial and general construction projects for the Foxboro Company, Polaroid, Mobil Oil, and others.

William R. Steele contributed the "Predictive Maintenance" section of Chapter 16. Mr. Steele is a consultant in vibration analysis and thermography. His company, Enviro-Management & Research, is headquartered in Arlington, Virginia.

James E. Armstrong contributed three chapters — on Deferred Maintenance, Libraries/Museums, and Manufacturing facilities, and wrote the primary and secondary schools section of the Educational Facilities chapter. Mr. Armstrong is currently the Property Manager for the Massachusetts Development Finance Agency at Fort Devens, the redevelopment authority charged with the economic development of this former military base. Previously, he was Director of Facilities and Operations for Colby Sawyer College; Manager of Facilities for the Boston Museum of Science; and Maintenance Supervisor for Anderson Power Products.

Charles J. Stuart, CPM is the author of "Reserve Funds." He founded The Replacement Reserve Report in 1984. The firm now has offices in New England and Florida and over 700 clients with capital plans, condition analysis, and reserve studies. The firm specializes in condominium associations and government-assisted housing, and the staff includes professional engineers and architects. Mr. Stuart is a Certified Property Manager designated by the Institute of Real Estate Management, with over 20 years of industry experience.

Pieter J. van der Have is the author of the Higher Education section of "Educational Facilities." He is currently the Director of Plant Operations for the University of Utah. In the course of his 31 years in the higher education facilities arena, Mr. van der Have has led staff responsible for departmental materials procurement systems, payroll and personnel, and budget preparation and tracking functions. He has also been involved in long-range campus planning, design, construction and value engineering. Active in APPA: The Association of Higher Education Facilities Officers for over 20 years, Mr. van der Have has served on the Board of Directors and is currently Vice President for Information Services for that organization.

Phillip J. DiChiara, CFM, CHE, is the author of two chapters in this book: "Health Care Facilities" and "The Maintenance Budget." He also made valuable contributions in his review and comments on the other chapters, and in sharing his insights on budgeting for facilities operations. Currently the Associate Vice President for Clinical and Support Services at Beth Israel Hospital in Boston, Mr. DiChiara has total administrative resonsibility for ten departments, over 500 employees, and an operating budget of $40 + million, including all Support Service departments. Previously he served as Vice President for Facility and Support Services for Fallon/Saint Vincent Healthcare System in Worcester, Massachusetts, and as Vice President of Lehigh Valley Hospital/Allentown Hospital and Healtheast in Allentown, Pennsylvania.

In addition to these authors, who defined this book as a unique and comprehensive resource, we are grateful for the vital input of our reviewers. Charles Buddenhagen, the Executive Director of Planning and Project Management for Georgetown University, reviewed the manuscript and provided valuable feedback to the authors and editors. He also wrote the Foreword. Further, Mr. Buddenhagen enlisted the assistance of the Maintenance Staff of the Facilities Management Division at Georgetown, whose hands-on experience made their review particularly valuable.

Roy Gilley III, AIA, reviewed and contributed valuable comments on "Historic Buildings." Renovation and restoration of existing buildings are among the specialties of Mr. Gilley's firm, Gilley Hinkel Architects in Bristol, Connecticut. Mr. Gilley is also active in this area through his work as Chairman of the Connecticut AIA (American Institute of Architects) Committee on Historic Resources, and participation in the AIA National

Committee on Historic Preservation, the National and Connecticut Trusts for Historic Preservation, and the Society for Preservation of New England Antiquities, and others.

This is Means' first reference book to incorporate the writings of so many contributors. Working with this group of professionals, so well-versed in their respective fields, has been enlightening and rewarding for us, as the book's editors. More importantly, we believe that this approach has enabled Means to deliver current coverage of practical information in a breadth that could not have been possible with one or two authors. We welcome comments from readers on this and all of our publications.

Mary P. Greene, Stephen Plotner, and Phillip Waier, Editors

Part 1

Facilities Operations Planning

Part 1

Introduction

Part 1 introduces maintenance planning and budgeting, in the context of the overall costs of owning and operating a facility. It also offers information on some of the available management tools to consider for improving the bottom line. Chapter 1, "The Building Proforma," presents the total financial picture, listing the cost components for operating a typical commercial facility, and demonstrates the process of financial analysis for property acquisition. By understanding the way in which various expenses are accounted for and covered, one gets a clearer picture of both the constraints and opportunities open to the enterprising and efficient facility or maintenance manager.

In Chapter 2, "Capital Planning," the facility manager may gain insight into what is required when major repairs or a need for new space become capital improvements, and an effective proposal must be prepared, reviewed, and prioritized.

A financial analysis may include consideration of outsourcing — contracting maintenance and repair work to outside contractors. The process of choosing the appropriate tasks to outsource, assessing the benefits of outsourcing, and evaluating contractors is the subject of Chapter 3. A similar analysis of potential benefits from a computerized maintenance management system (CMMS) is discussed in Chapter 5, after an overview of benchmarking in Chapter 4. A benchmarking study can be carried out before the maintenance plan and budget are created, to establish current performance levels and set goals for improvement that will be included in the new plan. Alternatively, a benchmarking study can follow the budget preparation process, since the data collected for budgeting (e.g., building condition, staff productivities, current maintenance practices, etc.) is essentially the same kind of information needed for the benchmarking study.

Chapter 6 deals with meeting codes and regulations in the course of practicing facility maintenance. These requirements have a very real associated expense, beyond the direct costs of performing regular maintenance and repair on building systems, and should be addressed in accounting for total maintenance costs. Codes and regulations can also drive major improvements, such as a fire protection system, which must be incorporated into capital planning.

With the financial background information in Part 1, readers will be prepared to take a "tour" of major building elements in Part 2, to discover where maintenance efforts, and dollars, are likely to be concentrated.

Chapter 1

The Building Proforma

William H. Rowe III, AIA, P.E.

Harry Truman is reported to have said, "A work of art is a lotta work." (If he didn't say it, he probably would have.) Well-designed, well-built, and well-operating buildings are a pleasure for the inhabitants and owners alike, and in the best of circumstances enrich the environment of the people who use them. The best buildings are the ones whose operations aren't noticed because they run so well. Buildings that are well-maintained, clean, functioning, and comfortable come close to the ideals expressed by Vitruvius in his *Ten Books of Architecture,* which strived for buildings that provide "beauty, commodity, and delight."

For anyone involved in operating buildings, there are many diverse items to deal with and many costs. Building owners, tenants, architects, engineers, real estate developers, brokers, banks and lending institutions, contractors, and building managers and operators must all deal with the costs of maintaining and repairing existing facilities. The skills required cross over all the building trades, from plumbers to painters, and over all the building services, from accountants to attorneys. The subject of this chapter is the factors that influence the cost of operating buildings. It is a "lotta work."

How Maintenance and Repair Fit into the Overall Cost Picture

It is important to understand the overall picture of how building costs are structured, and how facilities maintenance and repair costs affect and are affected by other costs. There are many costs associated with a building. They are called *line items* on the proforma, and include interest on the mortgage, annual property taxes, and operating costs. Clearly, the total annual cost of running a building is affected by many factors. As items such as taxes or interest increase, less money becomes available for other building expenses. A good understanding of the overall building finances is essential to the facility manager who seeks to provide adequate resources for proper building upkeep. Other line items may be a source of relief for stressed maintenance budgets.

Subsequent chapters review the factors that determine building maintenance costs, and then provide examples of how to apply maintenance management principles to specific building types. The information provided can be used to help property managers and building owners—before a building is placed on line—to develop building operating costs and to determine areas where alternate measures can result in more cost-effective maintenance. This information can help facility managers define cost items, identify areas where costs are likely to be expected, accurately estimate the cost of such work, and develop strategies to improve overall building performance.

Proforma — The Building as a Money-Making Machine

Buildings cost money to build (or buy) and to operate. In addition, they receive money, either in rent or in a budget to pay for upkeep. The financial analysis that compares the annual expenses with the annual revenue is called the *proforma* for the building or property. The proforma lays out, in financial terms, how the property will perform over time. Banks and investors require a proforma as part of the loan application to determine if the building net income will be able to pay for the loan. Public and nonprofit organizations use the proforma to determine the budget for properly operating a facility. The basic outline of a proforma for a small office building is shown in Figure 1.1.

The way to determine if a building is to be a financial success (in the case of a private building for profit), or at least to project the annual operating costs for budgeting (in the case of government or not for profit), is to develop a proforma that establishes both the expenses and the revenue stream. If your expenditures exceed your income, then your upkeep will be your downfall.

The proforma in Figure 1.1 is "run out" for four years. Normally it would be done for at least 10 years or the length of the mortgage, if that is longer. In this way, both the owner and lender can feel assured that there is enough money over the term of the investment to cover the cost of the note and maintain a successfully operating property. The net income at the end of each year is a measure of the profitability of the investment.

Line Items in the Proforma

The proforma chart in Figure 1.1 shows how to organize the financial picture. First, the income or revenue is listed. Each source of revenue is noted separately and estimated for each year. Then the expenses are listed again individually, with estimates of the expected costs for each year for which the proforma is "run out." Finally, the total revenues minus the total expenses for each year yield the annual profit or loss for the property. Following are more detailed descriptions of each of the line items in the proforma.

Revenue

Income-producing property receives revenue from the rents paid by tenants. The bulk of the revenue is from the major building function, such as offices or apartments. Often, however, there are additional income sources such as concession stands, roof antennas rented to communications or public safety organizations, parking, and other accessory uses. These are counted as part of the revenue stream.

In setting rents, or determining the income stream, the tenant negotiations usually take into account the fact that building expenses will rise on an annual basis. Operating expenses such as electric bills, taxes, and fuel are expected to increase. Lease agreements, therefore, usually have built-in escalator clauses that require the tenant to pay a percentage (sometimes more than 100%) of his or her pro-rated share of the annual increase in building operating expenses. Also, since inflation is also a part of current economic life, the rent may be subject to increases related to the cost of living. Rents are commonly broken out into a *base* portion and an *adjustable* portion. The base rent is usually set to the owner's fixed expenses such as interest on the property, expenses for the base (or first) year, and the

Outline – Building Proforma						
Annual Revenue	**Type**	**Yr. 1**	**Yr. 2**	**Yr. 3**	**Yr. 4**	**Description**
Rent	Dwelling					Annual rent for space. Define in lease the portion of utilities, taxes, cost of living, escalators, building operating costs, and other operating expenses shared by the tenant.
	Office					Allow for a 3-month period without rent when leases expire on commercial property.
	Parking					Income from parking vendor, or sum of daily receipts.
	Concessions					Rent from carts, sidewalk vendors, newspaper boxes, telephones.
	Antennas					Radio, TV, satellite, or other communications providing income.
Other						Additional revenue from running building outside normal business hours, temporary rentals of space, backcharges for services rendered to tenants outside lease provisions.
Revenue Total						
Annual Expenses						
Acquisition	Down payment					The amount of principal not provided by lender.
	Closing costs					Legal fees, title insurance, inspections, permits, sales tax, points to lender, bank fees, appraisal, and related costs.
	Principal					Varies each year depending on type of loan.
	Interest					Varies each year depending on type of loan.
	Planning					Early development cost to site investigation, design fees, legal opinions.
Capital Expenses	Equipment					Major repairs such as pointing, reroofing, new mechanical equipment, elevators.
	Tenant improvements					Cost of concessions to tenants to improve or occupy a space. Usually occurs at rollover.
Building Operating Costs	Leasing commissions					Fees paid to leasing broker as a percentage of collected rents.
	Management fees					Costs to hire management to run building, pay bills, hire staff to maintain equipment, and pay vendors for building operation and maintenance.
	Taxes					Real estate property taxes.
	Operating expenses					The direct cost of outside vendors for maintenance and repair.
	Ordinary repairs					Ongoing repairs and upkeep expenses separate from operating expenses.
Expenses Total						
Net Operating Income						

Figure 1.1

profit expected on the building. The adjustable portion includes all costs that may rise over the term of the lease. In the extreme case of a triple net (NNN) lease, the tenant pays the landlord the base rent plus annual increases for inflation and, separately, is responsible for all other expenses. Property tax, repairs to the property, and upkeep maintenance are all paid as if the tenant were the owner.

Public Buildings: For buildings operated by owners such as corporations or government agencies (which are not operated to generate income), the revenue stream is the annual operating budget. It is determined by first calculating the expense side of the proforma and then writing a request for funds to cover the anticipated expenses.

Public or not-for-profit property such as government buildings do not pay rent; instead, revenue is appropriated to pay for building operations. In most instances there is an account line item for building operations allotted to each department. Thus a school department or hospital complex would have funds to cover the expenses on several buildings. The line item for facilities maintenance and repair would appear on each year's budget for review.

Later in this chapter, the general costs for facilities maintenance and repair will be developed and fit into the overall building proforma. The following chapters of this book develop these costs in more detail.

Expenses

Acquisition Cost: The purchase, or capital, cost is paid at acquisition. The buyer will usually "put up" some of his own money in the form of a down payment. Down payments range from 5% to 25% of the purchase price. Lenders will have property assessed to make sure that it is worth the value of the purchase price so that in the event of a default, they can recover the value of their loan, at an auction if necessary. As a general rule, banks will limit the amount they are willing to lend based on a number of factors. For example, they may limit the amount of their loan to a percentage, say 80% of the assessed value. In such a situation, if the purchase price equals the assessed value, then the buyer must put up 20% in his/her own money to complete the sale, or find a lender for a second mortgage. This lender will take a "second position" — not an attractive place to be for an investor.

In addition to the down payment, the buyer must pay for closing costs. These include paying for the bank's assessor, title search, legal fees, points on the loan, and a variety of smaller costs associated with completing the deal.

The rest of the money for the acquisition is usually borrowed from a bank and paid back over time with interest. Each year there is a cost to the owner to pay for the acquisition in monthly payments to the bank to cover the principal and interest on the property. In some cases several banks and several different bank loans are needed. Figure 1.2, "Types of Loans for Real Estate Ventures" shows some of the more common methods for borrowing money for buildings. For new construction, one bank may be involved in the purchase of land, another in providing financing for the construction costs, and a third for the permanent financing. For income tax purposes, the interest is a deductible expense. The acquisition cost, or principal, is not deductible. However, the owner may depreciate the value of the property over a period of years using various methods as determined by the tax code. The period of depreciation has varied, over time, from 20 to 40 years, and is often revised by both the federal and state governments. When the period of depreciation is shorter, more of the principal can be deducted in the

Types of Loans for Real Estate Ventures

Terms (P = prime rate, M = average market rate, REIT = Real Estate Investment Trust)

Type of Loan	Description	Requirements	Typical Sources	Rates %	Coverage	Points	Terms (yrs.)	Amortization	Comments
A. Construction Loan Against Take Out	Finance project until C–Bullet Loan or other type replaces it.	Strong track record for similar projects and some personal security.	Commercial banks and some thrifts.	P+0.5 to P+2.0	Loan level tied to take out or 85/90% of costs.	0.5 to 1.5	Tied to completion.	Interest only.	Forward take outs usually limited to preleased buildings. Most loans are build to suit.
B. Open Ended Construction Loan	Finance project without other repayment source in place.	Used by experienced developers with full personal guarantee.	Commercial banks, some thrifts, and credit companies.	P+1.0 to P+2.0	Loan level tied to 70/90% of costs.	0.75 to 2.0	1–3 with additional miniperm option.	Interest only.	Difficult to arrange. Lenders prefer strong preleasing and tie to miniperm.
C. Bullet Loan/Permanent Loan	Medium term financing, mostly available on an immediate funding basis on fully leased buildings.	Good track record and quality building in good location.	Insurance companies and pension funds.	M+1.0 to M+2.0	1.2× to 1.3×	0 to 1	3 to 30	15 to 30	Higher rates for longer terms. Usually 65/75% LTV. High prepayment penalty. Limited/no personal guarantee.
D. Industrial Revenue Bonds	Tax exempt bonds for sites designated for development.	Commercial activities providing jobs and growth to depressed areas.	Government.	0.75P	1.2×	0	10	10	Programs require enactment of State and Federal programs.
E. Second Mortgage Loan	Used to finance a portion of acquisition or building upgrade.	Property value must support value of both first and second mortgages.	Credit companies and some banks.	M+1.5 to M+3.0	1.2× to 1.3× on combined loan.	1 to 3	2 to 10	Mostly interest only.	Prohibited in most C–Bullet Loans. Second lender at higher risk.
F. Wraparound Mortgage	Similar to E–Second Mortgage.	Similar to E–Second Mortgage.	Credit companies and some banks.	Blend	1.15× to 1.25× on combined loan.	1 to 2	2 to 10	None	Rate depends on first mortgage rate since lender charges a blended rate. Yield requirements on new money varies from (P+3) to (P+6). First mortgage must not exclude second mortgage.
G. Land and Land Development Loan	Finance land, acquisition, planning, design and infrastructure improvements.	Similar to B–Open Ended Construction Loan.	Some commercial banks and thrifts.	P+1.5 to P+3.0	N/A	1 to 3	1 to 3	Interest only.	Difficult to arrange.
H. Standby Loan	Available to repay construction loan if other source of repayment is not arranged.	Similar to B–Open Ended Construction Loan.	Credit companies and some insurance companies.	P+1.5 to P+3.5	1.2× to 1.25×	2 + 1 for each 6 month extension.	2 to 7	25 to 30	Lender prefers preleasing. Borrower prefers forward commitment from Bullet lender. Standby period 1 to 2 years.

Figure 1.2

9

Types of Loans for Real Estate Ventures

Type of Loan	Description	Requirements	Typical Sources	Terms (P = prime rate, M = average market rate, REIT = Real Estate Investment Trust)					Comments
				Rates %	Coverage	Points	Terms (yrs.)	Amortization	
I. Interim Loan or MiniPerm (Floating Rate)	Similar to C – Bullet Loan except granted on a floating rate basis.	Similar to C – Bullet Loan.	Credit companies, commercial banks, thrifts, and some Wall Street firms.	P + 0.5 to P + 2.0	1.2× to 1.5×	1 to 2	3 to 30	15 to 30	Lend on existing cash flow. Flexible prepayment.
J. Interim Loan or MiniPerm (Floating Rate)	Similar to C – Bullet Loan.	Similar to C – Bullet Loan.	Commercial banks, thrifts, some credit companies, and Wall Street firms.	M + 1 to M + 3	1.2× to 1.3×	1 to 2	3 to 40	15 to 30	Lend on existing cash flow. Moderate repayment restrictions. Periodic rate reviews possible.
K. Permanent Participating Loan	Same as C – Bullet Loan except lender takes a share of cash flow in exchange for more aggressive leverage.	Same as C – Bullet Loan.	Insurance companies and pension funds.	M + share of income	1.5× to 1.20×	0.5 to 1	10 to 15	Up to 30	IRR projections must be strong. Forward commitments possible if preleasing is in place. Participation in cash flow ranges from 25% to 75%.
L. Convertible Loan	Same as C – Bullet Loan except lender has the right to convert mortgage to ownership (50% +/–) in exchange for more aggressive rate.	Similar to C – Bullet Loan.	Pension funds.	M	1.1× to 1.15×	1	2 to 15	Up to 20	IRR projections must be good/strong. Equity portion usually requires preferential return plus 25%–75% share in ownership.
M. Debt/Equity Joint Venture	Finance 80%–90% of cost by a combination of debt and equity.	Similar to C – Bullet Loan.	Insurance companies, pension funds, and some credit companies.	M	1.1× to 1.15×	0.5 to 1	5 to 15	25 to 30	Debt and equity sources can be different. IRR projections must be good/strong. Equity portion usually requires preferential return plus 25%–75% share in ownership.
N. Land Sale/Leaseback	Financing technique for maximum leverage.	Similar to C – Bullet Loan.	Mostly REITs and real estate partnerships.	M	N/A	1	50 or more	N/A	Land is usually subordinated to leasehold first mortgage.
O. Building Sale/Leaseback	Off balance sheet financing, which allows lessee to control property.	Lease term that provides an overall rate of return that is acceptable.	Insurance companies, REITs, pension funds, and some off-shore investors.	Value is a function of capitalizing net operating income (NOI) at market rates, which may vary with the location of the property, lease potential, and available credit.					Lessee is usually of investment grade. Possibly leveraged with third party debt.

Figure 1.2 (cont.)

Types of Loans for Real Estate Ventures

Type of Loan	Description	Requirements	Typical Sources	Terms (P = prime rate, M = average market rate, REIT = Real Estate Investment Trust)					Comments
				Rates %	Coverage	Points	Terms (yrs.)	Amortization	
P. Outright Sale (Unleveraged)	An outright sale of property for cash to a medium- or long-term investor.	Must be investment-grade real estate with potential for superior appreciation appeal and/or high quality income stream.							Used for retail, multi-family, and industrial property. IRR targets should be good.
Q. Outright Sale (Leveraged)	Sale for cash combined with some level of debt financing of property to a medium- or long-term investor.	Similar to P – Outright Sale (Unleveraged).							Used for retail, multi-family, and industrial property. IRR targets should be good.
R. Syndication (Limited Partnerships)	The sale via a limited partnership vehicle of typically 99% of cash flow, depreciation and residuals if any.	Similar to C – Bullet Loan. Key issue is that the underlying debt ideally should be a fixed rate.	Investment bankers who sell partnership interests to individual investors and then purchase properties.	Varies with available programs such as low-income housing, historic or other tax credits, or government incentives.					Tax credits allocated to states are based on per capita limits and IRS code. Expensive and intricate financing strategy.

Figure 1.2 (cont.)

early years, which results in a lower net income for tax purposes and hence lower taxes. Owners prefer short or accelerated depreciation schedules.

The acquisition cost can include more than the actual cost of a building. It also involves:

Planning — Pre-purchase costs/due diligence

Site investigation and report, premarketing efforts, legal fees and negotiation fees, consultants for economic analysis, feasibility studies, and commissions for obtaining financing. Also included in this category would be resources to cover preparation for public hearings on any zoning, permitting, or conservation issues, and the cost of putting together the actual program and proforma packages used to entice investors.

Design

Architectural and engineering fees, site surveys, site assessment costs to determine hazardous wastes, and special consultants for environmental or design issues.

Construction

When possible, generally one lump sum; however, all projects should budget for costs beyond the initial contract sum for unknown conditions, design and programmatic changes, delays, government approvals, and related items.

Post-Construction

After the construction is completed there are some costs to be expected during the fit-up and shake-down period, despite the fact that most systems are usually covered in the first year's normal warranty period. Staff training, incidental furnishings and plants, charging fluids for fuel, and initial stores of building supplies are examples of such costs.

Capital Expenses and Building Operating Costs: In addition to acquisition costs, the annual capital expenses and building operating costs are determined to project total annual expenses.

Figure 1.3, "Building Cost Information to Review," contains a more complete list of items necessary to review when purchasing property. Items 1 – 6 are primarily associated with acquisition costs. Items 7 – 9 are used to determine capital improvement and building operating costs. The major categories of these expenses are:

1. *Title/Insurance/Real Estate Tax/Mortgages*

   These documents are examined to determine if the property is in fact owned by the seller. (The Brooklyn Bridge would not have been sold so many times if a title search had been performed!) They also identify the recorded mortgage holders who would be paid as part of the acquisition. A title search would reveal any land-use restrictions due to unmitigated or unresolved environmental problems, and would also reveal the presence of any outstanding liens to which the sale of the property would be subject and that would also need to be paid as part of any sale.

2. *Operations: Income and Expenses*

   The income and expense sheets form the basis of the data for the proforma. The bulk of contracts and leases pertaining to annual expenses can be found by reviewing these documents. Often a large error occurs, however, because each year there

Building Cost Information to Review

1. Title/Insurance/Real Estate Tax/Mortgages
   a. Original deeds, as recorded.
   b. Matters of record of which the title is subject, including trust and partnership agreements.
   c. Title insurance policy in effect (owner and lender).
   d. Insurance policies (property insurance, general liability, etc.).
   e. Real estate tax bills and correspondence on abatements and prior assessments and future assessments.
   f. Mortgage financing, including amendments, if any.
   g. Plot or site plan showing boundaries, easements (if any), and major ground features and topography.
   h. Bonds—performance/completion, labor and materials, and local government surety—presently in effect.
2. Operations—Income and Expenses
   a. Data referring to projected income and expenses.
   b. Existing leases and amendments, and projected leases already received, presently under negotiation and a standard lease.
   c. Chart of accounts to be used for the project operations and description of their accounting systems and their variations from the uniform system of accounts.
   d. Market studies conducted by or otherwise available to seller used in project development or any other market information, including initial marketing plan (by rate and occupancy group or sub-group).
   e. All office buildings deemed competitive, their rent and level of concessions.
   f. Pre-opening budget.
   g. Pre-opening service contracts.
   h. Pre-opening activities and schedule milestones.
   i. Resumés of proposed general manager, resident manager, and other key personnel.
   j. Proposed inventory of initial operating supplies.
   k. Proposed initial working capital budget.
   l. Permits, licenses, or pending applications that affect the operation of the project.
   m. Revenue histories and projections for all tenants (if any) paying percentage rents.
3. Management Contracts
   a. Existing or proposed property management, marketing and leasing contracts and associated fee structures.
   b. Other existing and/or proposed vendor or service contracts.
   c. Agreements affecting site and recreational amenities.
   d. Marketing, leasing, and other related agreements, and associated fee structures.
4. Master Plan Documents—All master plan related drawings and documents, including:
   a. Site plan and documents to verify planning board or similar agency approval.
   b. Documents related to clearance of environmental impact.
   c. Other master plan related documents, if any.
5. Zoning, Building Permits, Inspections, and Certificates of Occupancy
   a. Zoning:
      • Attorney's zoning opinion and certifications from architects and engineers regarding zoning.
      • Zoning applications, zoning analysis (height, setback, and average calculations), annual report, and decisions rendered.
      • Excerpts from P.U.D. zoning ordinance and internal design standards.
   b. Building permits:
      • Building permits for base building.
      • Building permits for infrastructure.
      • Other building department records.
      • Building permits for tenant space renovations.
      • Exceptions from Building and Zoning Code.
   c. Building inspections for base building and tenant work, including violations:
      • Base building
      • Elevators
      • Fire alarms
      • Electrical

Figure 1.3

- Plumbing
- Violations

d. Certificates of occupancy:
- Base building
- All tenant spaces

6. Building Documents (Contracts presently in effect in the building, all amendments and correspondence from contractors, architects, engineers, attorneys, brokers, and management.)

a. Design and construction contracts and correspondence for base building:
- Architectural and engineering design contracts for ongoing improvements.
- Interior design contracts.
- Other design contracts and/or consulting agreements.
- Performance and payment bonds.
- Insurance.
- Finishes, furnishings, and equipment, including handling and installation contracts and proposed inventory.
- Construction contracts for ongoing improvements:
  - Developer's agreement
  - Construction contract
  - List of subcontractors and costs
- Job meeting notes for ongoing improvements.
- Contractors' requisitions for payment.

b. Construction contracts and correspondence, including development agreement for tenant spaces.

c. Contracts for related work not included in general contract, including infrastructure:
- Landscaping
- Telephone
- Computer
- Interiors
- Furniture
- Infrastructure

d. Architects' and engineers' certifications:
- Certificate of substantial and final completion.
- Structural engineer's acceptance of work.
- Structural design criteria.
- Mechanical and electrical engineers' acceptance of the work.
- Release of liens from contractor for base building.
- Release of liens from contractors for tenant improvement work.
- Certificates by architects to owner, financial institutions, building commissioner, or building inspector.
- Certificates by engineers to owner, financial institutions, building commissioner or building inspector.
- Daily logs from on-site field representatives.

e. Building plans and specifications.

f. As-builts with addenda, bulletins, and change orders.

g. Testing reports:
- Soils compaction, concrete, compression and structural steel inspections.
- Freeze/thaw test procedures and results.

h. Fire safety plan.

i. Fire security system manual.

j. Acceptances by insurance carriers of fire suppression system.

k. Method and calculation of gross and net building areas.

l. Elevator control sequence.

m. Escalator control sequence.

n. Emergency elevator control sequence.

o. Emergency generator controls.

p. Roof guarantees.

Figure 1.3 (cont.)

q. Certification that building is not in flood plain and history of flooding on the site.
r. Survey of wetlands.
s. Site utility plan.
t. Building tenant standards and sample tenant agreement.

7. Building Operations and Maintenance
    a. Service contracts and warranties:
- HVAC
- Control systems
- Filters
- Pest control
- Alarm system testing
- Water treatment
- Radio service
- Artwork rental
- Sprinkler and standpipes
- Elevators
- Escalators
- Cleaning
- Plants
- Security
- Parking lot maintenance
- Landscape maintenance
- Window washing
- Roof guarantee
- Trash
- Water coolers
- Garage maintenance
- Plaza maintenance, other equipment/service
- Telephone
- Electric service
- Interior design service
- Automatic door equipment
- Towel and uniform cleaning

    b. Contracts for work in progress.
    c. Building management contract.
    d. List of warranty expiration dates.
    e. Maintenance schedules for equipment.
    f. Inspection records.
    g. List of stored materials.
    h. Energy usage.
    i. Breakdown of operation and maintenance budget.

8. Building/Capital Cost Data
    a. Operating capital and building cost since opening.
    b. Capital budgets for future years.
    c. Budget of committed capital items.
    d. Chart of accounts for capital items.
    e. Project budget and requisitions from lender.
    f. References for operating expenditures.

9. Miscellaneous
    a. Seller's package.
    b. Project appraisals conducted by owner.
    c. Promotional brochure for project.
    d. Brochures from architect, construction manager, and other key professionals involved with the project.

Figure 1.3 (cont.)

are many expenses for which written contracts are not available. Emergency expenses and smaller costs do occur and should be properly accounted.

3. *Management Contracts*

These contracts describe the basis of fees and expenses to management companies engaged to operate the building. They can include accounting services as well as the costs of managing the host of vendors that service the property for cleaning, elevator maintenance, HVAC repairs, and the like. See Figure 1.3, Section 7 for a full list.

4. *Master Plan Documents*

Master plans will assist building owners in predicting future costs for long-range planning, expansion, additions, environmental costs, or costs of extended public hearings and reviews.

5. *Zoning, Building Permits, Inspections, and Certificates of Occupancy*

A review of the zoning and building permits should satisfy the owner that the property is not in violation of zoning and that all required permits were applied for and obtained. If there are any violations, which are citations by the Building Department for deficiencies, they should be on record. Recorded violations or building permits not fully signed off represent exposure to costs for repair, professional services, and possible fines. If they are expected, they should be accounted for.

6. *Building Documents*

These contracts represent the cost of completing a construction project. Often a proforma is created during the construction period, as the construction lender is replaced by the permanent financing. The permanent lender will require that a proforma be completed showing the status of the construction activity,

---

Building Cost Information to Review (cont.)

e. Building surveys of comparable sites by seller.
f. Certificates (including architect's and general contractor's) of no asbestos or other hazardous waste.
g. Engineering certification of no hazardous waste.
h. Any environmental notification statement (ENS) or environmental information statement (EIS), if available, and/or other subsurface or groundwater studies, environmental studies or reports.
i. Tax analysis and tax statement.
j. Standard lease and tenant standards for improvement.
k. CPM report.
l. Letters, field reports, preliminary report, and list of information requested.
m. Copies of Engineer Inspection Reports.
n. Inspection Engineer qualifications.

Figure 1.3 (cont.)

percentage of completion including fees to architects and engineers, and evidence that tests have been completed and environmental and related public reports have been filed.

7. *Building Operations and Maintenance: Building Operating Costs*

Building operating costs are the expenses incurred in a year that do not generally have a useful life longer than that year. These costs are the primary focus of this book and are the costs in the proforma that represent the annual expenditures for operations, maintenance and repair. Each year money is spent to operate the building and to pay for taxes, supplies, cleaning, utilities, and all the other costs of operating and maintaining the building and its associated property. As previously shown in Figure 1.1, the major categories of building operating costs are:

Leasing Commissions

Over the term of a lease, the landlord will pay a commission to the broker who found and helped negotiate the lease. Terms of such commissions vary, but are currently in the range of 1% – 3%.

Management Fees

For any property there is a management cost. The entity that collects rents, contracts with vendors for services, prepares annual financial statements, and generally oversees the asset management will be engaged. On small properties, this is done by the owner, but it is still an expense to be accounted for on each property. The information for these costs was also discussed in Item 3.

Taxes

Real estate tax is a significant annual expense. It is assessed on an annual basis based on the tax rate and the assessed value of the property. The tax rate is set locally and varies from community to community.

Operating Expenses

Operating expenses consist of utilities, maintenance, and repair costs. They are normally paid annually from the operating budget. The basic difference is that maintenance is usually scheduled in advance, whereas repairs are for items that break down or that require emergency action, such as broken pipes or failed motors. Both qualify for the general description of ordinary repairs.

Maintenance Costs

Figure 1.3 lists some standard contracts that can be expected in a building. In many cases these costs are predictable and can even be established with vendors on an annual basis.

Service contracts:
- Alarm system testing
- Artwork rental
- Automatic door equipment
- Cleaning
- Control systems
- Electric service
- Elevators
- Escalators
- Filters
- Garage maintenance
- Parking lot maintenance

- HVAC
- Interior design service
- Pest control
- Plants
- Plaza maintenance, other equipment/service
- Landscape maintenance
- Radio service
- Roof guarantee
- Security
- Sprinkler and standpipes
- Telephone
- Towel and uniform cleaning
- Trash
- Water coolers
- Water treatment
- Window washing

Repair costs:

Costs of repair (not replacement, which would be a capital improvement cost), such as repair of leaks in a roof, repair of pump motors, tuning up of mechanical equipment, and ordinary repairs would fall into the category of operating expenses. For tax purposes, the full cost of these items can be deducted from the income in determining net taxable income.

8. *Building/Capital Cost Data*

Capital Expenses

Capital expenses, which are those expenses necessary to make major repairs, are paid in the year they occur. However, for tax purposes, the expense must be spread over several years, depending on the depreciation schedule permitted by the tax code. Capital costs cannot be expensed (deducted from the income in determining net profit) in the year when paid. These capital costs are depreciated over a longer term (i.e., 20 to 40 years). The term *capital expenses* applies to costs that will have a useful life of over one year. After a few years, major building components require upgrade — sometimes substantial upgrade.

Examples of capital costs include replacing a sprinkler system, upgrading elevators, changing cooling towers, waterproofing a plaza deck, retrofitting toilets, installing ramps and doors to comply with the requirements of the Americans with Disabilities Act (ADA), replacing building exterior caulking, re-roofing, repointing masonry, and upgrading the HVAC control system.

Many, if not most, capital expenditures are triggered by code requirements. Refer to Chapter 6 for a description of the codes and regulations that necessitate such expenditures.

Tenant Improvements

As an enticement to new tenants or to encourage existing tenants to renew their leases, there is an expectation that the building owner provide some form of tenant upgrade. Sometimes called *tenant concessions*, tenant upgrades are also part of any negotiations at the expiration of any lease. Agreed-upon sums may be allocated to the tenant for space improvements, such as new finishes, new walls, painting, and upgrading of electrical or mechanical systems. When such concessions are made, the landlord is basically financing the improvements over the term of the upcoming lease.

9. *Miscellaneous*

The documents listed in this category generally consist of background information that indicates both opportunities and problems associated with the property. In any thorough review of a site, the existence of certificates indicating the following conditions will give an overall sense of comfort and assurance that the costs and predictions have some rational basis:

- The property is free of hazardous materials.
- All taxes have been paid.
- Building appraisals are consistent with the offered purchase price.
- Architects and engineers have reported that the property is substantially in conformance with applicable codes.

Generally speaking, the operations and maintenance budget is one of the few expenses over which the owner or facility manager has significant control. Some expenses may be deferred, others eliminated. Some may be negotiated with tenants who will share the costs, and some represent a balance between large capital expenses that result in lower annual maintenance, or little or no capital expenses that result in high annual maintenance fees. A leaky roof may be able to be nursed along year after year, but at increasing expense as plaster falls or the structure is affected. Or it can be fixed early with a large capital expense that significantly lowers annual costs down the line.

Proforma Example

Suppose you are asked to make a proforma for a new building that is to be purchased on June 1, 1996. It will be renovated and then scheduled for occupancy on January 1, 1997. In 1996, there will be 7/12 of the normal full year to pay out in interest, and there will be no income from rents. Other revenues and expenses will also be adjusted for partial year and start-up.

Costs

The building will cost $475,000 to buy, but will require extensive renovation of $580,000 in construction costs and another $65,000 in soft costs for architects, engineers, attorneys, fees, surveys, and the like. A summary of the expected costs is shown in Figure 1.4.

From Figure 1.4 we can see that the owner intends to borrow the $992,500 and will pay from available cash the $95,000 down payment and half of the total planning or soft costs, $32,500.

Debt Service: Payments for Principal and Interest

We will assume that the project will be financed over 25 years and that the available interest rate is 10.75%. A note on these terms means that for every $1,000 borrowed, the owner will pay to the bank $9.63 per month, or $115.56 per year over the 25 years to amortize the loan. At the end of the term, $2,889 would have been paid back to the lender for the use of the original $1,000.

We will run the proforma out for the first ten years, sufficiently long to obtain an overall picture of how the building is expected to perform. Not all the money will need to be borrowed at acquisition, only the $380,000 portion for the purchase. Over the first year money will be "drawn down" from the construction note, and over the first two and one half years, money will be drawn down for the soft costs. This results in the debt service schedules shown in Figure 1.5.

Clearly the numbers in Figure 1.5 would be changed if separate banks were involved for the acquisition, construction note, and the permanent financing. Also, the cash picture would be altered if the monies were all borrowed at once and allowed to accumulate nonrisk money-market interest until they were expended.

Indexing Rates

For the proforma we will assume an annual inflation rate of 5% over the term of the mortgage, which will apply to expenses and income. Thus each year's expenses will increase over the base year by 5%.

Proforma Example – Acquisition Expense Summary					
Item	Amount	% Financed	Amount Financed	Owner Portion	Comments
Acquisition of Land and Building	$ 475,000	80	$380,000	$ 95,000	This is the down payment from the owner, paid at closing.
Construction Cost	$ 580,000	100	$580,000	0	
Soft Costs	$ 65,000	50	$ 32,500	$ 32,500	These costs will be spread over the first two years.
Total	$1,120,000		$992,500	$127,500	

Figure 1.4

Proforma Example – Debt Service Schedule						
Item	Amount Financed	Rate per $1,000	1996	1997	1998–2021	Comments
Acquisition	$380,000	115.56	$25,616	$ 43,913	$ 43,913	First year is only 7 months = 7/12 × $43,913.
Construction	$580,000	115.56	$22,342	$ 67,025	$ 67,025	First year construction activity begins in September (4/12 × $67,025).
Soft Costs	$ 32,500	115.56	$ 2,191	$ 3,756	$ 3,756	First year is only 7 months.
Totals	$992,500		$50,149	$114,694	$114,694	

Figure 1.5

Rent Renewal

For the particular time and place, the going rate for rent is expected to be $20 per square foot. In five years, the rent will increase at the rate of 5% per year, or

$$1.05 \times 1.05 \times 1.05 \times 1.05 \times 1.05 = 1.276$$

At the beginning of the sixth year, therefore, the base rent expected = $20 \times 1.276 = \$25.52/\text{S.F.}$

At the beginning of the eleventh year the base rent expected = $25.52 \times 1.276 = \$32.56/\text{S.F.}$

Parking Income

The project has 82 parking spaces. Assuming 80% occupancy at $5.00 per day, the daily income equals:

82 cars $\times$ \$5.00/car/day $\times$ 0.80 = \$328 per day $\times$ 200 working days/year = \$65,600 per year income. This can be increased each year by the inflation factor of 5%.

Building Areas

The building has a total gross floor area of 26,640 square feet and a total net floor area of 22,644 square feet. The difference is the floor area of stairs, elevator, and duct shafts. The area available for parking is 19,800 square feet, which for this project is expected to provide 82 parking spaces. All of the building area is subject to real estate taxes – there are no portions exempt.

Lease-up Period and Vacancy Rates

There are to be four spaces in the building, for Tenants A, B, C, and D. The table in Figure 1.6 explains the expected lease terms, areas, inflation, and percentage of expenses shared by each tenant. All the tenants are assumed to begin occupancy in January 1997 and have 5-year leases that are renewed in January of 2003 and 2008. Vacancy rate (or free months given as incentive to new tenants) and tenant improvement allocations are also noted.

It is expected that it will realistically take time to fully lease the building. Tenants are projected to sign five-year leases. Some tenants are expected to renew their leases. When they renew, the owner expects to provide tenant improvements amounting to $15 per square foot. For the tenants who do not roll over, a vacancy of three months will be built into the proforma with a larger allowance for tenant improvements of $25 per square foot to entice new tenants. A summary of the expected indexing rates, rents, and occupancy is shown in Figure 1.6.

Tenant improvements can represent a significant expenditure, and so they will also be financed over a 10-year term. For 10 years we will assume an annual rate of 9%. The Amortization Table in Figure 1.7 shows that for every $1,000 borrowed, the monthly payment equals $12.67 or $152.04 per year.

Vacancy rate is a measure of how successfully the building will be fully occupied. Conservative analysis allows for some period of vacancy for some tenant spaces.

Expenses

Acquisition Costs

The **down payment** of $95,000 has been calculated in Figure 1.4.

Closing costs are assumed for this example to be 1% of the purchase price, or 1% of $475,000, or $4,750.

Projected Base Rent Income and Tenant Improvement Expenses

Tenant	Area (Sq.Ft.)	Share of Expenses	Lease Year	Annual Base Rent ($/Yr.)	Occupancy Rate	Expected Base Rent ($/Yr.)			Tenant Improvements ($/Sq. Ft.)	Tenant Improvements		
						1997	2002	2007		First	Fifth	Tenth
A	5,000	0.22	1997	100,000	1	100,000			25	125,000		
			2002	127,600	1		127,600		15		75,000	
			2007	162,800	1			162,800	15			75,000
B	10,144	0.45	1997	202,880	1	202,880			25	253,600		
			2002	258,874	1		258,864		15		152,160	
			2007	330,289	1			330,289	15			152,160
C	5,000	0.22	1997	100,000	0.90	90,000			25	125,000		
			2002	127,600	0.90		114,840		25		125,000	
			2007	162,800	0.90			146,520	25			125,000
D	2,500	0.11	1997	50,000	0.90	45,000			25	62,500		
			2002	63,800	0.90		57,420		25		62,500	
			2007	81,400	0.90			73,260	25			62,500
			Total Base Rent			437,880	558,724	712,869	**Total Tenant Imp.**	566,100	414,660	414,660
									Index #	1	1.28	1.63
									Projected Tenant Imp.	566,100	529,100	675,000
						Annual Payments @ $152.04/1,000				86,069	80,444	102,627
Total	22,644	1.0										

The base 1997 rent for all leases is $20/S.F.
The base 2002 rent for all leases is $25.52/S.F.
The base 2007 rent for all leases is $32.56/S.F.
These base rents have been multiplied by the areas above to obtain the annual base rent.

Figure 1.6

Interest Rate	5 Years	10 Years	12 Years	15 Years	18 Years	20 Years	25 Years	30 Years
			Amortization Table					
			Monthly Payment—Multiply per $1,000					
5%	$18.87	$10.61	$ 9.25	$ 7.91	$ 7.04	$ 6.60	$5.85	$5.37
5-1/4	18.99	10.73	9.38	8.04	7.17	6.74	6.00	5.53
5-1/2	19.10	10.86	9.51	8.18	7.31	6.88	6.15	5.68
5-3/4	19.22	10.98	9.63	8.31	7.45	7.03	6.30	5.84
6	19.33	11.11	9.76	8.44	7.59	7.17	6.45	6.00
6-1/8	19.39	11.17	9.83	8.51	7.66	7.24	6.52	6.08
6-1/4	19.45	11.23	9.89	8.58	7.73	7.31	6.60	6.16
6-3/8	19.51	11.30	9.96	8.65	7.80	7.39	6.68	6.24
6-1/2	19.57	11.36	10.02	8.72	7.87	7.46	6.76	6.33
6-5/8	19.62	11.42	10.09	8.78	7.94	7.53	6.84	6.41
6-3/4	19.68	11.49	10.16	8.85	8.01	7.61	6.91	6.49
6-7/8	19.74	11.55	10.22	8.92	8.09	7.68	6.99	6.57
7	19.80	11.62	10.29	8.99	8.16	7.76	7.07	6.66
7-1/8	19.86	11.68	10.36	9.06	8.23	7.83	7.15	6.74
7-1/4	19.92	11.75	10.42	9.13	8.31	7.91	7.23	6.83
7-3/8	19.98	11.81	10.49	9.20	8.38	7.98	7.31	6.91
7-1/2	20.04	11.88	10.56	9.28	8.45	8.06	7.39	7.00
7-5/8	20.10	11.94	10.62	9.35	8.53	8.14	7.48	7.08
7-3/4	20.16	12.01	10.69	9.42	8.60	8.21	7.56	7.17
7-7/8	20.22	12.07	10.76	9.49	8.68	8.29	7.64	7.26
8	20.28	12.14	10.83	9.56	8.75	8.37	7.72	7.34
8-1/8	20.34	12.20	10.90	9.63	8.83	8.45	7.81	7.43
8-1/4	20.40	12.27	10.97	9.71	8.91	8.53	7.89	7.52
8-3/8	20.46	12.34	11.04	9.78	8.98	8.60	7.97	7.61
8-1/2	20.52	12.40	11.11	9.85	9.06	8.68	8.06	7.69
8-5/8	20.58	12.47	11.18	9.93	9.14	8.76	8.14	7.78
8-3/4	20.64	12.54	11.24	10.00	9.21	8.84	8.23	7.87
8-7/8	20.70	12.61	11.32	10.07	9.29	8.92	8.31	7.96
9	20.76	12.67	11.39	10.15	9.37	9.00	8.40	8.05
9-1/8	20.82	12.74	11.46	10.22	9.45	9.08	8.48	8.14
9-1/4	20.88	12.81	11.53	10.30	9.53	9.16	8.57	8.23
9-3/8	20.94	12.88	11.60	10.37	9.61	9.24	8.66	8.32
9-1/2	21.00	12.94	11.67	10.45	9.68	9.33	8.74	8.41
9-5/8	21.06	13.01	11.74	10.52	9.76	9.41	8.83	8.50
9-3/4	21.12	13.08	11.81	10.60	9.84	9.49	8.92	8.60
9-7/8	21.19	13.15	11.88	10.67	9.92	9.57	9.00	8.69
10	21.25	13.22	11.96	10.75	10.00	9.66	9.09	8.78
10-1/8	21.31	13.29	12.03	10.83	10.08	9.74	9.18	8.87
10-1/4	21.37	13.36	12.10	10.90	10.16	9.82	9.27	8.97
10-3/8	21.43	13.43	12.17	10.98	10.25	9.90	9.36	9.06
10-1/2	21.49	13.50	12.25	11.06	10.33	9.99	9.45	9.15
10-5/8	21.56	13.57	12.32	11.14	10.41	10.07	9.54	9.25
10-3/4	21.62	13.64	12.39	11.21	10.49	10.16	9.63	9.34
10-7/8	21.68	13.71	12.47	11.29	10.57	10.24	9.72	9.43

Figure 1.7

Interest Rate	Amortization Table (cont.) Monthly Payment—Multiply per $1,000							
	5 Years	10 Years	12 Years	15 Years	18 Years	20 Years	25 Years	30 Years
11	$21.74	$13.78	$12.54	$11.37	$10.66	$10.33	$ 9.81	$ 9.53
11-1/4	21.87	13.92	12.69	11.53	10.82	10.50	9.99	9.72
11-1/2	21.99	14.06	12.84	11.69	10.99	10.67	10.17	9.91
11-3/4	22.12	14.21	12.99	11.85	11.16	10.84	10.35	10.10
12	22.24	14.35	13.14	12.01	11.32	11.02	10.54	10.29
12-1/4	22.37	14.50	13.29	12.17	11.49	11.19	10.72	10.48
12-1/2	22.50	14.64	13.44	12.33	11.67	11.37	10.91	10.68
12-3/4	22.63	14.79	13.60	12.49	11.84	11.54	11.10	10.87
13	22.75	14.94	13.75	12.66	12.01	11.72	11.28	11.07
13-1/4	22.88	15.08	13.91	12.82	12.18	11.90	11.47	11.26
13-1/2	23.01	15.23	14.06	12.99	12.36	12.08	11.66	11.46
13-3/4	23.14	15.38	14.22	13.15	12.53	12.26	11.85	11.66
14	23.27	15.53	14.38	13.32	12.71	12.44	12.04	11.85
14-1/4	23.40	15.68	14.53	13.49	12.89	12.62	12.23	12.05
14-1/2	23.53	15.83	14.69	13.66	13.06	12.80	12.43	12.25
14-3/4	23.66	15.99	14.85	13.83	13.24	12.99	12.62	12.45
15	23.79	16.14	15.01	14.00	13.42	13.17	12.81	12.65

Figure 1.7 (cont.)

For simplicity, the **principal and interest payments** have been combined and are entered on the proforma chart from the Debt Service Schedule in Figure 1.5. For tax purposes, it would be necessary to split out the principal and interest portion each year since the interest portion is deductible, but the principal payments are not. The principal, or cost of the capital, may be depreciated over time in accordance with the rules for depreciation by the Internal Revenue Service.

The **planning costs** are assumed to be the soft costs of $65,000, spread over the first two years.

Capital Expenses

Equipment costs are assumed to be zero in the first year because of building warranties. We will set an initial expected cost for annual capitalization costs at 1% of the building value and allow this amount to escalate by double the index rate. Thus for the building initial value of $1,120,000, 1% means an allowance of $11,200 at the beginning of the second year with an increase of 10% per year thereafter, as the equipment has more wear and tear. This is a little risky, but not unheard of, since the building will be getting a full renovation.

Tenant improvement costs have been computed in the "Projected Base Rent Income and Tenant Improvement Expenses" chart in Figure 1.6.

Building Operating Costs

Leasing commissions are determined in accordance with the contract between the owner and broker. Payment can be a lump sum, or a percentage of the collected rents. We will assume a broker fee of 2% of the collected rents arranged by the broker.

Management fees such as broker fees are paid to the management firm based on the contract. The management fee may be a percentage of the value of the asset managed or a more detailed breakdown that accounts for the time and services provided. Since the sample building is relatively small, we will assume that the property is managed by part-time staff for a total cost of 0.5% of the revenues collected. This is based on the general rule that revenues and expenditures are approximately equal (obviously, somewhat more revenue is better!) and that work is necessary to both collect the revenue and provide for the services regarding expenses.

Property taxes are paid on the assessed value of the property. Currently, commercial property tax rates are in the range of $10 to $50 per $1,000 of assessed value each year. This property has an assumed tax value of $1,120,000. At an average rate of $30 per $1,000, this would mean a tax rate of 3%, or $33,600 per year. This amount will be indexed each year.

Operating expenses for scheduled upkeep items, such as janitorial services and the like listed in Figure 1.3, are often budgeted on a square foot basis. Generally, experience with a particular building type and historical costs are used as a guide. *Means Facilities Maintenance & Repair Cost Data* is also useful in establishing the annual expected operating costs. Benchmarking studies conducted by facilities organizations, such as IFMA, BOMA, and AFE (formerly AIPE), provide additional data for comparison. (See chapter 4,

"Benchmarking Facility Maintenance," and the Bibliography/ Recommended Resources.) We will assume an initial cost of $4.50/S.F., indexed annually at 5%. For the building area of 22,644 square feet, this means an initial annual expenditure of $101,898.

Ordinary repairs are the unscheduled costs of upkeep. Like capital expenditures, they occur randomly, but with an overall predictability. Because this building has planned for regular capital improvements and a strong maintenance budget, we will assign a relatively lower cost for repairs at $0.50/S.F. Thus the repair budget for the 22,644 square foot building will be $11,322, indexed annually at 5%.

We are now ready to fill in the proforma chart for the project (Figure 1.8).

Summary

The proforma (Figure 1.8) illustrates many factors useful to remember, and important to consider. The first year is always a net loss with considerable out-of-pocket expenses. Maintenance of the income stream (rents or annual appropriations) is critical to economic survival. For the example shown, approximately one-fourth of the gross revenue is dedicated to debt service, and one-fourth to operating costs. These are likely to be higher in many buildings.

After making allowances for predictable expenses there is only a small margin for error before poor weather, sloppy management, a drop in market revenue, or just bad luck can tip the proforma into a stream of net losses. Each year the annual budget needs to be examined critically and each line item in the operations and maintenance budget scrutinized for maximum effectiveness. The remaining chapters in this book review in detail how to make decisions on what items are neccessary, how to phase them, how much they will cost, what the impacts of alternatives are, and how to strategize property management issues for maximum effectiveness.

Familiarity with the principles outlined in this book should provide a reasonable level of confidence and basic knowledge about how to apply this information to successfully manage a facilities maintenance and repair program.

Outline – Building Proforma

Annual Revenue	Tenant	96	97	98	99	0	1	2	3	4	5	6
Rent	A	0	100,000	100,000	100,000	100,000	100,000	127,600	127,600	127,600	127,600	127,600
	B	0	202,880	202,880	202,880	202,880	202,880	258,864	258,864	258,864	258,864	258,864
	C	0	90,000	90,000	90,000	90,000	90,000	114,840	114,840	114,840	118,840	118,840
	D	0	45,000	45,000	45,000	45,000	45,000	57,420	57,420	57,420	57,420	57,420
	Parking	0	65,600	68,880	72,324	75,940	79,737	83,724	87,910	92,306	96,921	101,767
Revenue Total		0	503,480	506,760	510,204	513,820	517,617	642,448	646,634	651,030	655,645	660,491
Annual Expenses												
Acquisition	Down payment	127,500										
	Closing costs	4,750										
	Principal and interest	50,149	114,694	114,694	114,694	114,694	114,694	114,694	114,694	114,694	114,694	114,694
	Planning (soft costs)	16,250	16,250									
Capital Expenses	Equipment	0	11,200	12,320	13,552	14,907	16,398	18,038	19,842	21,826	24,009	26,410
	Tenant improvements		86,069	86,069	86,069	86,069	86,069	86,069	86,069	86,069	86,069	86,069
Building Operating Costs								80,444	80,444	80,444	80,444	80,444
	Leasing commissions	0	10,070	10,135	10,204	10,276	10,352	12,849	12,933	13,021	13,113	13,210
	Management fees	0	2,517	2,534	2,551	2,569	2,588	3,212	3,233	3,255	3,278	3,302
	Taxes		33,600	35,280	37,044	38,896	40,841	42,833	45,027	47,278	49,642	52,124
	Operating expenses		101,898	106,993	112,343	117,960	123,858	130,051	136,554	143,382	150,551	158,079
	Ordinary repairs		11,322	11,888	12,482	13,106	13,761	14,449	15,171	15,930	16,727	17,563
Expenses Total		198,649	387,620	379,913	388,969	398,477	408,561	502,689	513,967	525,899	538,527	551,895
Net Operating Expenses		-198,649	115,860	126,847	121,235	115,343	109,056	139,759	132,667	125,131	117,118	108,596

Rents are from the Projected Base Rent Income and Tenant Improvement charts.

Parking revenue is indexed by 5% each year. The above example shows a healthy profit. Real-life situations are not always so positive, and in any venture, be assured that both the buyer and seller will review different scenarios of the proforma in establishing the rents, interest rates, vacancy rates, index rates, and selling price. The debacle in the real estate market in the late 1980s showed us all how fragile some predictions can be.

Figure 1.8

Chapter 2

Capital Planning

Jack F. Probasco, Lisa H. Macklin, and David C. Marsh

Introduction

Capital planning is a key element to an organization's future. Outdated facilities can limit an organization in the delivery of their products and services. Few, if any, facilities are exempt from significant physical renovations, improvements, or new construction required to meet their functional expectations. Major renovations and new construction fall into the category of capital improvements, as do substantial repairs and equipment replacement, which would not be included in a normal maintenance and repair budget.

Because they are likely to be involved in one or all of these activities, it is important for facility managers and planners to be familiar with the way in which capital improvement projects are proposed, reviewed, and prioritized. Maintenance managers also need to have an understanding of the capital planning process in developing strategies for ongoing maintenance issues that may be best solved by a major renovation or improvement.

Planning for capital development in any organization can be a complex integration of space demands, financing, and the long range plans of the organization. An organization should prepare a capital plan that establishes the priority projects for the budget cycle. Input to the plan includes priority projects that were not funded in the last budget cycle and still are a priority for the organization, plus new projects that are physically and financially feasible. The number of worthy projects generally far exceeds available funds; hence funding priorities become an important part of this process.

The way in which an institution determines the priority, or ranking, of its capital project requests relates to the organization and its mission. Within institutions of higher education, capital project requests may be judged on such criteria as:

Academic Priority
- Relation to institution's mission
- Program excellence
- Research productivity
- Opportunities for cross-discipline collaboration

Physical Need
- Condition of existing space
- Unmet health and safety requirements

Financial Feasibility
- Cost sharing by requesting unit
- Net impact on operating costs
- Established funding principals

Physical Feasibility or Other Considerations
- Previous commitments
- Timing or staging issues
- Impact on other needs
- Linkages

While our experience is in planning higher education facilities, many of these concepts and procedures can be used by government and private organizations. For example, priorities for industry in ranking capital needs may include unit productivity, demand for product, and program excellence.

Project Genesis

The capital planning process begins with the perceived space needs of the units within the organization. The project is generally initiated by the unit assigned to the space or by the unit responsible for managing the space. A statement of need prepared by the requesting units should describe the type of space required, the justification for the space, and some concept of the capital and operating costs for the modifications or new construction. Two examples of capital need are provided below: one from a university and another from a hospital.

University Example

The following three potential capital needs requests may compete for capital funding:

The Department of Physics at the local university requests a building to replace their obsolete facility constructed nearly 50 years ago. It can no longer function as a research and teaching laboratory. The plumbing does not work properly, the building lacks central air conditioning, and the roof leaks. The condition of the building has a detrimental effect on the research that is being conducted in the building. The department believes that the building cannot function as a state of the art laboratory building even if it were renovated. A new building is estimated to cost $22 million.

A second request has come from the College of Business. The college has a donor who will provide $15 million for a new business building if the university will come up with matching funds to construct the facility. The College of Business has facilities that are only 25 years old. The facilities can be renovated for adaptive reuse should the university decide to fund half of the new building.

The College of Law has requested $10 million for an addition to its building to expand the library and add a student lounge and office space. The college's accreditation could be affected by the lack of proper library facilities. It claims to have only 60 percent of the stack space and reading areas required by the accrediting organization.

With these and many more requests for capital funds, how does a university develop a logical and fundable capital plan? What are the steps to be taken in analyzing the needs of each unit? What data are necessary for making decisions? Decisions made will certainly affect the quality of teaching and research at the university. The decisions may also have a negative impact on student enrollment, retention, and fees collected.

Hospital Example

A hospital has two major problems. One is that a new intensive care unit must be constructed. The current cubicles within the ICU have inadequate utility services and have only 75% of the required space. Carts and other equipment cannot be accommodated within each unit. In addition, the nurse's stations are not properly placed to serve the patients and the unit lacks space for private conversations between the family and the physicians.

Another equally important problem is that the Emergency Room must be replaced. The current facility lacks easy access from the highway. In addition to this access issue, the hospital has a security problem that could be improved by relocating the ER to the other side of the building. The condition of the current area has deteriorated with its constant use, and there are inadequate waiting areas for the friends and family of incoming patients.

Both capital projects will cost several million dollars. These major problems, along with the normal upkeep of the hospital facilities, are causing concern for the financial health of the hospital. A challenge facing planners is to determine how decisions will be made to keep the hospital competitive and responsive to the health care of its service area.

The process is basically the same whether the request comes from a university, a hospital, or a manufacturing plant. In today's competitive industrial setting there are always more requests than time and money can provide. There must be a system to determine the necessity of the project, the feasibility of the project, and the funding priority within the organization.

Establishment of Need

The requesting unit generally has the responsibility for establishing the need. This unit should provide a summary of project requirements describing the project in sufficient detail so that an analysis of need and a preliminary cost estimate can be determined. The size and complexities of a project will determine the level of detail. In general, the following information should be provided:

- Project description
- Summary of space requirements
- Relationship to unit's mission, customers, market, etc.
- Justification and verification of need
- Estimated costs
- Critical dates

The space requirements should identify the special needs, such as environmental conditions, utilities, etc., to assist in identifying the capital and operating costs for this project. In addition, the amount and type of space must be verified as being essential. Many capital requests are just wish lists or something that would be nice to have.

Once the need for the project is established and justified at the appropriate levels, the project moves to the management assessment stage.

Management's Assessment of Project Need (Concept Approval)

Based on a preliminary description, management must assess that the project is consistent with the unit's mission, goals, and objectives, and that the project is conceptually feasible. For major projects this "judgment call" is a critical milestone, since a decision to proceed will commit significant resources (either internal, such as staff time, or external, such as consultants) for further feasibility planning. The key question is, "Should the organization commit funds to determine if the project is financially and physically feasible within the present economic and political environments?"

Approval at this point usually presumes the release of funds to prepare a more detailed description of the project and the resources required for its implementation. Once management has determined that this project is needed and the concept is sound, a feasibility study is undertaken.

Project Feasibility

A project feasibility study begins with the materials developed as part of the unit's need assessment and management evaluation. The organization must determine the physical and financial feasibility of the project before proceeding.

The feasibility study must define the project scope (size), develop space plans, evaluate potential sites, develop cost estimates and funding sources, and establish a project schedule. The outcome of this study is a statement of project requirements.

The feasibility study should provide an in-depth analysis of the following topics.

Space Needs Calculation
- Are space planning guidelines available to assist in determining space requirements? (National standards have been used for years in higher education to determine space requirements.)
- How does the proposed space request compare to space available at peer institutions or competitors?
- Are the space requests realistic and justifiable?
- Are the data used in the space calculations (staff, clients, etc.) in line with management's expectations and funding capabilities?

Figure 2.1 shows sample space requirements calculations for a university feasibility study.

Project Alternatives
- Is there an alternate solution for solving this space problem?
- What is the quality of the existing facilities?
- Do the existing facilities need to be remodeled, or are new facilities required?
- Are there facilities available elsewhere for this intended use?
- Is the space currently available within the organization?
- If space is available, how much remodeling is necessary for the space to be usable?
- If leasing is an option, is the lease rate within the budget?
- Should new space be constructed? Is it in the budget, or should a temporary solution be implemented until funds are available?

Reassignment and/or Adaptive Reuse of Existing Facilities
- Will the existing space be vacated, and can it be used by other units?
- Can this space request be satisfied through the use of existing space?

SPACE REQUIREMENTS CALCULATION

CHEMISTRY

Space Type		Calculated ASF	Current ASF
Office	Exec Admin	220	175
	A/P	1,400	775
	Clerical	240	335
	Clerical Student	105	n/a
	Faculty	2,380	1,675
	Grad Student	1,235	1,097
	Visiting Emeritus	200	n/a
		867	569
	Office Total	6,647	4,626
Research, Teaching & Discretionary	General Funds Research	9,900	9,942
	Sponsored Research	19,050	17,439
	Research Total	28,950	27,381
	Scheduled Teach Labs	8,200	7,156
	Discretionary	880	561
Subtotal		44,677	39,724
Special Use	General Storage	200	0
	Unscheduled Labs	4,532	4,532
Grand Total		49,409	44,256

Current Space by Building

Miller Chemistry Bldg.	44,256 ASF

Current Staff

Personnel Category	Current FTE	Current Office Avg. (ASF)	Projected FTE
Exec Admin	1.00	175	1.00
A/P	8.00	97	10.00
Clerical	2.00	168	2.00
Clerical Students	1.52	n/a	1.75
Faculty	15.50	53	17.00
Grad Students	17.50	63	19.00

Summary of Current ASF

Exec Admin	Admin Office	175
A/P	Admin Office	775
Clerical	Clerical Office	335
Faculty	Faculty Office	1,675
Grad Students	Grad Office	1,097
Service	Office Service	569
Discretionary	Conference	561
GF Research	Faculty Office	783
GF Research	Office Lab	94
GF Research	Res Lab Serv	2,093
GF Research	Research Lab	6,324
GF Research	Staff Office	648
Sp Research	Faculty Office	775
Sp Research	Genl Office	341
Sp Research	Res Lab Serv	3,815
Sp Research	Research Lab	11,890
Sp Research	Special	618
Teach Lab	Scheduled Lab	7,156
Unsch Lab	Special	2,378
Unsch Lab	Teach Lab Serv	2,154
Total		44,256

Figure 2.1

Facility Condition

- Can the facilities be renovated to meet the needs of the requesting unit?
- If the project is a renovation or replacement, what is the condition of the existing facilities?

Site Issues

- If the project is a new facility, has a site been identified? Is the site clear for construction?
- What are the requirements of the users that may influence the site location?
- What affinities to other units are necessary?
- What are the physical requirements, such as access to roads, location to necessary utilities, etc.?
- What site criteria are necessary to meet the unit's need for program affinity or customer access?
- Is the site compatible with the long-range plan of the organization?

Special Facility Needs

- Are there special structural needs, such as high bay areas and floor loading?
- What are the utility requirements within the facility (HVAC, electric ventilation, plumbing, etc.)?

Space Quality for Requested Facilities

- Is there a need for high-tech facilities requiring expensive laboratory space?
- Are there unique architectural features that will be required?
- Does the location dictate that the building exterior fit into the surrounding aesthetics?

Project Cost Estimates

- What is the cost per square foot?
- Can a comparison be made to similar types of facilities?

Impact on Operating Budgets

- If additional space is provided, what effects will this have on the unit's operating costs? Who pays for the additional costs?
- Will the project reduce the operating costs? For example, a renovation project may reduce the maintenance costs.

Linkages to Other Projects

- What effect will this project have on other projects?
- Is there a domino effect where another building will need to be renovated or demolished?

Project Schedule and Schedule Constraints

- What is the timing for funding?
- What effect will funding delays have on this project?

Possible Funding Sources

- Are there potential donors to the project?
- Will it be necessary to bond the design and construction?
- Will revenue generation be used to fund the facility?

Figure 2.2 shows the functional relationships in capital planning.

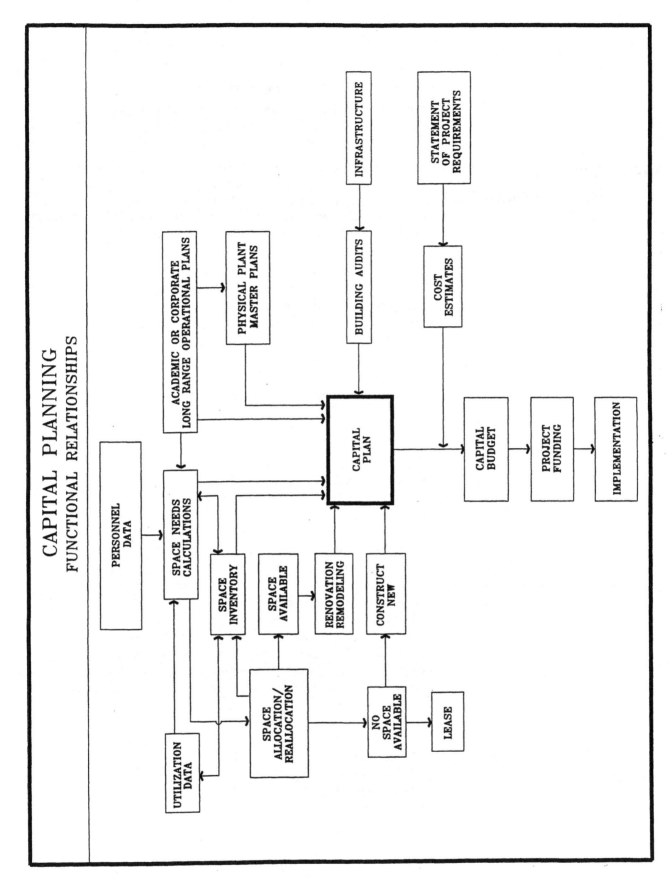

CAPITAL PLANNING
FUNCTIONAL RELATIONSHIPS

Figure 2.2

The result of this study will produce three possible outcomes:

1. The project is physically and financially feasible within the capital planning period.
2. The project is not physically feasible for the upcoming budget cycle, but is feasible for a later date (two to six years).
3. The project is not physically or financially feasible.

The project feasibility study is completed with a statement of project requirements.

Statement of Project Requirements

A Statement of Project Requirements (SPR) is developed as part of the project feasibility study and, if the project is approved and funded, it becomes the basis for the Program of Requirements. The SPR defines the scope of work, determines space needs, establishes the preliminary project budget, and investigates siting and other relevant planning issues to determine that the project is both physically and financially feasible.

Generally, the SPR should include the following:

- Project overview — Brief description of the project
- Characteristics of the program
- Space analysis — Determination of space needs for the unit, a space plan, and space to be released by the unit
- Description of facility needs
- Site issues:
  - Required affinities
  - Desired affinities
  - Description of possible sites
  - Recommended site
- For all sites identified provide:
  - Locations of utilities available
  - Topographic information
  - Parking
  - Pedestrian ways
  - General utilities requirements
  - Development objectives, including access and circulation, open space, etc.
- General and special needs of the project:
  - General or special environmental conditions
  - Specialized HVAC and lighting and power needs
  - End-use utilities (steam, water, power, communications, etc.)
- Quality — A statement describing the desired quality level of the new or remodeled space with respect to aesthetics, materials, and systems
- Project schedule
- Financial needs:
  - Project costs
  - Design costs
  - Relocation costs
  - Operating costs
- Source of funds

Project Funding and Prioritization

Once the project has been defined by the SPR, a determination must be made that identifies the source of funds within the organization's priorities and within the funding structure. Criteria for determining the funding priority needs to be established. For example:

- Health and safety issues
- Making the units competitive within the marketplace
- Obsolescence of existing facilities that cannot meet the expectations and mission of the users
- Impact on operating funds
- Previous commitment

While funding capital construction is never an easy task, there are several approaches available:

- Bonding (most common method)
- Current fund balances (pay-as-you-go)
- Reserve (see Chapter 20)
- Lease purchase

In the case of institutions of higher education, there are other sources such as contributions (especially those with associated naming arrangements as well as local, state and federal government funds).

Timing

In most organizations there are always a number of capital requirements. Many projects are either in competition for the same funds or dependent on another project to be completed prior to beginning the one under consideration. Two key factors to consider are:

- The domino effect, whereby the construction of one project will vacate space for use by another unit.
- Financing must be available when this project is needed.

In cases where the construction of one project will vacate space for use by another unit, problems can arise unless the project schedule and funding are carefully coordinated. It is recommended that a project schedule (such as the one shown in Figure 2.3) be developed early in the planning process. Scheduling conflicts and funding problems often can be detected by means of a Gantt chart.

Capital Plan

Once all the information is collected, the organization must determine if the project has high organizational commitment, high physical need, and high financial feasibility.

The project must be integrated into the organization's total capital plan. It is critical that priorities for funding and implementation be established, and that the total annual or biennial funding expectations be determined. Once the plan is in place and funding is available, the work of implementing the capital plan begins.

Conclusion

In summary, capital planning can be an intense process of information gathering, reviewing, analyzing, planning, and budgeting. Proposals that units submit to an organization are reviewed for feasibility, transformed into projects, and given priorities—or rejected. In this process, feasibility reviews ensure that appropriate values and agreed upon criteria establish the direction for long-range planning and build on the Mission and Vision Statements of the organization.

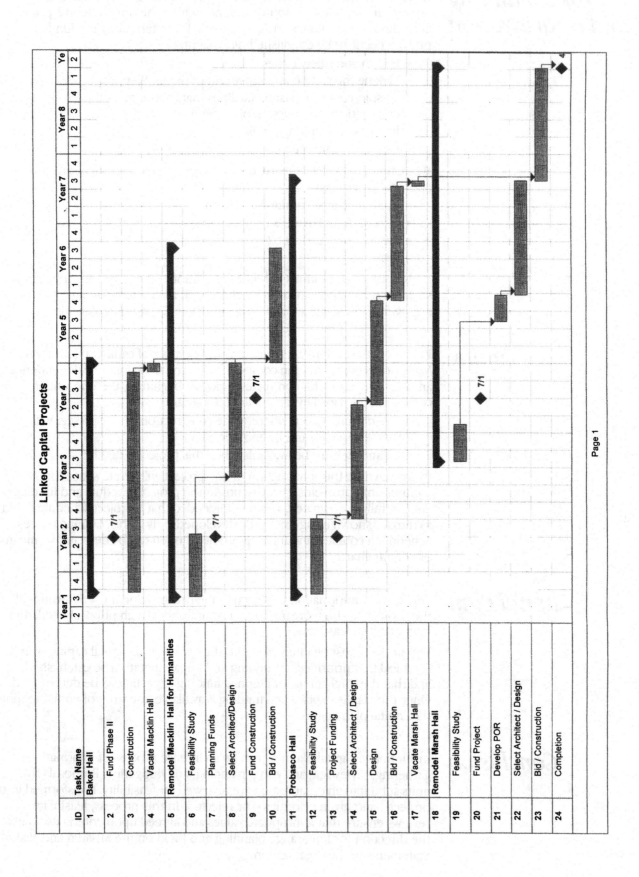

Figure 2.3

Chapter 3

Maintenance Outsourcing

Mark Peacock and Christopher Park, AIA Assoc.

Today's competitive environment is creating a need for change in every industry. As a result, companies are structuring their organizations by identifying the activities that bring them competitive advantage in the marketplace. Once these core activities are identified, companies can focus on ways to reduce the costs of these activities and to increase organizational flexibility (the ability to quickly respond to changing economic conditions). Non-core activities are subject to intense scrutiny, often viewed as distractions or drains on scarce capital. While these activities are necessary, they do not improve the company's competitive position. Today's top performing organizations have demanded that non-core service providers maximize value to the company, optimize capabilities, and increase service levels. Facility management is often viewed as a singular non-core activity, yet facility managers recognize that organizational success can be critically dependent on activities provided by the facility management group.

Outsourcing is often suggested as a way to quickly address the demands of facility management. Results, however, have been mixed. Some companies identify too few or even the wrong activities to outsource. An organization may identify an activity for outsourcing that is already performed cost-effectively. Others may experience a cultural backlash from outsourcing or choose an inadequate outsourcing partner. Even more harmful is offering core activities (those that provide organizations a competitive advantage) to outside vendors because of the perceived cost savings. Understanding the potential benefits and corresponding risks related to outsourcing all or part of the facility management function is one key to an organization's success.

A Structured Approach to Outsourcing Evaluation

It is important to use a structured approach when evaluating the potential outsourcing of some or all of the facility management and maintenance functions. An analytical approach removes emotion from the decision and provides objective data for appropriate decision-making. The evaluation process should consist of clearly defined activities that provide a step-by-step methodology for completing the analysis. Figure 3.1 outlines six distinct steps, and includes Change Management and Information Technology threads that run throughout the approach. The critical decision point is illustrated in the fifth step—the decision to contract with a vendor or to reengineer internal processes to meet or exceed the external benchmarks identified through the Request for Proposals (RFP) process. The steps outlined in Figure 3.1 are described in detail in the following sections.

Evaluating Outsource Candidates

The first question often asked in an outsourcing evaluation is, "What tasks should be performed internally and what tasks should be outsourced?" The three primary drivers for retaining services within the organization are illustrated in Figure 3.2 and described below.

- **Cost**

    A rational organization will not pay an outsource firm more than it is already spending for comparable service. However, companies often fail to account for all of their expenses, including

Structured Approach to Outsourcing Evaluation

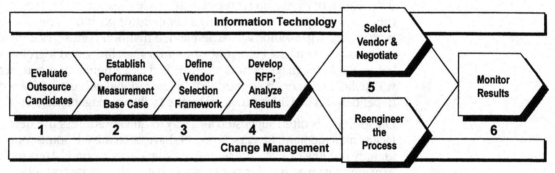

Figure 3.1

Three Service Retention Drivers

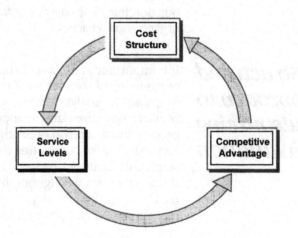

Figure 3.2

management, administrative, and opportunity costs. If the analysis is incomplete, activities that are candidates for outsourcing may be overlooked.

- **Service and Performance Levels**
  Some activities may be too complex for outsourcing. For example, research and development facilities often have unique maintenance and security requirements for clean labs and other specialized spaces within a facility. There may be too few vendors available to provide the required combination of service and performance. Many organizations cannot find a vendor because of overspecification. Internal providers for certain complex facilities, such as labs or medical facilities, define overly complex services and performance levels, often increasing the scope and cost of outsourcing proposals unnecessarily. Key operations staff should be involved in the specification of service levels, especially for technically complex facilities. An objective, realistic view of service and performance levels can help to avoid this complication.

- **Competitive Advantage**
  In general, core activities that provide an organization with a competitive advantage should not be outsourced. For example, some organizations utilize automated facility planning methods to respond to changes in the market by moving the workforce quickly and efficiently. This flexibility and responsiveness creates a competitive advantage. However, if an outsourcing market exists for a specific activity, the organization should analyze the *sustainability* of the current advantage. The number of well-established facility management service providers is an indicator that facility management activities are often not core competencies and do not provide a sustainable competitive advantage. In the facility planning example noted above, the advantage may not be sustainable because more and more outsource firms are providing automated planning tools.

The evaluation of outsource candidates consists of five discrete steps, illustrated in Figure 3.3:
- Identify key business processes (e.g., strategic facility planning, security, moves and changes).
- Define critical dimensions for each process (e.g., specialized knowledge of business, required response time, required service level, unique technology).
- Apply a weighting factor to each critical dimension.
- Score each critical dimension to measure core competency support.
- Force-rank all business processes by weighted score to generate a priority list.

The business processes that score high in the evaluation process should be retained internally; those that score lower are candidates for outsourcing.

Establishing the Performance Measurement Base Case
The next step of the approach is to define a performance measurement base case, gather the performance data, and analyze the results. Some performance measurements may already exist — response times, quality satisfaction ratings, etc. — but it is important to reclassify these measures for easy comparison with benchmark data. The base case analysis should focus on three areas — organization structure, work activities, and costs.

Documentation of the organization's structure is readily available or easily produced. The description should include not only names, job titles, and reporting relationships, but also length of service, salary grade and rate, and other key descriptors that may be subject to comparison during the RFP analysis.

Next, the business process should be broken down into discrete work activities and key tasks within each activity. An example of the appropriate level of detail appears in Figure 3.4. Often, this data does not exist in usable form or is difficult to gather and analyze in the form required. Independent analysis of day-to-day activities using accounting, payroll, or other data can usually simplify and speed up the data gathering and analysis process.

Once the organizational structure, activities, and key tasks are identified, the cost of each activity must be analyzed to establish the base case. Because the largest single component of facility management and maintenance activities is salary and benefits, the labor cost for each activity and task needs to be calculated. This requires a restatement of typical budget categories into costs directly associated with a particular activity, known as *activity-based costing.* A comparison of a typical budget and an activity-based cost structure is provided in Figure 3.5. Since vendor RFPs will be provided with pricing categorized by activity, it is important to restate internal costs in the same format. While labor costs are the largest single component, other costs include materials (including inventory and carrying costs), information systems, training and education, and others.

Five Steps to Evaluate Outsource Candidates

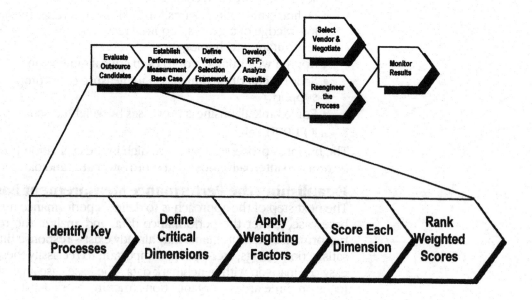

Figure 3.3

Work Activity and Task Framework

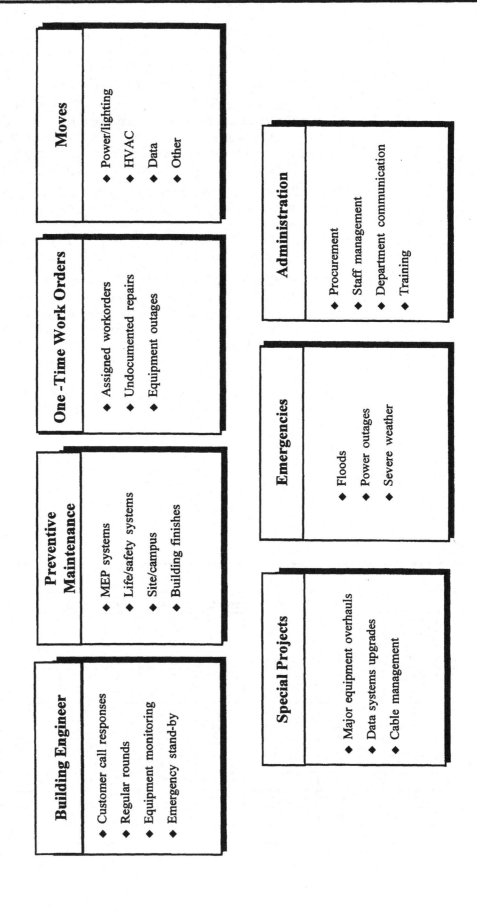

Building Engineer
- Customer call responses
- Regular rounds
- Equipment monitoring
- Emergency stand-by

Preventive Maintenance
- MEP systems
- Life/safety systems
- Site/campus
- Building finishes

One-Time Work Orders
- Assigned workorders
- Undocumented repairs
- Equipment outages

Moves
- Power/lighting
- HVAC
- Data
- Other

Special Projects
- Major equipment overhauls
- Data systems upgrades
- Cable management

Emergencies
- Floods
- Power outages
- Severe weather

Administration
- Procurement
- Staff management
- Department communication
- Training

Figure 3.4

While establishing the base cost structure is an important part of the analysis, it does not describe the whole picture. Other performance measures need to be identified to support the conclusions. Larger organizations can benefit from an *internal benchmarking* process where similar measures are tallied across different facilities or regions. Examples of other performance measures are listed in Figure 3.6.

One useful performance measure is *staffing per square foot,* which varies by trade, facility type, and organization. For example, the International Facility Management Association (IFMA) *Benchmarks II, Research Report #13,* provides detailed data on maintenance staffing. Figure 3.7 provides an example of ranges of staffing per square foot by key trade.

After the cost structure and performance measures are produced, it is important to validate and modify them based on discussions with internal customers (i.e., building occupants). Mismatches between service levels and customer expectations are a good place to begin modifying business processes and are the first area to review responses from vendor RFPs.

Defining the Vendor Selection Framework

The most important step in outsourcing is selecting the right vendor. The vendor's capabilities, experience, and strategic directions need to be matched to the organization's requirements if the relationship is to meet expectations. The first step in achieving the right match is to extensively survey the market for potential vendors in target or closely-allied businesses. The list of vendors should be exhaustive, based on industry data, trade journals, previous experience, and discussions with colleagues inside and outside of the organization. Potential vendors must be screened

								Budget vs. Activity-Based Costing Comparison	
Engineering	**One-Time Work Orders**	**Emergency**	**Landscaping**	**Moves**	**Planning**	**Administrative**	**Total Amount**	**Budget Category**
$228,000	$28,500	$28,500	$57,000	$ 85,500	$114,000	$28,500	$ 570,000	Wages
32,600	4,075	4,075	8,150	12,225	16,300	4,075	81,500	Employment Taxes
98,400	12,300	12,300	24,600	36,900	49,200	12,300	246,000	Benefits
8,800	1,100	1,100	2,200	3,300	4,400	1,100	22,000	401K
					2,000	6,000	8,000	Travel
18,000	2,000				4,000	4,000	28,000	Supplies
					2,500	16,000	18,500	Computer Equipment
5,500	5,500	1,000	1,000				13,000	Uniforms
2,000	4,000	2,000					8,000	Small Tools
						5,000	5,000	Telecommunications
$393,300	$57,475	$48,975	$92,950	$137,925	$192,400	$76,975	**$1,000,000**	**Activity Totals**

Figure 3.5

Example of Performance Measures

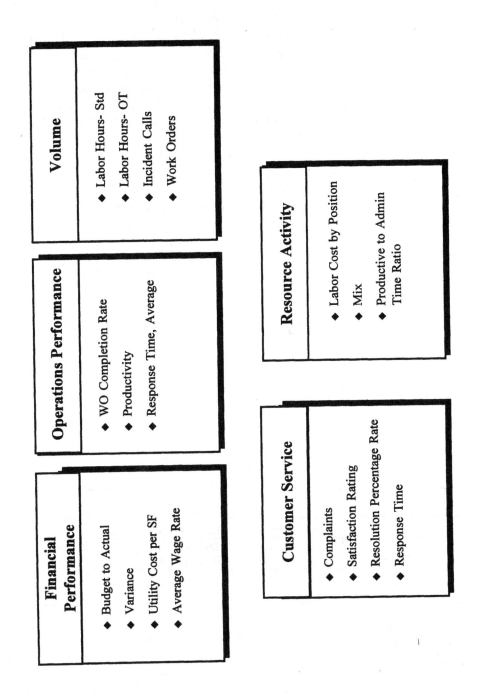

Figure 3.6

through successively finer levels of scrutiny, and may come from a variety of core disciplines, including real estate, engineering, construction, design, and management firms.

The results of the first two steps in the methodology — Evaluating Candidates for Outsourcing and Establishing the Performance Measurement Base Case — are combined with vendor data to yield the selection framework. Figure 3.8 illustrates one example of a summary analysis that defines the issues influencing vendor selection.

The diagram in Figure 3.8 illustrates two critical dimensions of vendor selection — Availability of Outside Vendors and Degree of Core Relevance — and the relative locations of all processes under consideration in relation to the axes. The availability of outside vendors can be measured by the market research required to develop the list of potential vendors, while the degree of core relevance is measured by the force-rank scoring process completed during evaluation.

The diagram categorizes business processes by the two dimensions, yielding a strategy for pursuing opportunities for improvement:

- Processes that are highly relevant to the core business and have limited outside expertise available should continue to be provided internally, although they should be analyzed to verify adequate performance.
- Processes that are highly relevant but have extensive outside expertise available should be controlled (led) internally but could be leveraged with outside resources through less expensive or more efficient labor structures.
- Processes that are low in their measure of core relevance and have limited outside expertise should be candidates for partner

Staffing by Trade for a Range of Facility Sizes					
Trade	100,000 S.F.	100,001 to 200,000 S.F.	200,001 to 500,000 S.F.	500,001 to 1,000,000 S.F.	1,000,000 S.F.
Overall	5.6	6.0	21.9	34.7	91.9
Carpenters	1.2	1.7	2.1	4.6	6.7
Electricians	1.2	1.5	1.9	3.8	12.5
Painters	1.1	1.6	1.6	2.0	3.6
Mechanics	2.5	3.2	3.5	7.8	22.4
Helpers	1.3	1.7	2.7	3.4	5.0
Temp or Contract	1.4	3.5	6.7	14.5	15.5
Janitors	6.2	8.8	13.2	31.7	53.1
Movers	1.5	3.2	4.4	5.0	6.1
Locksmiths	1.0	1.3	1.6	1.7	2.0

Source: *Benchmarks II, Research Report #13*, International Facility Management Association

Figure 3.7

arrangements of some type, since switching vendors is limited by market availability. Partner arrangements are often characterized by greater risk- and reward-sharing, longer agreements, and open disclosure of cost structures and other business issues.

- Processes that are low in relevance but have extensive service providers available can be straight contract arrangements, since the services are commodities and price competition is keen.

The diagram in Figure 3.8 also illustrates two other factors to consider:

- The relative size of the circles is indicative of the cost structure for each process. By presenting data in this way, management can quickly ascertain which processes are the most costly, and hence the most likely to improve profitability by a reduction in cost.
- The arrows indicate the direction of the business process over time with regard to its relevancy and the availability of outside expertise. Contracts should be evaluated over time as business processes move along in these two dimensions.

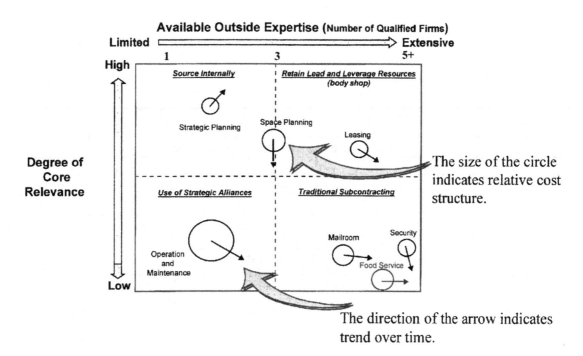

Summary Vendor Selection Matrix

Figure 3.8

Finally, the development of the selection framework includes defining and weighting vendor selection criteria. Criteria should be defined for key categories—vendor fundamentals, cost, performance measurement targets, and so on.

Developing the RFP and Analyzing the Responses

There are a number of issues to consider when developing a Request for Proposal (RFP) for outsourcing services. The first is whether to go to the market with an *open* or a *blind* RFP. Open RFPs are characterized by disclosure of the client name, location, and other key information to potential vendors, while blind RFPs use an independent third party to keep the client identity confidential. Risks and benefits of each approach are summarized in Figure 3.9. Since the decision to outsource or internally improve has not yet been made, a blind RFP can be an important data-gathering tool for benchmarking purposes. Once the blind vs. open decision has been made, the RFP can be structured using standard formats. A typical RFP includes the following sections:

Summary and Introduction. The summary and introduction present a brief overview of the organization's objectives and the services requested. By clearly stating needs and expectations upfront, the organization allows the vendor to make an informed decision about a response without studying the entire document. Other logistical information should be included such as appropriate locations, description of the facility, sizes, key dates, equipment, and other issues. This section should also include a review and decision timeline, details about walk-throughs or facility tours, and a discussion of selection criteria and the selection process.

Scope of Work. The scope of work should clearly and completely describe all of the work involved. It should describe activities and tasks that are candidates for outsourcing and performance levels required for each, as well as related activities that are outside of

Open vs. Blind RFP: Risks and Benefits		
	Risks	**Benefits**
Open RFP	• Marketplace is aware of potential contract. • Staff awareness creates uncertainty and concern.	• Vendors have opportunity to gather extensive information, making responses more complete. • Marketplace awareness may increase competition.
Blind RFP	• Lack of information gathering makes responses less complete. • Vendors have negotiating room because of uncertainty.	• Staff awareness of potential outsource is minimized and controlled. • Objective third party is involved to make independent assessment of responses.

Figure 3.9

the scope of work. This section should use as much quantitative data as possible without providing vendors with information they can use during contract negotiation. Cost and head-count data should be omitted. Lengthy discussions on how processes are completed should be avoided; you are asking the vendors to tell you how they will provide the services, not to provide them the same way you do.

Response and Proposal Formats. To facilitate review and comparison of various responses, the vendor should be provided with a definition of the format required for responses, including specific forms or examples for the candidate to complete. Pricing plans and staffing sheets are two examples; others include staff experience and background forms, and customer reference forms.

Reviewing RFPs: The Variance Analysis

After responses have been received, the vendor proposals can be compared to one another and measured against the performance base case defined earlier. An important component of reviewing the responses includes *variance analysis*, which requires a detailed review of each vendor proposal. The variance analysis calculates the independent effects of each cost component in the overall cost structure. While types of variance analyses will vary with processes being reviewed, typical analyses include:

Loaded Rate Variance: the difference between the organization's benefit-loaded payroll costs and those of respondent vendors. (This assumes that the facility manager has access to key compensation and benefits cost data, a necessity for the variance analysis.)

Base Rate Variance: the difference between the organization's base payroll rates and those of respondent vendors.

Performance Variance: the difference between the number of hours required for internal staff to complete an activity vs. performance targets set by vendors.

Personnel Mix Variance: differences between internal seniority mix, trades and skillsets and those of the vendor.

Head-count Variance: differences between a vendor's proposed staffing plan and the current model, by position level.

An example of a variance analysis completed for Rate Variance is provided in Figure 3.10. The variance analysis provides critical insight into what is driving a variance in cost between internal activities and outsourced activities. The information is key in the decision process to outsource or retain facility services.

Vendor responses to RFPs should be scored against the selection criteria established earlier. In addition, qualitative issues should be taken into account — timeliness of responses, interest level, similar qualifications, team "chemistry," and others. These factors play an important role in the eventual working relationship that may develop. Figure 3.11 illustrates one tool that is helpful in summarizing comparisons of vendor responses. The matrix defines selection criteria along one axis and vendor candidates along the other and provides a quick visual representation of analysis results.

Key Decision Point: Outsource or Reengineer?

Once vendor responses have been compiled, analyzed, and summarized, the organization can make an informed decision about creating internal change to meet external performance standards. The RFP process provides key data to structure an internal change or reengineering effort and provides the first critical component – *the change imperative*. Often, reengineering efforts fail because organizations are not completely committed to productive change. The RFP analysis provides external validation of the need for change, and provides targets for performance and goal-setting. Combined with external benchmarking analyses, the external RFP data can be used to communicate the need for change to management and staff.

The decision to reengineer or outsource carries inherent risks and benefits, some of which are illustrated in Figure 3.12. Fortunately, neither answer is permanent. If internal change is selected as a means to improve performance, goals can be measured on a regular basis. If expectations are not met, the outsourcing alternative can still be pursued. It is more difficult, however, to outsource functions initially and then recover them later – the rebuilding process can be time-consuming, costly, and result in decreased service levels. Hence firms often begin their efforts by moving toward internal change, but they keep the outsourcing alternative alive as a tool to promote continued improvement.

Selecting Outsourcing

In the event that outsourcing is selected as the means for change, the contract negotiation process and choice of contract types are important in ensuring that performance targets are identified, addressed, and met. There are some critical success factors that are important to consider when entering vendor negotiations:

Understand strengths and weaknesses. Vendors know more about the marketplace for outsource services than facility or maintenance managers do and have written many favorable outsource contracts. However, your

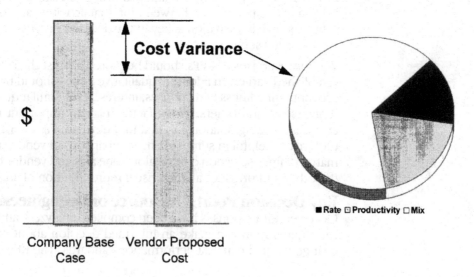

Figure 3.10

Example Vendor Comparison Matrix

Priority of Criteria — Low ➝ High

	General Business & Financial Strength	Services Approach and Flexibility	Technology Platform & Direction	Location	Proposal Completeness & Effectiveness	Price
Vendor 1	Neutral	Neutral	Neutral	Neutral	Neutral	Good
Vendor 2	Good	Neutral	Neutral	Neutral	Neutral	Neutral
Vendor 3	Poor	Neutral	Neutral	Good	Poor	Good
Vendor 4	Neutral	Neutral	Neutral	Neutral	Neutral	Poor
Vendor 5	Neutral	Neutral	Neutral	Neutral	Neutral	Poor
Vendor 6	Good	Neutral	Neutral	Neutral	Poor	Poor
Vendor 7	Good	Neutral	Good	Neutral	Good	Neutral
Vendor 8	Poor	Neutral	Poor	Good	Neutral	Poor
Vendor 9	Neutral	Neutral	Neutral	Neutral	Good	Good

Legend: ● Good ◐ Neutral ○ Poor

Figure 3.11

Outsourcing vs. Reengineering Risks and Benefits		
	Risks	**Benefits**
Outsourcing	• Loss of control. • Cost escalation over time. • Performance degradation over time. • Inability to reinternalize key processes.	• Initial cost reductions. • Leverage of high-cost staff. • Some transfer of risk to vendor.
Reengineering	• Requires significant management involvement and commitment. • May require investment in training, systems, etc. • Change imperative must be clear and convincing.	• Control is retained while service is improved. • Cost reductions available, short-term (quick hits) and long-term (strategic change).

Figure 3.12

understanding of the organization, its goals and mission, and working environment provide the balancing advantage. Use the experience of other parts of the organization that may have used outsourcing previously. In particular, the information systems organization may have learned a great deal if mainframe systems were outsourced within the last decade.

Play the "internal change" card. Let the potential vendor know that internal change is a viable option, which will be pursued if satisfactory terms are not negotiated. Be open and honest with vendors, because they are making a substantial investment in the response and negotiating process. Negotiate in good faith.

Tie compensation to measurable performance targets. Ensure that the contract clearly defines specific quantitative and qualitative goals and that the vendor's compensation is tied to both. Vendors should be willing to go at risk for at least some of the compensation package, but may expect that exceeding goals improves the picture for them. The contract should specify the means by which performance will be measured and results interpreted.

Selecting Reengineering

In the event that internal change is used to improve performance, many of the same critical success factors apply. Management and staff should know that outsourcing remains a viable alternative, although this approach requires caution. Compensation structures should be revised to reflect an expectation of exceeding performance targets.

Reengineering as a method for change is well-documented and references are readily available. The change methodology should be developed to match the requirement for change (dramatic or incremental) and the appetite for change (change-averse or ready and willing to change).

Monitoring the Results

One of the most important considerations in outsourcing or internal change is the development, implementation, and ongoing use of performance measurement and monitoring methods and systems. Recent surveys have illustrated that only about 50% of facility managers have formal performance monitoring methods and systems in place. Those that are in existence focus on typical measures (budget, actual against budget, etc.). To succeed and provide valuable feedback to the change process, specific performance metrics should be measured on an ongoing basis. The base case model defined earlier presents an excellent basis for ongoing measurement and, in fact, development of the base case acts as the initial definition for performance monitoring systems. Chapter 5 includes discussion of technical systems that provide maintenance management performance reporting.

Change Management

In either the outsource or the reengineering case, *change management* is an important aspect of eventual success, and is often misunderstood. Figure 3.13 provides examples of "change levers" — methods and tools by which positive change can be communicated and shared with the organization. Change is inherently difficult and creates uncertainty; it is up to the team implementing the change to moderate the uncertainty. While change management activities cannot eliminate the *fear* associated with change, they help reduce the associated *uncertainty*.

One of the most important elements of the change management program, however it is defined, is the approach for assisting displaced employees. While many organizations have specific human resources policies in force to provide support, those organizations that do not should consider providing some benefits if a reduction in staff is necessary:

Outplacement Assistance. Assist displaced employees in finding alternative work arrangements by providing space in the facility, resumé writing and production support, telephone support, and outplacement counseling. Many firms offer packages of these services.

Retraining/Reeducation. Provide the opportunity for displaced employees to learn new skills or trades in their efforts to find alternative work. Computer and technical skills are becoming increasingly important.

Internal Placement Support. If the reduction is part of a larger change initiative at the organization, it may be possible to relocate displaced staff in other facilities or sites.

Vendor Direct Hire. Displaced employees are hired by outsource vendors as a condition of the contract for a specified period of time.

Change Management 'Levers'

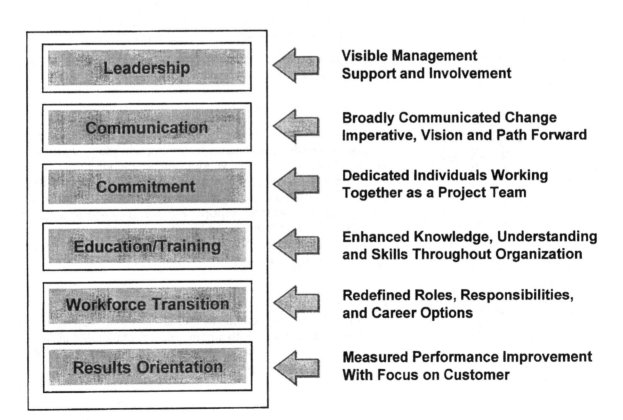

Leadership	Visible Management Support and Involvement
Communication	Broadly Communicated Change Imperative, Vision and Path Forward
Commitment	Dedicated Individuals Working Together as a Project Team
Education/Training	Enhanced Knowledge, Understanding and Skills Throughout Organization
Workforce Transition	Redefined Roles, Responsibilities, and Career Options
Results Orientation	Measured Performance Improvement With Focus on Customer

Figure 3.13

Another key element of change management is changing the focus of the internal organization to customer service instead of a support function. Facility management and maintenance organizations increasingly recognize the benefit of treating other employees as internal customers, and including those internal customers in the change process works to everyone's benefit. Performance measures should include a substantial focus on qualitative and quantitative customer perceptions and satisfaction.

Management Involvement

One of the key threads running throughout the outsourcing evaluation methodology is the importance of management involvement. Very often, the initial direction for considering outsourcing as an alternative is generated by management-level performance or other concerns. The methodology presented here provides an opportunity to meet or exceed management expectations while still retaining the possibility of providing change internally rather than using the marketplace as a change tool. Senior management often considers outsourcing as an easy, quick cost-reduction opportunity, and, when presented with the alternative of meeting performance and cost targets using internal staff, are open to such suggestions. Some key points to consider when involving management in the process include:

Communicate completely, openly, and often. Leave it to management to decide how much communication is too much. Be frank about the process and share concerns and observations frequently.

Consider management's point of view. Understand what drove the initial request to consider outsourcing as an alternative, and try to assess the situation from their perspective.

Try to understand or identify unspoken drivers or goals. Often there are unspoken issues or "drivers" to consider, especially when outsourcing is suggested as an alternative.

Demand input into the process. Outsourcing is, above all, a cooperative effort and should be done *with* an organization, not *to* it.

Information Technology

The increased availability and lower costs of processing-intensive facility management and maintenance technology should be an integral part of the change process. Outsource vendors are price-competitive because, among other reasons, they have the use and benefit of new technologies. Examples of such uses include automated problem reporting and work assignment, performance reporting, space inventory and management, and cost chargeback monitoring and reporting. The change process should include a detailed technology component, since increasing automation helps reduce manual costs.

Conclusion

Outsourcing certain facility management and maintenance functions can provide tremendous benefits by increasing service levels, decreasing costs, or both. However, the outsourcing process, if handled in the manner described above, can also yield performance targets that enable an internal change process to succeed. The decision on whether to outsource certain functions or reengineer those functions should be based on extensive and objective data gathering and analysis, and a reasoned approach to the issues. Management involvement, change management, and information technology are critical components of the success of either change methodology.

Chapter 4

Benchmarking Facility Maintenance

Mary P. Greene

To create a maintenance program, estimates and budget, a substantial amount of information must first be collected—about the facility itself, and regarding the methods used to manage its maintenance. This same data can be fed into benchmarking studies—whether they are conducted internally, or for the purpose of comparing the organization's maintenance performance to that of others in similar facilities. Whether benchmarking is done prior to the planning and budgeting process using existing data, or afterward, using data gathered for that task, it is a valuable exercise for assessing and improving the performance of a facility's maintenance functions.

Benchmarking, as defined in *Webster's Ninth New Collegiate Dictionary,* is "a point of reference from which measurement may be made," or "something that serves as a standard by which others may be measured." The practice of benchmarking in facility management seeks and describes the industry's best examples of methods, products or processes in order to establish a standard of quality, productivity or operating costs. An organization assesses its own performance against that standard, with the goals of identifying its deficiencies and implementing plans to reach or exceed the defined level of superior performance.

Making the Decision to Benchmark

Benchmarking requires an organization's commitment to an improvement plan, and to the installation of specific procedures that will bring the organization to its desired level. Senior management must be willing to devote adequate resources to this task. Fortunately, the case for pursuing a benchmarking study is increasingly easy to make. More and more organizations see this process as an essential component in ensuring their competitive advantage. Many have found that the improvement of even one practice as a result of benchmarking has more than offset their investment in the study. Reports of these results in other organizations can be useful in presenting the need for benchmarking to management. If it can be shown that the information gathered in a benchmarking study will be clear and useful, will help improve services and implement positive change, and can be shared with others in the field, then the organization should be a good candidate for benchmarking.

The facility's staff must also be committed to the benchmarking process, in the sense that they must be willing to change procedures or the organizational culture or structure, where required, to make identified improvements. Changes need not occur overnight or on a grand scale; benchmarking studies may be performed one department at a time, with data collected, results analyzed, and improvement plans implemented before going on to the next department or function.

How Is Benchmarking Performed?

Benchmarking involves gathering and quantifying objective data about the facility. That data is a first step toward improved staff and facility performance and has value apart from its use in the benchmarking process. The collected data is then compared to similar information gathered from other organizations, or from an organization's own "member" facilities or campuses at other locations. A major advantage of benchmarking is the element of competition spawned by a performance comparison of various buildings or campuses within an organization. Each group seeks to outperform the others, and everyone's level is raised.

Benchmarking can also be done using an established standard in the form of published costs such as *Means Facilities Maintenance & Repair Cost Data*. This publication defines national average productivity standards against which a maintenance department can be measured. This approach challenges the maintenance staff to meet these standards or to document circumstances that prevent their attaining this productivity.

While the basic process will be similar for many organizations, there are, of course, issues specific to the type, age and location of each facility that must be addressed. Special conditions, such as 24-hour use of hospital facilities, may also affect the performance of maintenance and repair as well as other functions. Maintenance of manufacturing or retail facilities may be affected by their own unique performance requirements.

Facilities organizations such as the International Facility Managers Association (IFMA) and the Association of Facilities Engineers (formerly the American Institute of Plant Engineers) have performed benchmarking surveys and published their findings for members. (See the Bibliography/Recommended Resources at the back of this book.) IFMA's *Benchmarks II, Research Report #13* is the result of that organization's 1993 benchmarking program. This report contains benchmarks based on survey responses from more than 300 professional members in the United States and Canada. Some of its measurements include maintenance, janitorial, utility, and support, as well as churn rate and cost of moves, square footage per employee, and other indicators. IFMA has also published a workbook that explains the benchmarking process and takes readers through ten steps, from planning a benchmarking study, to analyzing collected data, to deciding on appropriate actions, and finally, steps to take for continuous improvement. IFMA's database of survey responses is available, along with a list of the organizations that contributed their data. IFMA's 1996 survey is expected to be published in 1997 in a subsequent benchmarking report. In addition, IFMA holds "Best Practices" forums, and publishes the proceedings and compilations of member information from a "Best Practices" questionnaire. These publications are available by contacting IFMA's research department. (See the Appendix, a directory of facility associations.)

APPA: The Association of Higher Education Facilities Officers also conducts benchmarking studies. This organization has collected data through their biennial "Comparative Costs and Staffing Survey" for over 15 years. APPA's new and evolving "Strategic Assessment Model", developed in partnership with American Management Systems, Inc., is designed to provide a graphic illustration of the range and depth of facilities activities, and a road map for improvement. The model is based on a core set of benchmarks and a process for equalizing them for adaptation to institutions of higher learning, regardless of size, funding source or mission. APPA also plans to provide best practice process information as they continue to refine their Strategic Assessment Model.

Figures 4.1 and 4.2 are excerpts from APPA's Strategic Assessment Model survey. Figure 4.1 requests actual amounts or values. These are used to develop the distribution matrix for determining appropriate factors or percentages for each level of four core components. Figure 4.2 reflects the current values for each item in each level. The values/percentages are moving targets and will change with the data being submitted by the participating institutions.

The National Association of College and Business Officers (NACUBO) launched a benchmarking project in 1991. It covers 25 core functional areas, including facilities management. The study is intended to provide data for internal budget development; operations enhancement, such as business process reeingineering; total quality management; cost reduction; or service improvement efforts. NACUBO plans to offer the benchmarking survey as an ongoing service, refining the benchmarks and increasing the number of participants in the database. More than 280 institutions have participated to date.

Any of the above organizations can be contacted for more information at the addresses and telephone numbers provided in the Appendix.

Another resource is *Maintenance Management Audit,* by Applied Management Engineering, PC and Harvey H. Kaiser. This workbook-format publication allows internal staff or consultants to review and analyze current maintenance management performance; then develop corrective actions, with procedures for monitoring the outcome of these corrective actions. Figures 4.3 and 4.4 are sample pages (a Guideline Checklist and an Effectiveness Rating Worksheet) from that publication.

Tools for Benchmarking

Listed below are some of the basic tools or methods for getting a benchmarking study under way:

- Gathering key participants to list areas where benchmarking can be most valuable.
- Listing areas where there are quality concerns, competitive weaknesses, and/or excessive costs.
- Using a questionnaire, developed in-house and designed in consideration of future information exchange partnerships. The questionnaire should profile the facility so it can be matched accurately for comparison with others. Processes also need to be addressed. Costs or "metrics" should be covered as well. Questions should be asked in such a way that the answers can be organized in a useful database. All potentially ambiguous terms, such as "overhead," should be clearly defined. (For example, overhead should include specified components.) Questionnaires can be distributed in hard copy form, or on an in-house computer network.
- Using benchmarking guides or workbooks from associations.
- Participating in association forums for discussion of facility management techniques in other organizations.

Strategic Assessment Model

APPA SURVEY
Please fill in the value for each separate benchmark indicator
for your college or university. Use actual data from FY 1995.

FINANCIAL PERSPECTIVE	AMOUNT or VALUE
Capital Replacement Value (CRV) $	$
Operations and Maintenance (from IPEDS) $	$
Renewal $	$
Plant Adaptation $	$
Deferred Maintenance $	$
Facility Operating Expenditures $	$
Gross Institutional Expenditure (GIE) $	$
Gross Square Feet	
Basic Labor Rate ($/hour)	$
Overhead Rate ($/hour)	$
Comments or Additional Financial Indicators:	

INTERNAL PROCESSES	VALUE OR %
Total Job-Related Injury	
Total Available Work Hours (Excluding vacation, sick, holiday, and other leave with pay)	
Total Energy Consumption (BTUs)	
Total GSF Served	
% Task/Standards	
Comments or Additional Process Benchmarks:	

INNOVATION & LEARNING	$ AMOUNT
Total Training Costs $	$
Total Personnel Compensation Costs (include fringe benefits)	$
Comments or Additional Innovation and Learning Benchmarks:	

CUSTOMER SATISFACTION	% or VALUE
% Services Measured	
% Stakeholder Measured	
Continuous Improvement (Circle one Value)	Yes: full use / Yes: nearly full use / Some: mixed results / No: A little use / No: No use for continuous improvement
Comments or Additional Customer Satisfaction Benchmarks:	

Name: _____

Institution: _____

PLEASE MAIL OR FAX THE RESULTS TO:
Steve Glazner
APPA
1446 Duke Street
Alexandria, VA 22314-3492   Fax: 703-549-2772

Figure 4.1

STRATEGIC ASSESSMENT MODEL – APPA BENCHMARKING SURVEY

Level	Strategy	Financial Perspective		Internal Processes		Innovation & Learning		Customer Satisfaction	
5	Interdependent Planning and Budget	Renewal $/CRV Plant Adaptation $/CRV Maintenance $/CRV Deferred Maintenance $/CRV Facility $/GIE $ Cost/Square Foot Load	> 4% > 2% > 6.5% 0 - 5% > 15% > $8.00 all costs, work hrs reduced	Safety Energy % Task/Stds	≤ 2% ASHRAE > 90%	Trng Costs	> 5%	% Services Measured % Stakeholder Measured Continuous Improvement	100% 100% Yes
4	Integrated Strategic Plan	Renewal $/CRV Plant Adaptation $/CRV Maintenance $/CRV Deferred Maintenance $/CRV Facility $/GIE $ Cost/Square Foot Load	3 - < 4% 1.5 - < 2% 5 - < 6.5% > 5 - 10% 12 - < 15% $6.00 - $8.00 all labor & overhead	Safety Energy % Task/Stds	10% 3xASHRAE 60 - 90%	Trng Costs	5%	% Services Measured % Stakeholder Measured Continuous Improvement	75% 75% Yes
3	Clearly Defined Goals and Objectives	Renewal $/CRV Plant Adaptation $/CRV Maintenance $/CRV Deferred Maintenance $/CRV Facility $/GIE $ Cost/Square Foot Load	2 - < 3% 1 - < 1.5% 4.5 - < 5% > 10 - 15% 10 - < 12% $4.00 - $6.00 labor, fringes, vehicles, small tools, & uniforms	Safety Energy % Task/Stds	15% 5xASHRAE 30 - 60%	Trng Costs	4%	% Services Measured % Stakeholder Measured Continuous Imp'mt	50% 50% Some
2	Mission Communicated and Effective	Renewal $/CRV Plant Adaptation $/CRV Maintenance $/CRV Deferred Maintenance $/CRV Facility $/GIE $ Cost/Square Foot Load	1 - < 2% 0.5 - < 1% 4 - < 4.5% > 15 - 20% 8 - <10% $3.00 - $4.00 labor rate plus fringe benefits	Safety Energy % Task/Stds	20% 6xASHRAE 10 - 30%	Trng Costs	3%	% Services Measured % Stakeholder Measured Continuous Improvement	25% 25% No
1	Mission Statement	Renewal $/CRV Plant Adaptation $/CRV Maintenance $/CRV Deferred Maintenance $/CRV Facility $/GIE $ Cost/Square Foot Load	< 1% < 0.5% < 4% > 20% < 8% < $3.00 basic labor rate	Safety Energy % Task/Stds	≥ 25% > 6 ASHRAE ≤ 10%	Trng Costs	0 - 2%	% Services Measured % Stakeholder Measured Continuous Improvement	0% 0% No

	Additional Financial Indicators	Additional Process Benchmarks	Additional Innovation & Learning Benchmarks	Additional Customer Satisfaction Benchmarks
Please write in any recommendations for additional benchmark indicators you wish to be considered (optional)				

Please mail or fax the results to:

Steve Glazner, APPA
1446 Duke Street
Alexandria, Virginia 22314-3492
Fax Number: 703-549-2772

Name: _____

Institution: _____

DUE DATE: December 1, 1995

Figure 4.2

A. Guideline Checklists

C-6 Backlog of Deferred Maintenance and Repair

Backlog Reporting

- The backlog of deferred maintenance and repair consists of the accumulated total of facility condition deficiencies which have been deferred, primarily due to lack of funds in annual budget cycles. Has the backlog of deferred maintenance and repair been compiled during the past 12 months? _____

Inspections

- Was this amount developed as the result of on-site inspections of all significant facilities during the past two years? Is this amount updated at least annually by adding inflation cost to uncompleted backlog, deleting completed work, and adding newly found deficiencies? _____

Backlog Analysis

- Are the past year's backlog amounts and trends recorded and reported? Are these amounts and trends categorized to provide more effective evaluations, for example, by type of facility, cost center, location? _____

Backlog Reduction Plan

- What is the percentage of backlog as it relates to current replacement value? Is the overall condition of facilities considered good, fair or poor? Has a target backlog amount been established and is there a formal plan for reducing backlog to the target level? _____

Comments

Figure 4.3

B. Effectiveness Rating Worksheet

B. Workload Identification		RATING	TARGET
1. Facilities Inventory	5 x __		20
Records not prepared or outdated	0		
Current facility list — general data incomplete	1		
Current facility list — general data complete	1		
Current facility list — detail data complete	1		
Active Updating Procedure	1		
2. Facility Condition Inspection	9 x __		36
No structured function	0		
Scope of inspection established and personnel available	1		
Schedule established and inspections accomplished	1		
Inspections accomplished $\pm$ 10% of schedule	1		
Generate major portion of O&M-funded M&R	1		
3. Work Request Procedure	4 x __		16
Procedure not documented or outdated	0		
Procedure documented and distributed	1		
Authorization controlled	1		
Structured processing	1		
Status feedback provided to requestor	1		
4. Equipment Inventory	5 x __		20
Records not prepared or outdated	0		
Current equipment list — detail data incomplete	1		
Current equipment list — detail data complete	1		
Current equipment list — computerized	1		
Active updating procedure	1		
5. Preventive Maintenance (Equipment)	6 x __		24
No structured procedure	0		
Procedure established, includes hours, frequency, schedule	1		
Work Orders issued and personnel assigned	1		
PM accomplished $\pm$ 10% of schedule	1		
Equipment records maintained, reviewed	1		

Figure 4.4

- Using consultants who can formulate a benchmarking study tailored to your organization, identify possible information partners, and help collect the data and analyze the results.

Considerations for Maintenance Benchmarking

Maintenance is a likely first target for benchmarking, as this facility function in often well documented, allowing comparison with the same function in other organizations. It may also be easier to share information on maintenance with others than is the case for other activities, since maintenance procedures are not typically seen as trade secrets.

A maintenance benchmarking study needs to define issues specific to its type of facility. Appropriate performance measurements for staff and equipment need to be specified. It is also important to identify factors, such as the age of a building or its components, that contribute to variations in performance level. The level and effectiveness of predictive maintenance (e.g., the collection and analysis of data to predict future equipment failure or identify defective components) and preventive maintenance (e.g., routine inspections, maintenance and improvements) will very likely play a major role in a maintenance department's overall performance.

Some of the areas for measure, comparison, and improvement in maintenance include:

- Type of industry the facility serves.
- Square footage of space maintained. (This may be divided into gross area and rentable and/or usable area.)
- Annual maintenance budget (total, and per square foot of usable or rentable area).
- Number of facility employees.
- Square footage per employee.
- Number of maintenance employees.
- Ratio of workers to supervisors.
- Ratio of maintenance workers to total facility employees.
- Budget for outsourced (contracted) services.
- Elements of work (e.g., janitorial) that are contracted.
- Status of maintenance and repair records.
- Status of maintenance supply records.
- Hours expended on monitoring and inspecting equipment.
- Scheduled vs. actual labor-hours required to complete predictive and preventive maintenance tasks.
- Ratio of planned to preventive and emergency maintenance and repair tasks, in terms of quantity and cost.
- Ratio of repair vs. preventive maintenance.
- Ratio of exterior maintenance costs to usable or rentable square foot of area.
- Breakdown of maintenance tasks by specialty (e.g., what percentages are electrical, HVAC, security, painting, carpentry, etc.), and use of in-house vs. contracted services for each.
- Definition of trade responsibilities (e.g., what tasks are carpenters, mechanics, electricians, etc. responsible for?).
- Janitorial costs per usable square foot.
- CMMS system: Is one in use and what does it cover?
- Staff training: Is it regularly budgeted and scheduled?
- How work is prioritized. The kinds of tasks that are categorized as "critical," "important," and "routine".
- Labor-hours expended on standard repair tasks.

- Administrative labor cost.
- Cost of worker tools.
- Cost of vehicles.
- Utility costs per square foot.
- Trash removal costs/revenue from recycling.
- Life Safety (maintenance of life and safety equipment, alarms, signage and exit doors).

Conclusion

Once areas for improvement have been identified, goals need to be formulated, corrective actions taken, and an ongoing program established to ensure that progress will be made and measured. There may still be some restrictions imposed by the facility owner on actions that can realistically be taken to improve the facility's "score" against the performance of leaders in the field. Facilities planning departments are often involved in these issues, and can be helpful in presenting a case for improvements.

It must also be remembered that benchmarking is an ongoing process. An organization's standards may (and should) change as staff improve both performance levels and measurement techniques. Data gathering for benchmarking studies can also uncover heretofore unexplored areas for improvement. Further, an organization might develop new goals through strategic planning which require new benchmarks.

Benchmarking is a proactive approach to improvement, whether measured in terms of higher quality of product or service, increased productivity, or lower costs. It can be incorporated into an overall improvement effort, along with other methods such as total quality management and reengineering. It is designed to keep an organization current with regard to technological and other applicable solutions for facility management. If this task is approached with the materials and information such as those available from industry associations, benchmarking offers the additional benefit of information sharing with a number of colleagues in the field, and the development of partnerships that will foster continuous improvement.

Chapter 5

Computerized Maintenance Management Systems

Chris Keller and Dan Hounsell

Few people would dispute the importance of computers in business today. Computerization has led to increased efficiencies throughout the business world. Downsizing in the nineties has forced many facilities departments to do more with less, making tools for efficiency ever more essential. Many organizations have implemented computerization using a combination of spreadsheets, scheduling and word processing software to support the maintenance function. Today, Computerized Maintenance Management Systems (CMMS) are going far beyond these basic functions and integrating facility management into the company's MIS mainstream. Specialized CMMS software not only helps schedule preventive maintenance (PM) and repair activities, but also supports the following related tasks:

- Maintenance of parts inventory
- Maintenance of equipment data and repair histories
- Identification and allocation of resources
- Calculation and reporting of costs
- Management of the deferred maintenance backlog

The success of CMMS has been documented by all sorts of facility types, including government agencies, hospitals, manufacturing plants, and retail malls. A recent study by the A.T. Kearney, Inc. consulting firm, published in *Industry Week* magazine, found that the payback period for an investment in a CMMS can be as short as 14.5 months. The major advantages identified in the study are:

- Improved preventive maintenance program
- Enhanced equipment uptime through equipment repair history analysis
- Improved planning and scheduling
- Improved inventory control
- Reduced maintenance costs
- Timely work performance and cost reporting

The anticipated purchase of a CMMS follows the same analytical process as any other significant business purchase:

- Analyzing the needs
- Selecting a system
- Selling the system to management
- Implementing the technology

Analyzing CMMS Needs

Maintenance departments are as different as the people who manage and staff them. As a result, there are vast differences in the way each selects a Computerized Maintenance Management System (CMMS). Despite the differences, each must go through one common process — analyzing department facility needs to match them to the power and capabilities of a CMMS.

A needs analysis is really two processes: integrating long-term plans and auditing operations. Long-term plans result from the first stage of a CMMS implementation — setting goals and objectives. The goals may be as simple as bringing all buildings on line, or as complex as integrating CMMS with Building Automation Systems (BAS) or Computer-Aided Facility Management (CAFM) systems. The audit process is divided into four areas: technology, work flow, paper flow, and resources. These audits provide an analysis of existing conditions and help determine future requirements for functionality and procedures.

Audits also identify current procedures of your organization that are inefficient — especially those based on resources. For example, say one or more versions of two different spreadsheet software packages are being used in your organization. Determine if these can be replaced by one version of one package or possibly by internal CMMS software. The audit process lets the company re-evaluate work procedures based on tasks, not resources.

Technology Audit

A technology audit provides a list of current systems (hardware and software) that can be used to run the CMMS or that will integrate with it. This audit may include workstations, networks, printers, modems, mainframes, PCs, building management control systems and bar coding systems, as well as communications systems — phones, walkie-talkies, pagers, E-mail, V-mail, fax machines, and video conferencing.

First, provide a complete specification of all hardware and software, current and proposed. The version number, installation date, and cost of the components will help in evaluating and maintaining the CMMS. Also, identify the users — title, function, and skill level — of each component, as well as the primary task each component performs.

One aspect of a technology audit that will tie into the other three audits is putting together an inventory that includes tracking all information in the department on maintenance tasks. You need not provide a comprehensive inventory of descriptions and counts of every item, but items should be identified, described, and categorized.

The audit's purpose is to determine what the CMMS will manage now and in the future, as well as to identify items maintained in separate systems that may integrate with the CMMS. Items that may be included in this list are floor plans, construction documents, equipment leases, and time sheets, as well as management and documentation of the budgeting process, housekeeping, equipment maintenance, grounds maintenance, building leases, property leases, utilities, furniture, phones, computers, vehicles, hazardous waste, asbestos and parts inventory.

The next three audits offer different ways of analyzing the same process being considered for automation through a CMMS:

- The work flow audit analyzes the methods, procedures, and processes used to accomplish the various functions being considered for automation.

- The paper flow audit tracks paperwork associated with work flow.
- The resource audit helps determine the level of expertise available to implement and maintain the CMMS.

They address four questions:

- Current functions. What are the primary maintenance functions?
- Methodology. How are the primary maintenance functions accomplished?
- Desired functions. What functions should be accomplished, but are not?
- Future requirements. What are the future plans for the department?

Work Flow Audit

This audit comprises a list of functions that may be automated and can be written based on goals and objectives and the technology audit.

Construct a work flow diagram for each function, such as on-demand work, preventive maintenance, and project work orders. Identify the function; the task required to complete the function; the title, skill level, and wage rate of personnel for each task; time required by each person for each step of the task; required information for each task; a list of decisions and/or approvals to be made; and the next task in the process.

Ask these questions — "Is this item being done efficiently? What is its purpose?" — for all items identified in the diagram.

The diagram also should show how the functions link to other systems in the company. For example, accounting does not need access to individual work orders, but may want to know the costs associated with maintaining a building or property. Accounting may also need a list of assets. Maintenance costs also may be needed to analyze true costs of equipment for figuring economic value added. Cost data also may be used by department managers in budgeting, and by the strategic planning committee in long-range planning activities.

Also, create work flow diagrams for future tasks. Many maintenance departments are trying to improve their efficiency, which entails changing functions or resources, or both. The processes will affect the work flow diagrams either by modifying, adding to, or eliminating the existing diagrams. These diagrams can help determine if suggested changes are valuable, as well as help determine future CMMS requirements.

Paper Flow Audit

The paper flow audit defines or provides examples of reports and descriptions of non-standard reports or desired reports. If a report is useful, the format, data, and calculations necessary to generate the report must be identified. This information will be used in the system specification step of the implementation process. While generating the diagrams, a paper trail can be identified.

There are two kinds of paper generated in a work flow. The first consists of reports internal to the process. The second is informational reports used outside of accomplishing the task described in the work flow diagram. An example of an internal report is a work order. An example of an external report is a cost analysis of monthly work orders for one building.

The paper flow diagram will look similar to the work flow and should identify all reports needed for each function and associated tasks. The paper flow diagram may help to reduce the amount of paper required, since many of the reports may be kept entirely within the system.

Resource Audit

The resource audit describes work flow processes that you would like to do differently. Your goal should be to determine what is not working so that you can identify and make the necessary changes/modifications. In this step, you should also determine tasks that you would like to do, but cannot currently accomplish due to a lack of resources.

To determine the optimal system layout, one must first assess the technical skills of staff, because the computer system requires not only a knowledge of the maintenance tasks, but also a technical understanding of the CMMS system. For example, if technicians are very computer literate, they can enter information, and collect and update work orders directly into the computer. This saves everyone a lot of paperwork and time. It requires terminals distributed throughout the locations for easy access. The alternative is to use the computer system to generate the paper work order, which is then given to the technician. The technician then comes back and completes the paper work order, turning it over to a data entry person.

Another key step is generating an organizational chart. This chart should show the breakdown structure of the maintenance department and should include a description of their primary duties and functions as they pertain to the system. It also may be helpful to expand the chart beyond just the system, to include other departments besides the maintenance department, since this system will not typically be isolated. It may start off that way, but the future trend is to integrate with other systems.

Technological enhancements in computers and software are enabling computer systems integration, in the form of Computer-Aided Facility Management (CAFM) systems and Building Automation Systems (BAS).

A detailed organizational chart will help establish the criteria and parameters for the type of system you would like. One decision you need to make is whether you want to actively plan integration into CAFM systems and building automation systems. If so, make sure the system is flexible enough to accommodate your needs.

Selecting a CMMS

The Computerized Maintenance Management System (CMMS) market has matured to the point that the top five software packages deliver 75%–90% of the features required by most maintenance departments.

The key to selecting a CMMS is to look beyond the special features and see how the software can be made to provide the remaining 10%–25% of the functions that your facility's maintenance requires. The anticipated functional goals of the CMMS should be the primary determinant in evaluating what, if any, CMMS package is appropriate for the operating functions of your organization and/or department.

Initial Issues

The capability of software to meet functional requirements is still a basic criterion, but it does not necessarily guarantee a system's success; most systems deliver the minimum functional requirements. Software upgrades appear every year, with a major upgrade every three to four years. If the program does not quite have a feature that is sought, it probably will appear in the next upgrade, or the next. The ability of software to change with users' needs is what enables CMMS manufacturers to remain in the software business.

The first questions to be answered during the process of selecting a CMMS are:

- Does it provide all of the required and most of the desired functions?
- How often is it updated?
- What is the one-, three- and five-year software development plan?
- What support structure is available? Manufacturer support, vendor support, user groups, consultants?
- Is the manufacturer financially stable and investing in new technologies?
- What position does the software occupy in the marketplace? How many installations? How many years has the product been available?
- Can the system be cost justified?

One Key Feature: Adaptability

Once these questions have been answered and the basic criteria are met, there is one key element to focus on: *adaptability*. Can the program be economically modified to provide the required missing features? Can it be adjusted to provide the 10%–25% of missing features? Can it be formed and reformed as the needs change? One reason that most programs provide only 75%–90% of the features necessary to manage the performance of maintenance functions is that the required features change as an organization evolves. The organization of a company changes faster than the upgrade cycle of CMMS software. This results in a permanent lag between an organization's needs and the software's capability.

The phrase, "It's impossible to be all things to all customers," remains true. Software manufacturers haven't discovered how to provide Just In Time (JIT) software. The only way to come close is to customize a system. As a result, customization plays an important role in handling a task that is not available.

One Key, Three Issues

Three issues make adaptability the one key feature in selecting a CMMS:

- The need to modify the program to meet required and desired functional requirements
- The missing and elusive 10%–25% of the functions
- Integration with other corporate systems of information management.

The first two issues are critical to the short- and mid-range success of a CMMS. The third issue affects long-term use of the system. Integration with existing and future systems is a topic currently on the leading edge of technology. Many instances still exist where one software package does not "communicate" with another software package.

Before you select a CMMS package, make sure that the intended interface of the CMMS system with your existing software is qualified. While interoperability is possible, industry standards are not quite in place. Most programs are moving toward corporate-wide integration.

The current term for this integration is Computer-Integrated Facilities Management (CIFM®). While this concept is not new, it is just now becoming technologically feasible. Having the appropriate hardware and software to achieve CIFM is only part of the answer. Currently, the missing ingredient in implementing a CIFM system is industry standards.

Interoperability

Standards for interoperability—the smooth interface between software—have been and are in the process of being developed. These standards allow CIFM to become a reality. Personal computers (PCs) based on the Intel chip using Windows® as the operating system is the current dominant system standard.

The American Society of Heating, Refrigeration and Air-Conditioning Engineering, Inc. (ASHRAE) has created the BACnet standard for automated building control systems to exchange data. BACnet applies not only to the sensors and control programs but also to the systems that might link to them, such as CMMS and CAFM systems. BACnet is emerging as a standardized data communication protocol for facilities automation operations.

Another standard under development is the Industry Alliance for Interoperability (IAI). The IAI is a non-profit organization founded by Autodesk, Archibus/FM, AT&T, Carrier, Lawrence Berkeley Laboratory, HOK, Honeywell, JB&B, Primavera, Softdesk, Timberline, and Tishman. As defined by its founding members, the IAI is an association of leading architectural/engineering/construction (AEC) industry companies that includes manufacturers, design firms, construction companies, building owners, and software companies. It is a standard for the exchange of information between all disciplines throughout the life cycle of a building.

The IAI has created Industry Foundation Classes (IFC), which expand on the most recent object-oriented software technological developments to create intelligent objects for the building industry. This will in turn dramatically increase the sharing and collaboration of information of the AEC marketplace.

As a building progresses through its life cycle, the owner and his or her vendors collect information from building projects. Sources contributing to this information pool may include the initial design process, the construction process, ongoing maintenance, renovations, and upgrades. Typically, the information is collected, summarized, and filed. The only surviving piece of information, if any, is a summary report. With the progression of the building's life cycle, information collected in a previous project is often recollected, sometimes at great expense. The primary focus of interoperability is to collect and store information so that each successive project can use data previously collected.

Selling the System to Management

Maintenance departments looking to install or upgrade a CMMS often stress the many impressive features of the software. In reality, however, these features offer little more than promises written in the wind—and in manufacturers' sales brochures. The challenge for a maintenance manager requesting funds for a CMMS is to translate grand promises and intangible benefits into concrete data and real-world facts that facility executives can use to make a decision.

The question that a facility executive might pose to a maintenance manager requesting funds to buy a CMMS can be phrased this way: "Given the fact that we're cutting the fat from every other phase of facility operations, why should we spend thousands of dollars on a CMMS?" With lofty concepts such as return on investment (ROI) and life-cycle costing dancing in their heads, maintenance managers set out to cost-justify the request. That task, however, is not as overwhelming as it might have once been. Facility executives today are more receptive than in years past to the need to computerize facility operations. As a result of this shift in favor of automation, the task of cost-justifying has changed from a simple yes/no proposal to a more complex proposal that turns on why, how, and how much.

Starting the Process

Maintenance managers will need to begin the cost-justification process by focusing on the software that offers the most power in the right places.

Since maintenance management is becoming more data-intensive each day, a CMMS offers data management power that is crucial to efficiency and success. According to Thomas Westerkamp, a facilities maintenance consultant and author of *Maintenance Manager's Standard Manual*, a CMMS provides users with:

- Rapid information storage and retrieval, such as a record of recent repairs on a chiller that has just failed
- Rapid calculation, such as the total value of inventory on hand
- The ability to handle large amounts of information, such as data relating to hundreds of pieces of equipment, thousands of work orders, and large inventory of spare parts and equipment

When this power is applied to maintenance management, it can translate into the following potential benefits:

- Lower labor costs
- Lower material and equipment costs
- More efficient preventive and predictive maintenance
- More effective work scheduling
- More accurate cost projections and budget

One word of caution: Don't expect a software system to solve organizational and performance problems in a maintenance department. The system can only work with what it's given, and if what's available is in a state of disrepair, the system will do little to fix the problems. As one CMMS consultant put it, "If you purchase a software system to clean up a mess, all you'll end up with is a computerized mess."

Gathering Data

This is perhaps the most formidable part of the cost-justification process, largely because it involves the most work. The entire case for cost justification will depend on the maintenance manager's ability to prove that a CMMS will help make the department more efficient and, as a result, more effective. But before projecting future performance, a base of data is needed from which to project. The data that will best make a case for a CMMS is:

- The amount of time and money currently spent on labor for regular, routine maintenance tasks
- The amount spent on equipment and materials for such work
- Whether the number of work orders for such tasks is rising or falling
- The amount of maintenance-related work contracted to outside services

Cara Rodgers, a computer consultant with Graphic Systems, Inc. in Cambridge, Massachusetts, recommends using data relating to routine maintenance tasks for two reasons. First, these duties form the bulk of the work performed during a normal work day. Emergency maintenance might be more costly on occasion, but such projects are difficult to predict in terms of duration or cost and, as a result, are less reliable for purposes of cost justification. Second, decision-makers are likely to have an easier time understanding the reasoning behind a spending request if they can understand and identify with the tasks being described.

Assessing Costs, Obvious and Otherwise

Instead of a list of promises and hopes, facility executives need a realistic assessment of costs relating to the purchase, set-up, and implementation of a CMMS. The most obvious of these costs include the purchase price, installation charges, and hardware needed to operate and enhance the system.

Many other costs lie further down the road, but pointing out these costs prior to purchase will help present a more realistic picture of total costs. Among the hidden costs:

- Training for new CMMS users
- Ongoing technical support
- Computer disks, paper, and related supplies
- Replacement parts and general system maintenance

Perhaps the largest hidden cost arises from compiling the information contained on work orders. This information—work requested, equipment specifications, workers assigned, materials used, and time required, among other items—forms the database on which the CMMS relies for most of its operations. For example, when a department uses a CMMS to track annual maintenance on a chiller, to predict how much to budget for repairs on the unit next year, to schedule technicians to perform the work, and to supply needed parts and materials, it does so by analyzing the chiller's repair history. This information, contained in the computer database, comes from past work orders. As much as 50%–60% of the total cost of the CMMS is likely to arise from entering equipment and labor data from work orders, organizing this information, and maintaining it for future use.

Quantifying the Benefits

A report on the condition of maintenance in the facility, a list of problems that contribute to this condition, and a request for funds to solve the problems through computerization will not go far unless the maintenance manager can put the problems in terms that facility decision-makers can understand; namely, the savings and overall financial benefits that the software system can produce.

Figures on savings and efficiency will vary based on a number of factors, such as the configuration and capabilities of the package purchased, the size of the facility, and the ability of managers and staff to adapt to and use the software effectively. Rodgers offers the following figures in two important areas that maintenance managers can use as projections in cost-justifying a CMMS:

- **Labor.** A CMMS is likely to result in savings of 10%–20% of current spending on labor. Technicians who know that management is tracking their work closely are likely to work more productively than they might otherwise. As a result, they perform more work and need less overtime to complete projects. In some cases, increased worker productivity might allow the facility to keep fewer people on staff or to bring outsourced work back in-house.
- **Materials.** A fully functioning CMMS can cut maintenance material costs by 10%–15%. Savings come primarily through more accurate cost estimating and improved reordering and material tracking.

The final figure that will weigh heavily in a CMMS purchase decision is the payback period. Facility executives will want to know—and a maintenance manager must be able to provide—information on when the facility will begin to see a return on its investment. Coming up with a reliable figure can be difficult. It is best to be conservative in estimating the payback, and be ready to defend it.

The typical payback period for many CMMS packages is around three years, according to Rodgers, a figure that may vary widely depending on the software, the facility, and the users' ability to maximize the software's power. Also, the payback time might be longer if the facility has an existing system that requires the conversion of work history data for use with the new system.

As a final note on payback, one consultant offered this scenario: When facility executives request information on ROI, a maintenance manager might consider countering by discussing what is being lost from the lack of computerization. The cost of buying and implementing a CMMS might be steep, but the cost of doing nothing or too little might be even steeper.

Making the Pitch

Maintenance managers should structure and present a request for a major purchase, such as a CMMS, so that it delivers as much relevant information as possible, with a minimum of hearsay, clutter, and double-talk. In a phrase, "Get to the point."

Software consultants offer a range of ideas for packaging and presenting a CMMS purchase recommendation, which are summarized below:

- *Keep it short.* Be as brief as possible in making the initial proposal or recommendation while still covering the key points. Be prepared, however, to provide more information to follow-up questions.
- *Use graphics.* Charts, diagrams, tables, and even photos, when used strategically with a written report, often make the case better than words alone. A chart showing potential cost savings from a CMMS, such as the one shown, in Figure 5.1, helps to sell the system in terms of its financial benefits.
- *Sell solutions.* Decision-makers, like everyone, want to know how a potential purchase will solve a problem, head off a disaster and, importantly, create additional benefits and profits. Framing the purchase request in terms of how the system will solve department and facility problems puts the request in the best light.
- *Look at the big picture.* Maintenance software today interfaces more easily with microprocessor-controlled equipment and other software systems within facilities, and a growing number of CAFM systems now offer their own building operations and maintenance module. The maintenance mission now is an integral part of facilities operations. Talk about it that way, especially in cost-justifying a CMMS purchase. Facility executives who can see potential benefits of a CMMS outside the maintenance department are likely to find more reasons to support the purchase.

A CMMS is no cure-all for problems and concerns facing the commercial and institutional maintenance mission, but it offers the maintenance department its best shot at harnessing technology to solve problems, streamline operations and contribute to the overall efficiency of the facility.

These benefits, as well as others that will occur as users take advantage of the power of the system, far outweigh any research and preparation that maintenance managers must carry out as part of the cost-justification process.

Implementing CMMS Technology

CMMS technology offers organizations tremendous power. To get such centralized information systems started requires careful planning. Successful implementations consist of three aspects:

- Enlisting management support
- Unifying the desire to make such a system succeed in the facility
- Developing a detailed, realistic action plan for implementing the system

First, the continued support of upper management is crucial to the success of CMMS and CAFM installations. Management needs to understand and agree with the benefits and direction of this system. Following such a conceptual agreement, management plays a critical role in allocating financial and managerial resources to bring the system on site. Management needs to approve funds to buy the appropriate hardware and software, but also must authorize people to lead and make decisions for the project.

Second is unifying the company behind the idea of adopting a centralized information system. System users and executives all must commit to understanding and accepting this new technology. Shared information systems rely fundamentally on the cooperation, use, and support of all levels of an organization.

Savings Realized with CMMS		
Savings Category	**% Saved**	**Amount Saved**
Lower equipment downtime		
Increased maintenance labor efficiency		
Overall maintenance cost reduction		
Improved supervisor effectiveness		
Improved parts and materials availability		
Lower preventive maintenance costs		
Lower parts and materials inventory		
Lower purchasing costs for maintenance parts and materials		
Reduced outside maintenance contractor costs		

Adapted with permission from *CMMS User Handbook,* Thomas Marketing Information Center

Figure 5.1

The third and most critical aspect is the creation of the plan of action for implementing the system. The implementation plan consists of five distinct steps:

- Customizing and installing the system
- Creating the database
- Training personnel
- Making the transition to using the system
- Maintaining the system

Customizing and Installing the System

Once a system has been selected, customization and installation issues arise. The implementation plan sets the direction for how hardware and software will be adapted to meet the facility's particular parameters. Provision for the documentation of custom features is set at this point. Hardware and software are customized to meet the criteria in the system specification and are prepared for actual installation. The plan then outlines the installation process and determines the interface of the CMMS/CAFM system with the existing computer system.

Creating the Database

This is the heart of the CMMS and establishes the initial bank of information files. A great deal of data entry time is needed, for example, to inventory equipment or enter PM procedures. To accommodate this need, the implementation plan must dictate a procedure for data entry into the system and also should prescribe how data in other computer formats will be converted or translated for inclusion in the CMMS.

The creation and connection of the databases is the single most labor-intensive aspect of getting the system running. Note that this step, therefore, will be one of the cost centers for the system.

Training and Transition

Once the system is fully defined, the focus turns to training and transition to the system. Making users comfortable with the system can make or break the viability of CMMS. Structured training sessions are usually recommended because they can demystify the system and introduce proper methods and procedures to those who will use it. Explaining features fully, from hardware to customized software, helps users understand and learn the system. Good training, coupled with company commitment to the computerization, can greatly ease a facility's transition to the system.

Maintaining the System

The last consideration in the implementation plan is the CMMS maintenance. The plan must establish a routine for:

- Updating system databases
- Making backup records
- Archiving old files

Specifying how a consultant can update the databases is critical, as is assigning relevant departments to update each database — or field in the database. This assignment ensures that data is accurate and up-to-date.

If the data is not kept current and accurate, many of the advantages of using a centralized system are lost. Similarly, predefined procedures for keeping the hardware and software current protect the continuing usefulness of a CMMS. Creating a user group and CMMS steering committee can further reinforce the commitment to succeed with the system. Computer technology is continually changing. Consequently, flexibility in a CMMS, and in a facility itself, can help ensure the system's success. Planning implementation carefully and supporting the system's use fully can make the difference between productivity and disappointment.

Conclusion

The future of CMMS technology is without limit. In fact, strong demand already is driving both equipment manufacturers and automated building operations companies to find an interface with CMMS/CAFM software. Ultimately, the power to drive the direction for the future development of these systems and the accompanying technologies rests with the facilities. When facilities commit themselves to employing these systems as an integral part of their business practice, the possibilities of the computer revolution become realities.

Based on excerpts from articles appearing in Maintenance Solutions magazine.

Chapter 6

Codes and Regulations

William H. Rowe III, AIA, PE

In the past, expenses for building maintenance were based primarily on the need to maintain and repair buildings properly over their useful life. With the enactment of building codes in the 1920s, building owners were required to follow applicable regulations regarding the types and extent of work on buildings, and fund associated costs.

More recently, environmental and worker safety codes have also been enacted that have increased the cost of operating and maintaining buildings. The significance of these costs is twofold. First, expenditures are required to conform to the code requirements. Second, changes to the building are often required, which reduce the available space for program functions, thereby reducing the space available for producing income. This chapter reviews basic code issues and requirements and provides information that can be used to assess the likely costs associated with code-required improvements.

The Role of Codes in Maintenance

Government codes and regulations have an impact on the day-to-day operations of a facility. In many cases they are the principal trigger for capital improvement and maintenance and repair expenditures. In the past, expenditures for building improvements were most often the result of building wear and tear or the market-driven need to attract tenants. While these needs continue to exist, codes and regulations have also become a significant force behind both building capital improvements and operating costs.

Codes and regulations have been traditionally viewed as applying only to new construction or renovation when it occurs. There has, however, been a trend toward retroactive legislation, which requires capital expenditures for buildings that were previously accepted by governing authorities when built. While in the past "preexisting non-conforming uses" have been allowed to remain, existing buildings are now required, in many areas, to outlay monies for certain improvements. In the areas of public safety, the environment, and public accommodation (such as the Americans with Disabilities Act), owners have been required to retrofit buildings to conform to recently enacted regulations.

Codes have also influenced the operations and maintenance budgets, particularly in the area of worker safety. The Occupational Safety and Health Act (OSHAct) requires that significant steps be taken to minimize injury to workers in areas of employment. For facility managers who work with construction trades where accidents are likely, this has meant an increase in costs to provide for worker safety. Although sometimes costly, these measures should provide a long-term benefit in the form of increased productivity and fewer Worker's Compensation claims.

Which Codes Apply to Operating Facilities?

For facility managers, the codes that have the greatest day-to-day significance are those that apply to renovating and maintaining existing buildings. In each locality it is important to compile a list and copies of the relevant applicable codes that apply to each building. The Bibliography/Recommended Resources contains an extensive list of some of the more significant codes and regulations currently in effect. Be aware that each year certain codes are updated and national standards are adopted. Information on zoning, historic districts, and environmental laws should be kept current as well.

As we have seen in Chapter 1, the costs for operating existing buildings include both capital improvements and operating expenses for maintenance and repair. Below is a breakdown of these items and a discussion of how to determine the applicable codes to follow for each.

Capital Improvements

For existing buildings, capital improvement work is most often contracted out to others. Work for capital improvements is described more fully in Chapters 1, 2, and 19, and includes items such as:

- Replacement of equipment (e.g., elevators, cooling towers)
- Upgrade of common areas (e.g., lobbies, toilets)
- Tenant improvements

These improvements are subject to the applicable codes (including those that apply to new buildings) such as:

- State Building Codes
  - Plumbing code
  - Elevator code
  - Fire code
- NFPA (National Fire Protection Association) Codes
- ADA Regulations

Codes, which are acts of the legislature, typically incorporate by reference or adopt without restrictions recognized national standards. Some are listed below. Code organizations for specific building elements can be identified through manufacturers of related products.

National Codes

BOCA	Building Officials and Code Administrators International
UBC	Uniform Building Code
SBCCI	Southern Building Code Congress International

Standards

AIA	American Institute of Architects
ASHRAE	American Society of Heating, Refrigeration and Air Conditioning Engineers
ACI	American Concrete Institute
ASTM	American Society for Testing and Materials
AISC	American Institute of Steel Construction
ANSI	American National Standards Institute
ARI	Air Conditioning and Refrigeration Institute

AWS	American Welding Society
JCAHO	Joint Commission on Accreditation of Health Care Organizations
NPA	National Plywood Association
SMACNA	Sheet Metal and Air Conditioning Contractors' National Association
UL	Underwriters Laboratories

Operating Expenses, Maintenance, and Repair

Work in this category is often completed by a combination of outside vendors and in-house staff. The codes just listed for capital expenses apply. In addition, contractors, the building owner, and maintenance staff are further subject to the OSHAct, including the "lock-out tag out," and confined spaces regulations, reviewed later in this chapter. OSHAct also applies to capital improvements. Responsibility for compliance rests primarily with the employer, a role the owner or facility manager often fills for maintenance and repair work.

Renovation Costs as a Trigger for Additional Work Required by Codes

When a building owner initiates renovative work, a capital expense or a large repair such as fixing a roof may be combined with several other projects. It is very important for a facility manager to know the extent to which a particular request for work from one department or administrator will trigger other costs, so the overall budget is controlled. Knowing the total cost picture is essential when improvements are planned. Failure to accurately predict project costs, with their associated implications, can be disastrous.

The following discussion covers some code-related factors that could affect the cost picture, and are therefore key to understanding the costs of the work beyond those directly connected with the specific project requested.

ADA

Removal of Barriers

On July 26, 1991, the Department of Justice, Office of the Attorney General, passed the Final Rule on the Americans with Disabilities Act, which prohibits discrimination on the basis of disability by private entities and requires that "places of public accommodation and commercial facilities" be designed and constructed to be readily accessible and usable by persons with disabilities.

It is now required that existing places of public accommodation remove physical and communications barriers where such changes are "readily achievable" and can be accomplished without much difficulty or expense. In addition, alterations and new construction in places of public accommodation and commercial facilities must make those areas accessible "to the maximum extent feasible." The Attorney General is charged with the investigation and prosecution of violations.

Cost implications to owners can be extensive. Capital improvement programs need to take into consideration the ADA requirements. The implications to existing facilities can be divided into two large categories as described below:

1. One requirement of ADA is that architectural and communication barriers that are structural in nature be removed in existing facilities where removal is "readily achievable and able to be carried out without much difficulty or expense." Examples of ways of achieving that accessibility are:

- Install ramps
- Install curb cuts at sidewalks and entrances
- Reposition shelves
- Rearrange tables, chairs, vending machines, display racks, and other furniture
- Reposition telephones
- Add raised markings on elevator controls
- Install flashing alarms
- Widen doors (to tenant spaces on rollover)
- Eliminate turnstile or provide an alternative accessible path
- Install accessible door hardware
- Install grab bars in toilet stalls
- Rearrange toilet partitions to increase maneuvering space
- Insulate lavatory pipes under sinks
- Install raised toilet seat
- Install full length mirror
- Reposition paper towel dispenser
- Create designated accessible parking spaces
- Install accessible paper cup dispenser at existing inaccessible water fountain
- Remove high-pile, low-density carpet
- Install vehicle hand controls

2. A priority of compliance is urged (not required) and includes:
   - Access to a place of public accommodation from sidewalk, parking, or transportation
   - Access to those areas where goods and services are made available to the public
   - Access to rest room facilities
   - Any other measures to provide access

The removal of barriers is meant to be readily achievable and take into consideration both cost and the difficulty in achieving the recommendations. Alternate proposals for achieving the end result without the structural removal of the barrier are allowed. For example, valet parking might be used for access to a garage when ramping or other constraints would be unreasonably expensive or technically infeasible.

New Construction and Alterations: New construction and alterations to provide access are not required by law, but when they are undertaken for other reasons, the existing element, common area or element of the alteration, as well as the route to the area, toilets, water fountains and telephones serving it, must be made accessible "to the maximum extent feasible" (28 Code of Federal Regulation Part 36 Subpart 36.402). This applies to physical alteration of property begun after January 26, 1992.

Alteration is defined as a "change to a place of public accommodation or commercial facility that affects or could affect the usability of the building or facility." Alterations include but are not limited to remodeling, renovation, rehabilitation, reconstruction, historic restoration, changes in structural parts, or rearrangement in plan configuration of walls and full height partitions. Normal maintenance, reroofing, painting or wallpapering, asbestos removal, and changes to mechanical and electrical systems are not alterations unless they affect the usability of the building. When a primary function area is being altered, the accessibility provisions are required only to the extent that the added costs are not disproportionate to the overall cost of the alteration. An accessibility alteration is disproportionate to the overall alteration when the cost exceeds 20% of the cost of the alteration to the primary function area. When costs are disproportionate, the path of travel shall be made accessible to the extent that it can be without incurring disproportionality. An example is a planned alteration costing $100,000, which triggers $30,000 in accessibility costs. The owner may defer 10% of the accessibility work, provided that the accessibility work that is done achieves a path of travel.

Performing a Self-Study Audit: The capital improvement programs of facilities should be evaluated according to the impact of ADA requirements. A survey and implementation program should be established if not already completed. Although "self-evaluation" is not required by the ruling, the Department of Justice (which enforces the law) recommends the development of an implementation plan that "if appropriately designed and diligently executed" will be considered evidence of a good faith effort at removal of barriers. Such self-evaluations include a survey of existing conditions by property managers and a plan and cost to achieve compliance. In addition, proposed architectural plans as well as tenant space need to be examined for compliance with ADA.

Under ADA, buildings that serve the public should have had low-cost or no-cost improvements made already. Proper signage, clearing of aisles for wheelchairs, simple ramps, and the like are expected to be in place. Where any improvements are being made (except those to repair building wear and tear), the buildings are expected to be made to conform to ADA. There are some dramatic examples of buildings with an expected capital improvement cost of $50,000 for actual tenant improvements, where a ramp, accessible toilets, elevator or lift, and accessible doors and hardware required by ADA or other local mandates as a part of this work, can actually increase the total project cost by an additional $100,000.

The ADA does allow for exemptions to compliance for certain ordinary repairs, but many states do not. The ADA also allows that to the extent that repairs for accessibility exceed 25% of the programmed improvements, the portion over 25% may be deferred until other program expenses are implemented. Again, many states do not permit such an exemption.

The effect of some of these compliance requirements is that facility managers often must choose between not making needed improvements because they will trigger ADA or other expenses they cannot afford, or being susceptible to a suit from the Department of Justice or a state agency for violating accessibility rights, with the resulting fines and litigation costs.

Asbestos

If there is a chance that asbestos will be encountered by the workers implementing the intended capital improvement, it must be abated. Asbestos may be present in insulation around an old boiler or its piping; or the piping itself may be asbestos-insulated and be buried in walls. Asbestos fibers may be present in certain older suspended ceiling tiles, in older vinyl composite floor tiles, and on the undersides of metal deck and around steel beams for fire insulation. The asbestos may have, over time, fallen off onto other surfaces where the fibers look like dust, which must be removed prior to construction workers entering the space.

During asbestos abatement the space usually must be vacated of inhabitants. Temporary quarters must be found for users of the space while the abatement is under way, even if they might not need to evacuate the space for the programmed capital improvement work that follows.

While asbestos is an undisputed health hazard and the effects of asbestos are devastating, most of the known cases of asbestos-related illness occurred at a time when asbestos workers worked in environments where the fibers floated freely. In mines or when asbestos was sprayed onto beams, workers without the benefit of respirators breathed in an environment that was a "snowstorm" of asbestos fibers. Pipe insulators "breathed" asbestos fibers for years in their work. In today's buildings the effects are less pronounced. Several countries, including England, have stepped back from the strict regulations in asbestos removal, preferring to recognize that limited exposure of a day or two with precautions and proper equipment is preferable to wholesale removal of asbestos with larger dangers resulting from its disturbance. Nevertheless, U.S. code requirements must be adhered to, and asbestos-related renovation and repair costs accounted for.

Oil Tanks and Site Assessment

Buried oil tanks can leak, especially if they are old. Older industrial, mill, or manufacturing sites may have "hot spots" of hazardous or toxic chemicals in the ground. In any project, particularly one that involves excavating on land previously used, there is a potential exposure to environmental costs of removing unwanted materials. If a boiler is being replaced, its fuel oil tank leak may be discovered or tested. Contaminated soil may be discovered at the location of the new emergency generator, or the old transformer may show it contains PCBs. Dry wells may be excavated, only to show that chemicals were dumped into them. While the probability of such expenses occurring on a particular site is fortunately low, the actual costs when they are encountered are often staggering.

HVAC

When a capital improvement is undertaken, there is often the need to provide for or upgrade ventilation in spaces as they are renovated. Many older buildings do not provide proper ventilation or supply enough quantities of fresh outdoor air to interior rooms. Proper exhausting of toilets is another frequently unexpected expense in a building renovation. Restaurants at ground levels often have poorly built exhaust ducts from stoves, including less than smoke-tight construction or proper hoods and exhaust air capacities. In some environments ducts may have accumulated yeasts, mildew, spores, or bacteria, which should be cleansed as the equipment is uncovered by the work and becomes more accessible. In high-rise buildings, any repair is likely to initiate a discussion of introducing smoke evacuation or additional tie-ins to the fire alarm system for control of building fires.

Plumbing

Plumbing fixtures are replaced when broken. In some states, if a plumbing fixture is replaced, it must conform to ADA. Replacing a broken fixture with a new one that meets ADA requirements may, in some cases, not be enough; it may actually be necessary to gut and expand the toilet room to conform to ADA, also redesigning the adjoining rooms in the process.

Backflow Prevention

In many buildings there is a cross connection between fire protection (sprinkler and standpipe) systems and domestic water. There are also many boilers and make-up water systems to mechanical equipment that are not separated from the domestic water supply. In any renovation that involves plumbing, most plumbing codes will require that if not already provided, a properly designed backflow preventor be installed to prevent contaminated water from mechanical systems from being siphoned into the domestic water supply.

Fire Protection

Sprinkler systems in buildings are viewed by fire departments as among the most effective devices for preventing the spread of fire and for saving lives. They are required in many types of buildings. Hospitals, high-rise buildings, basements, boiler rooms, storage rooms, and many more spaces are required to have sprinkler systems. In any renovation of a commercial or public facility, it is likely that the issue of installing sprinkler systems will be discussed. In some communities, any substantial renovation of more than 7,500 square feet must be sprinklered. When required, it means more than installing the sprinkler pipes. Ceilings must be removed and replaced, often triggering a need for new lighting. Spaces must be evacuated by users so the work can proceed. Walls and often carpets are affected by the work, or it is decided that since the space is empty and the ceilings are being replaced, it is a good opportunity to renovate the entire space. Finally, when a sprinkler system is installed, the fire service line in the street is often not of adequate pressure, in which case a fire pump and emergency generator must be installed to run it in the event of a power failure. A fire pump and the emergency power are often not allowed to be in the same room as the electric equipment or the mechanical room. Consequently, a major project involving new ceilings, lights, stairwell standpipes, emergency generator and fire pump with an exhaust to the roof and installation of a fuel tank can all result from what begins as a fairly modest building improvement project.

Electrical Power

Demands for electrical power in the United States have grown exponentially. In any renovation, electrical expenditures outside the program area may be required. Demands for power to computers, lighting, and air conditioning have stressed the capacity of many building electrical systems. Energy codes have demanded a reduction in energy levels, many of which have been carried out by facility managers in cooperation with utility companies who often have actually paid for the improvements. When new work is done, it is often the case that all electrical work above a suspended ceiling must be removed because it is not properly rated for a return air plenum, the most common type of ceiling used today.

Fire Alarm Systems

One of the fastest growing areas of concern (and code requirements) in the building industry is in fire alarm systems. Any air handler that supplies more than 2,500 cfm of air should be tied to a fire alarm system. Smoke detectors are being required in more types of spaces. Thus when a renovation is initiated, it is likely that the fire alarm system will need to be expanded. In many cases, older fire alarm panels are at capacity in terms of the number of points available, or have parts that are difficult to replace, as systems have been rapidly updated from electric to electronic circuitry. Thus a simple repair may trigger updating a significant portion of the fire alarm system, including better interactive communication with the fire department.

One of the bright spots on the horizon is the recent development of "wireless" fire alarm and other building communication systems. These are installed as individual devices that alert a central computer, which then controls elevators, HVAC systems, electric and other fire alarm devices, and communicates with the fire department. Once fully developed and accepted, these systems are expected to provide improved building performance at an acceptably low cost.

Soft Costs: Architectural/Engineering Fees

Almost all projects involving capital improvements require an architect or engineer to see that the work is properly designed and will comply with applicable codes and regulations. For owners with a long-term interest in properties such as hospitals, colleges, government institutions, and the like, the value of professional services has long been understood. For owners of market-driven property, working with professionals also has significant benefits, including the increased value on resale for property that can demonstrate a history of well-designed improvements by recognized architects and engineers.

Design professionals are also valuable in making presentations to the building commissioner, variance board, zoning board, and in public hearings, town meetings, citizens group forums, and to boards of directors and trustees. Many projects need to be explained to a host of agencies while keeping the central theme and project scope intact, and architects are skilled in such matters.

Project Representation During Construction

Perhaps one of the most valuable and least understood costs is proper review of the work during construction. It is enormously valuable for an owner to provide for a project representative during construction, preferably the design architect or engineer, who can: make timely decisions as the work proceeds; respond to changes in the work necessitated by existing conditions or to changes required by the owner; provide input for designs initiated by the architect; and provide a quick response to contractors' questions so the work is profitable for the contractor and expeditiously completed for the owner.

Environmental Agency Expenses

Projects that are close to wetlands or coastal areas, that can impact wildlife, or that are large and may affect existing social structures are subject to an environmental assessment. In addition, projects in historic or zoning overlay districts are required to be reviewed by historic commissions, zoning boards, planning development councils, or similar local or state agencies. The costs for such reviews may involve industrial hygienists, environmental assessment firms, and attorneys specializing in site analysis, as well as significant design costs as the project undergoes a series of rounds of review by agencies having jurisdiction. Expect delays of about a year for such reviews to be completed even on ordinary projects.

Building Codes

Virtually every building in a city, town, or municipality is governed by a building code. Building codes are minimum standards written to protect public health, safety, and welfare. Typically, building codes establish minimum design requirements for all building systems, including building enclosures, roofs, windows, plumbing, electrical and HVAC systems, and the components most frequently updated by building operators. The owner is legally obligated to establish and install building systems that conform to the governing code.

Building codes regulate design, methods of construction, quality of materials, and building use for the structures within a political jurisdiction. The local code may be state- or city-drafted, or it may be adapted from one of several recognized national standards. The Building Officials and Code Administrators (BOCA), the Uniform Building Code (UBC), and the Southern Building Code Congress International (SBCCI) are some of the several national organizations that have established a model building code.

Issues such as fire protection, material quality, and structural integrity are addressed in detail in each building code to ensure the public's safety. Virtually every aspect of any capital improvement or tenant improvement—from the design, to the materials, to the methods of installation and quality of workmanship—is regulated by building codes.

Code Structure

In addition to the codes, which are actually enabling acts passed by the legislature, there are reference standards produced by a large number of trade organizations, such as the NFPA (National Fire Protection Association), which are often incorporated into the model codes. Committees in these organizations update the standards regularly. A state code may stipulate, for example, that the provisions of the latest ACI (American Concrete Institute) standards on concrete construction, or ASHRAE (American Society of Heating, Refrigeration and Air Conditioning Engineers) standards on ventilation apply to the state code. Thus many standards, industry guidelines, and reference regulations are made a part of most building codes.

Since all these standards (there are approximately 44,000 in the United States) are revised periodically, just determining which editions of which standard apply to a particular community at a particular time involves considerable effort.

For a code to apply, it must be enacted by the related legislature or governing body. There are federal codes for federal property, state codes for public and private property within a state, local city and town codes for particular localities, and separate codes for certain authorities such as bridge and tunnel, subway, transportation, and airport authorities.

In addition to the codes, there are laws that affect building design, whether incorporated into the code directly or by reference. At the federal level, Congress has passed several pieces of legislation that affect buildings, some of which include:

- The Americans with Disabilities Act (ADA)
- The Comprehensive National Policy Act
- The Clean Air Act (CAA)
- The Federal Water Pollution Control Act (FWPCA)
- The Occupational Safety and Health Act (OSHAct)

Some of these acts affect only federal property; others are standards for the nation that all buildings are mandated to follow. There are also local laws on topics such as elevator upgrading, structural certifications of the exterior facade and stairs, and retroactive legislation for fire alarms, sprinklerization, and stairwell pressurization, which are part of the rules governing buildings even if they are not incorporated into the codes.

States adopt or write building codes for themselves. Some states have written entire codes, while others have adopted the model codes that are written by private organizations such as BOCA, SBCCI, NFPA, or the UBC. States that adopt model codes may also amend certain sections where they wish to change particular provisions.

Not all states require that the state code be used in all its communities. In some areas, each community or selected cities and towns have the authority to pass legislation that includes the right to have their own building code. Determining what the code regulations are in a particular community and which referenced standards apply is sometimes difficult. For each location, the governing authority needs to be determined. It may be the federal government if it is a federal installation, the state building code, or a local building code. The local fire official may also be designated as having jurisdiction for certain matters. The referenced standards may not agree with each other or with the code that incorporates them. Finally, there may be situations that are not covered by the code, or where a strict interpretation of the code would lead to a condition that is difficult or impossible to obtain. Where the resolution of a conflict is necessary, the designer, usually a registered architect or professional engineer, recommends a solution and then reviews it with the local officials for acceptance. When necessary, the solution may require a variance to formalize its acceptance by the community. There is a growing awareness that the regulatory process is complicated, and efforts toward code harmonization are being made. Overall building performance is being examined as an alternative to compliance with a series of individual requirements.

As currently written, most codes can prohibit occupancy or use in the most minor circumstances. One missing plug, one grille that is not yet installed, or one poorly-set thermostat can cause a delay in the acceptance and occupancy of an entire building. Efforts are under way to establish an overall rating system for building performance with a method that allows a building to be utilized when it meets the overall standards for safety.

Licensed Designers and Contractors

Much of the work done in buildings falls into the category of ordinary repairs and is generally exempt from codes and permit regulations. However, capital improvements and work that must be done by licensed trades will generally require that a permit be "pulled."

Building codes generally require that construction of new buildings, or substantial renovations to buildings other than one- or two-family residences, be designed by registered architects or licensed professional engineers who stamp their drawings. Architects and engineers are licensed in each state by the respective Division of Registration. In addition to architects and engineers, other professionals such as landscape architects and land surveyors may be licensed. Certain building trades are also typically licensed; for example, electricians, plumbers and gasfitters, and sanitarians.

While many trades such as those listed above are licensed, many others are not. For example, sheet metal workers, painters, tile setters, ceiling installers, caulkers, insulators, control wiring specialists, and communication wiring specialists are not regulated by most building codes. The anecdotal evidence is that selection of contractors by experience and reputation is the best way to ensure quality work.

Depending on how a state is organized, one or more departments may be involved in building construction. Departments of Public Safety, Health, Administration and Finance, Economic Affairs, Corrections, Education, Environmental Protection, Mental Health, Mental Retardation, the Architectural Barriers Commission, and the State Historical Commission will likely have codes, rules, regulations, or standards to follow for their areas of interest.

The building code is usually enforced by a state Department of Public Safety or its equivalent, which regulates certain codes including:

- Building Regulations (Building Code)
- Fire Prevention (Fire Prevention Regulations)
- Elevators (Elevator Code)
- Sprinkler Fitters and Pipefitters

- Plumbing Code
- Electric Code

Codes for Life Safety vs. Public Policy

The rules that govern buildings are divided roughly along two basic lines: those that govern health and safety issues and those that govern public policy. The provisions that specify proper ventilation, smoke exhaust in an emergency, and adequate heat all fall into the general category of health and safety. Issues such as energy conservation, ADA, or the incorporation of trade standards into the codes reflect public policy.

Because under current law the codes determine the systems installed in a building, those systems that are mandated by code have a significant commercial advantage. Consequently there is a natural tendency for the code organizations to be pressured to include particular equipment by suppliers or organizations. Professional societies and the open nature of the process have helped to guard against this tendency.

Codes provide a significant benefit for licensed construction trades and professionals such as architects and engineers. For one thing, they act as a shield against frivolous lawsuits. Since codes are generally regarded as representing a level of quality and standard of acceptance, buildings that conform to codes are generally considered to be designed properly. Hence a contractor, architect, or engineer who meets the code requirements is generally assumed to have acted responsibly. As codes become more complicated and in some instances contradictory, systems will need to be developed to determine levels of compliance. Some areas that codes currently mandate may be replaced with more overall performance-driven criteria, much like the FSES (Fire Safety Evaluation System) available from BOCA and NFPA. There needs to be a distinction between life safety, public policy, and workmanship of construction issues in the design of buildings.

Building Permits

For work other than ordinary repairs, most jurisdictions require that a building permit be obtained. It is generally necessary for a community at large to determine that adequate sewer and water are available, that the fire department can respond adequately in the event of a fire, that the uses of the property meet zoning and historical restrictions, and that the provisions of the governing building code are met before construction is allowed to proceed.

The Building Commissioner or local Inspector of Buildings usually enforces the Building Code. Other agencies may also play a role in the review process. For example, the Department of Public Health may review plans for hospitals and further require a separate approval from the Department of Public Safety. Other codes may have little review. For example, currently there is no review agency for ADA.

Designing to Meet Code Requirements: The Role of Architects and Engineers

The designer is responsible for designing the building to meet code requirements and comply with local by-laws, such as parking requirements, permissible sign sizes and locations, and zoning restrictions. The design drawings are reviewed by the authorities having jurisdiction. Each code specifies the requirements for the design drawings. Typically, drawings submitted to the building official for review are expected to contain the following:

- Accurate location and dimensions of all means of egress from fire.
- An occupancy schedule of persons in all occupiable spaces.

- Method and amount of ventilation and sanitation.
- Methods of fire stopping.
- Schedule and details indicating trim and finishes.
- Seal of architect or engineer if building is 35,000 cubic feet.
- Site plan showing size and location of new construction, existing structures, distances from lot lines, street grades, and finished grades.
- Engineering details and calculations stamped by a P.E. of structural, mechanical, and electrical work as determined by a building official.
- Size, section, and relative locations of all structural members with design loads.
- Type of construction and fire-resistive rating of all structural elements.
- Building exterior envelope, U values, R values, and lighting levels.

Once the drawings are approved, the design architect or engineer is expected to review the shop drawings, procedures for code-required quality control materials, and the critical design components for general compliance with the design requirements.

General Code Requirements

From the above list of information required on drawings, it can be seen that health and safety considerations are foremost. There are two basic principles of health and safety: containment of a fire and provision of a means of escape or egress. Containment of fire is primarily obtained by providing floors and walls of sufficient fire rating to contain a fire and building systems that will detect and help limit its spread. Facility managers need to have a working knowledge of some of the basic principles of general life safety considerations, listed below:

Architectural Layout

Egress
There should generally be two ways out of a building or floor, so that if one is blocked by fire, the alternate route can be taken.

Fire Ratings
Walls that define corridors and stairs are "fire-rated" so as to contain a fire for a limited period of time, typically one or two hours. This allows occupants to evacuate and fire fighters to enter the building with some measure of protection. Fire walls must have their integrity maintained. Holes are limited; openings in them must be protected; doors must have door closers.

Fire Separation
Certain portions of a building must be separated from others by fire-rated construction. Different uses, such as office, retail, public assembly, and mercantile, must be protected from one another by fire-rated walls and floors.

Fire and Smoke Dampers
Duct penetrations through fire-rated walls or floors must be protected by dampers, which will close in a fire to block transmission of fire and/or smoke and maintain the fire integrity of the wall.

Structural Requirements

Overall Strength

A building is expected to be strong enough to withstand normally occurring forces of a 25- to 100-year frequency without collapsing or experiencing excessive deflection. Dead loads from permanent building weight, live loads from people and equipment, snow, wind, and earthquakes are all considered in the arrangement and sizing of structural systems and members.

Plumbing

Requirements include the proper number of toilet fixtures for the population, with some consideration for additional fixtures for women's facilities (who often provide care for the young and older people), as well as drinking fountains, drains, and proper maintenance of clean water for domestic use, with proper design of sanitation systems.

Heating and Air Conditioning

Basic considerations are adequate heat in winter to maintain a healthy environment and air conditioning for certain spaces where inhabitants are fragile, such as hospitals.

Ventilation

Since building spaces are enclosed, they need to be vented with fresh air to replenish oxygen and maintain odor control. In addition, proper ventilation helps to maintain a healthful environment and good indoor air quality. Because it is costly to condition outside air for heating, cooling, or humidity control, there are economic pressures to establish the minimum ventilation requirements for building spaces. Proper ventilation is associated with indoor air quality (IAQ), which is discussed later in this chapter and in Chapter 11, "Mechanical Systems."

Electrical Systems

Electrical systems must meet code considerations for safety. Proper wire size and fuse or circuit breaker protection are fundamental. Adequate numbers of receptacles, and adequately sized equipment such as transformers and switchgear, are also required. The placement of switches and panel boards, and the types of devices in certain areas, such as GFI (ground fault interrupt) receptacles in wet areas, are defined.

The electrical considerations also involve proper lighting and energy efficient devices, such as the use of fluorescent light rather than incandescent wherever possible.

Fire Alarm Systems

Detection of incipient or ongoing fires is an essential part of building life safety. In residences, local alarms that wake occupants are common and essential. In larger buildings, the placement of smoke detectors, heat detectors, duct smoke detectors, pull boxes, horns, lights, and automatic alarms to fire departments for response are used more and more.

All of the above systems are affected in any building work and can result in expenditures that are associated with building maintenance and repair.

Environmental Regulations

Over the past decade a growing understanding of the effect of building systems on the environment has resulted in legislation aimed at better protecting the environment. The thrust of this legislation is to protect the natural environment and control pollution. The following is a brief summary of the major pieces of legislation that currently affect HVAC systems.

Federal Water Pollution Control Act (FWPCA)

(Amended by the Clean Water Act of 1977)
Public Law 92-500
Safe Drinking Water Act (SDWA)
Public Law 93-523

These laws govern the control and record keeping regarding the discharging of waste materials into a building's sewer or drainage system. Hazardous wastes from plumbing and HVAC systems that are governed by this act include laboratory wastes, acid wastes, process water, chemically treated water, oils, and equipment coolants.

Occupational Safety and Health Act of 1970 (OSHAct)

This legislation is best summarized by quoting the "general duty clause," which states that "...each employer shall furnish to each of his employees employment and a place of employment which are free from recognized hazards that are causing or are likely to cause death or serious physical harm to his employees." OSHAct outlines the procedures and penalties that are involved in compliance. Workplaces are sources of possible health hazards. The presence of toxic chemicals, the use of machinery, or working in an environment that is stressful to the body because of excess noise can all pose real dangers to individuals if not properly controlled. Exposure to harmful effects in the workplace generally are expected to be "as low as reasonably achievable" (known by the acronym ALARA) and applies to all workplace environments. For managers of facilities maintenance and repair, the OSHAct regulations have had an impact on cost as well as overall procedures.

Figure 6.1 is a partial list of known possible hazards in the work environment.

OSHAct has specific requirements for a wide variety of on-site activities that may expose individuals to hazardous materials. Some of these requirements are included in Figure 6.2. Each of the specific OSHAct provisions is written to require that an employer assess for potential

Partial List of Possible Hazards in the Work Environment		
Chemical	**Equipment**	**Environment**
Toxic waste	Tools	Fire
Asbestos	Machinery	Ozone
Lead	Barricades	CFCs
Cadmium		Heat stress
Mercury		Confined spaces
Cleaners		Noise
Carbon monoxide		Radiation
Chlorine		
Solvents—VOCs		
Pesticides		

Figure 6.1

dangers, provide information to workers on the hazards likely to occur, provide adequate signage to alert persons to locations where hazardous materials or conditions exist, provide protective equipment or procedures to minimize the dangers, and have in place emergency and evacuation procedures in the event of an accident or injury.

The act has contributed to an increase in overall safety in the workplace. Initially, facility managers absorbed some significant costs to establish procedures and implement systems to comply. There is an overall awareness that some of the current administrative procedures that have been required are burdensome and ineffective and that the system and authority of OSHAct inspectors needs some improvement. Facility managers are nonetheless increasingly aware of both the responsibilities and costs of providing improved safety for employees in the workplace.

Figure 6.2 is an overview of those aspects of OSHAct that are of primary significance to building managers, and that can influence the costs or procedures that are a part of facilities' operations.

CFCs and HCFCs

Clean Air Act of 1970 (CAA)
(Amendments, Reauthorized in 1990)
Public Law 101-549

The CAA is divided into 11 titles; of these, titles IV, V, VI, & VII are directly related to HVAC systems operation and maintenance.

Title IV focuses on a reduction in acid rain and mandates a general reduction in the emissions of sulphur dioxide (SO_2) and nitrogen oxides (NO). Acid rain is formed when sulphur dioxide and nitrogen oxides react with atmospheric chemicals to form acids, such as sulfuric and nitric. SO_2 and NO are products of combustion from automobiles, incinerators, power plants, and boilers that burn high sulphur content hydrocarbons.

Title V mandates the permitting and certification of personnel who work with CFC and HCFC refrigerants and the associated equipment, the permitting of recovery and recycling equipment, and the permitting of all major sources of air pollution. Fines are set at a minimum of $10,000 per day per violation.

Title VI pertains to stratospheric ozone protection and schedules the phaseout of CFC and HCFC refrigerants, as well as halons, carbontetrachloride (CCL_4), and methylchloroform (CH_3CCL_3). CFC and HCFC environmental considerations are discussed more extensively in the next section. Until these chemicals have been fully phased out and substitutes become available, special taxes and surcharges are being levied on these compounds and the monies used to fund research into alternative chemicals and environmental research.

Title VII pertains to the enforcement of the CAA and the consequences of noncomformance, also discussed further in the following section.

Refrigerant Applications
Refrigeration systems typically utilize refrigerants as a fluid that circulates between the evaporator, compressor, condenser, and expansion valve. Refrigerants have a low boiling point at atmospheric pressure, which makes them useful and effective in providing cooling.

The vast number of refrigeration machines currently use either CFCs or HCFCs as a refrigerant. CFCs are chlorofluorocarbons and HCFCs are hydrochlorofluorocarbons, and both are commonly used in air conditioning equipment and chillers. Other fluids, such as glycol or brine, are used in

Outline Summary of Selected OSHAct Regulations		
Regulation	**Title**	**Outline of Compliance Requirements**
1910.25 1910.26	Portable Wood Ladders Portable Metal Ladders	Applies to ladders used in construction. Uniformity of steps and overall length for specific trades, such as painters and masons, are specified.
1910.66	Powered Platforms for Building Maintenance	Applies to units permanently installed and dedicated to interior or exterior building maintenance (construction scaffolding is covered elsewhere). Building owners are responsible for determining that equipment meets OSHAct standards before each use and providing assurance of this to employers of workers who will use the equipment. The criteria and methods of assurance, an emergency plan, and review by a professional engineer with expertise in these matters are required.
1910.67	Vehicle Mounted Elevating and Rotating Work Platforms	General safety requirements established to assure that such units are stabilized, workers are safe, and hydraulic and electrical components are in proper working order.
1910.95	Occupational Noise Exposure	Implement a hearing conservation program; maintenance shops and construction areas. Provide the following: • Audiometric testing of employees • 8 hour time weighted average 85 decibels (DBA) • Written notification of hazards • Personal protective equipment
1910.96 1910.97	Ionizing Radiation Nonionizing Radiation	Prevent overexposure to radiation such as X-rays, ultraviolet or infrared light in restricted areas. Certain facilities such as hospitals have radiation safety officers, who identify potential sources of radiation and train staff on proper procedures for managing the location, movement, and disposal of radioactive materials. Also, procedures for monitoring the levels of radioactivity, signage, provision of restricted areas, evacuation procedures, and information on health effects are provided.
1910.101 1910.102 1910.103 1910.104	Compressed Gases Acetylene Hydrogen Bulk Oxygen Systems	Procedures for storing and handling compressed gases in cylinders such as hydrogen, oxygen, acetylene, nitrous oxide are discussed. Additional information available from the Compressed Gas Association.
1910.11	Flammable and Combustible Liquids	Regulates bulk, container, and portable tank storage of flammable and combustible liquids. Requirements for design and construction of flammable containers and rooms, storage requirements, office areas, maintenance storage, and ventilation standards are provided. Emphasis on containers of liquids (including aerosols, gasoline, paints, thinners) not exceeding 60 gallons. Liquids such as paints, oils and varnishes used for painting or maintenance, but kept for less than 30 days, are exempt from the regulations.
1910.12	Hazardous Waste Operations and Emergency Response	Standards for clean-up of existing hazardous waste sites and emergency response to release of hazardous substances. Facilities with hazardous materials present are required to have policies and procedures in place to respond to emergencies involving release of hazardous materials.

Figure 6.2

Outline Summary of Selected OSHAct Regulations (cont.)		
Regulation	**Title**	**Outline of Compliance Requirements**
1910.132	Personal Protective Equipment	Employers are required to assure that personal protective equipment (PPE) for workers is provided, used and maintained in a safe and sanitary condition. Such equipment shall be used to minimize dangers from:
1910.133	Eye and Face Protection	
1910.134	Respiratory Protection	• Impact
1910.135	Head Protection	• Penetration
1910.136	Foot Protection	• Compression – rollover
1910.137	Electrical Protection	• Chemicals
1910.138	Hand Protection	• Heat
		• Harmful dust
		• Light (optical) radiation
		The employer shall conduct an assessment of the workplace – usually a walkthrough to observe the following potential sources of hazards:
		• Motion or movement in personnel that could result in collision with stationary objects.
		• High temperatures that can result in burns, eye injury, or ignition of equipment.
		• Chemical exposures.
		• Harmful dust.
		• Light radiation; i.e., lasers, welding, brazing, cutting, furnaces, high intensity lights, and similar sources.
		• Falling objects or potential for dropping objects.
		• Sharp objects that might pierce the feet or cut the hands.
		• Rolling or pinching objects that could crush the feet.
		• Layout of the workplace and location of co-workers.
		• Electrical hazards.
		The employer should review past injury and accident data to identify problem areas. PPE appropriate to the anticipated hazards (e.g., to head, hand, foot, eye) are to be provided.
1910.14	General Sanitation	Proper sanitation in toilet facilities, vermin control, waste disposal, showers, food handling, eating and drinking areas should be provided. Food consumption in toxic areas is prohibited, and there are requirements for washing facilities and lavatories.
1910.14	Safety Color Code for Marking Physical Hazards	Proper marking of physical hazards is established: red for danger, fire protection, and stop; yellow for caution and for marking falling and tripping physical hazards.
1910.15	Accident Prevention Signs and Tags	Specifications for accident prevention signs are required at specific hazards, especially where failure to do so may lead to accidental injury. The basic categories of signs are as follows: • Danger • Caution • Safety instruction • Biological hazards • Accident prevention Signs must contain a signal word and a major message and must be readable at five (5) feet or greater as warranted.

Figure 6.2 (cont.)

Outline Summary of Selected OSHAct Regulations (cont.)		
Regulation	**Title**	**Outline of Compliance Requirements**
1910.15	Permit-Required Confined Spaces	Applies to certain spaces such as manholes, crawl spaces, vaults, and similar spaces not commonly occupied and also where workers are subject to dangers of hazardous air, collapse, or other significant dangers. In particular, a space is considered a "confined" space if it: • is not ordinarily inhabited by people and is large enough for an individual to enter. • has limited or restricted entry or egress. • is not designated for continuous employee occupancy. A space is considered a "Permit-Required Confined Space" if the above conditions exist and if further the space: • contains or has potential to contain a hazardous atmosphere. • contains a material substance with the potential to engulf the entrant. • has an internal configuration such that an entrant could be trapped by inwardly conveying walls or by a floor that slopes downward and tapers to a smaller cross section. Or: • contains any other recognized serious safety or health hazard. When a permit-required confined space is identified, the following procedures are required: • Hazard identification—the identity and severity of each hazard in the space must be determined and characterized. • Hazard control—procedures and practices that provide for safe entry into the space must be established and implemented. • Permit system—a written permit must be prepared, issued, and implemented. Upon completion of the task, physical barriers such as caution tape, and posting of signs and barriers, prevent unauthorized persons from entering. • Employee training—employees who enter the space, serve as standby attendants, or issue permits must have completed the confined spaces training program. • Equipment—appropriate equipment such as air sampling devices, retrieval equipment, respirators, and ventilation blowers must be provided. • Rescue/emergency procedures for personnel and equipment must be established and implemented. • External hazard protection—physical barriers, "men working" signs, and similar devices to control potential hazards posed by pedestrians and vehicles must be provided. • Written plan—above items must be incorporated with implementation.
1910.15	The Control of Hazardous Energy (lockout/tagout)	Requires employers to establish a program that will enable workers to disable equipment, particularly when the unexpected energization of the equipment or release of stored energy could cause injury to persons working on the equipment. It applies specifically to the control of energy during servicing. The following are required: • Written program. • Employee training and certification tagout devices. • Periodic inspection of procedures. • Coordination with outside contractors and shift/personnel changes. • Lockout or tagout devices, which are affixed to equipment in conjunction with a written program of maintenance procedures and training sessions, and allow workers to minimize dangers of bodily harm from electrical or mechanical activation of machinery and equipment.

Figure 6.2 (cont.)

Outline Summary of Selected OSHAct Regulations (cont.)		
Regulation	**Title**	**Outline of Compliance Requirements**
1910.16	Fire Brigades	Requires that facilities establish a written policy that recognizes the existence of a fire brigade for dealing with interim fires. A training program, PPE, emergency response plans, and respiratory protectors are to be provided.
1910.157	Portable Fire Extinguishers	Provides for the placement, use, maintenance, and testing of portable fire extinguishers: • Inspect visually monthly. • Travel distance for class A and D fire extinguishers is 75 feet or less. • Travel distance for class B fire extinguishers is 50 feet or less.
1910.158- 1910.163	Fire Extinguishing Systems	Provide training annually. Provides for specifics on fire-related systems.
1910.176 1910.178 1910.179 1910.184	Handling Materials — General Powered Industrial Trucks Overhead and Gantry Cranes Slings	Provides for handling of materials; related to secure storage, housekeeping, and clearance limits to prevent damage to materials, spills and the like. Operators of power trucks who move products are required to undergo training.
1910.21	General Requirements for All Machines	Requires that guards be affixed to machines such as grinders, portable power tools, and power saws. In some cases, manufacturers of equipment may need to be consulted to determine how to retrofit older equipment.
1910.21	Woodworking Machinery Requirements	Provides for safety-related procedures in carpentry shops or other work areas where wood is cut or finished.
1910.22	Abrasive Wheel Machinery	Abrasive wheel machinery may be used only if properly guarded when the wheel is greater than 2 inches in diameter. Specific requirements are provided for flanges, work rests, guard exposure angles, and bench and floor stands.
1910.22	Mechanical Power Transmission Apparatus	Standards on motor-driven equipment such as supply and exhaust fans and compressors; deals with both direct and fan belt drives, the condition of the pulleys, and the drive mechanisms. Levels of required illumination and standards for space clearance and restricted access are provided.
1910.24	Guarding of Portable Powered Tools	Requires that equipment be properly guarded, inspected, maintained, and handled. Generally: • Identify the types and applications of tools. • Identify specific compliance requirements. • Ensure compliance with manufacturers' specifications and requirements.
1910.252	Welding and Welding Operator Oxygen Fuel Gas Welding and Cutting Arc Welding and Cutting Resistance Welding	Provides for specific regulations regarding welding by employees. Covered topics include oxygen-fuel, arc and electric resistance welding. General requirements include specifications on: • Fire prevention and protection • Ventilation • Confined spaces • Personal protection equipment (PPE) • Signage • Health hazard training • Gas storage • Arc shielding

Figure 6.2 (cont.)

Outline Summary of Selected OSHAct Regulations (cont.)		
Regulation	**Title**	**Outline of Compliance Requirements**
1910.301-1910.399	Electrical	Provides a comprehensive set of standards regarding electrical systems. Topics include initial design, installation, and maintenance of systems. Much of the standard is devoted to building construction and renovation. For every major replacement, modification, repair or rehabilitation, the standards include regulations on: • Examination, installation, and use of equipment • Splices • Arcing parts • Marking • Identification of disconnecting means • Guarding of live parts • Protection of conductors and equipment • Location in or on premises • Arcing or suddenly moving parts • 2-wire DC and AC systems to be grounded • AC systems 50 to 1,000 volts not required to be grounded • Grounding connections • Grounding path • Fixed equipment required to be grounded • Grounding of equipment connected by cord or plug • Methods of grounding fixed equipment • Flexible cords and cables, uses and splices • Hazardous locations • Entrance and access to workspace (over 600 volts) • Circuit breakers operated vertically • Circuit breakers used as switches • Grounding of systems of 1,000 volts or more supplying portable or mobile equipment • Switching series capacitors over 600 volts • Warning signs for elevators and escalators • Electronically controlled irrigation machines • Ground fault interrupters for fountains • Physical protection of conductors over 600 volts
1910.1	Air Contaminants	OSHAct regulates exposure to over 700 compounds. Examples include solvents, caustic gases, pesticides, asbestos, mercury, antineoplastic drugs, waste anesthetic gases, and welding fumes. These may be present as gas, mist, vapor, particle, or fume and may or may not be discernible. Therefore, management is responsible for alerting workers to potential hazards. Plumbers working on laboratory sinks, and sheet metal workers repairing exhaust ductwork from a toxic gas facility or an infectious disease room, are examples of simple tasks that may have hazardous substances associated with them. MSDS sheets are used to assess the materials that may be encountered.
1910.1	Inorganic Arsenic	Regulates exposure to inorganic arsenic.

Figure 6.2 (cont.)

Outline Summary of Selected OSHAct Regulations (cont.)		
Regulation	**Title**	**Outline of Compliance Requirements**
1910.1	Lead	Exposure to lead is regulated. Baseline sampling is required in potentially contaminated areas. If lead levels greater than 30 micrograms per cubic meter of air are detected, intervention is required. This includes training, blood tests, and sampling.
1910.1	Bloodborn Pathogens	Healthcare workers who deal directly with patients are most at risk. Hospital workers who come in contact with medical wastes that spill into the environment, repair work in patient rooms, waste drain systems, research facilities in HIV and HBV, and similar risks are required to be notified of the potential for such risk and provided with PPE and policies and procedures for dealing with the substances and avoiding harm.
1910.1	Ethylene Oxide	Provides expanded regulations for this highly toxic gas used in the sterilization of hospital equipment. Health hazard training, monitoring, respiratory protection, emergency evaluation, and PPE standards are regulated.
1910.1	Formaldehyde	Provides expanded regulations for handling exposure. The principal areas of concern are training, PPE, and monitoring during cleanup.
1910.12	Hazard Communication	Requires that information on hazardous materials be made available and is the main focus of OSHAct. The Material Safety Data Sheets (MSDS) provide essential information for use of products. The basic contents of a hazardous communication program will include the following: • General information on health and physical hazard determination • Hazardous chemical lists • Labels and other warning signs • MSDS sheets • Employee information and training • Contractor's responsibilities • List of facility contacts and responsibilities • Methods of MSDS procurements • MSDS locations in facility • Training programs • Program effectiveness audits All hazardous materials must be located and inventoried at each site to determine: • Product name • Manufacturer • Container size • Container type • Number of containers • Location within facility The above information is required to be kept current, usually with a data management system and verified with periodic inventories. Training programs on the hazardous materials (hazmat) must be in place and shall provide for the following: • Location of facility's MSDS • Location of facility's hazard communication program • Location of hazmat inventory lists • Designated coordinators of the facility and their responsibilities • Labeling • Methods to detect the presence of a chemical (odor, appearance, color) • Emergency treatment • Discussion of air sampling data

Figure 6.2 (cont.)

Outline Summary of Selected OSHAct Regulations (cont.)		
Regulation	**Title**	**Outline of Compliance Requirements**
1910.12		• Potential physical effects
		• Potential acute chronic health effects
		• PPE, work practices, spill/emergency actions
1926.1 through 1926.1148	Safety and Health Regulations for Construction	This part of the code details the requirements to be met for all construction related activity. The sections covered are divided into the following subparts: • General • General Interpretations • General Safety and Health Provisions • Occupational Health and Environmental Controls • Personal Protective and Life Saving Equipment • Fire Protection and Prevention • Signs, Signals, and Barricades • Materials Handling, Storage, Use, and Disposal • Tools—Hand and Power • Welding and Cutting • Electrical • Scaffolding • Floor and Wall Openings • Cranes, Derricks, Hoists, Elevators, and Conveyors • Motor Vehicles, Mechanized Equipment, and Marine Operations • Excavations • Concrete and Masonry Construction • Street Erection • Underground Construction, Caissons, Cofferdams, and Compressed Air • Demolition • Blasting and Use of Explosives • Power Transmission and Distribution • Rollover Protective Structures, Overhead Protection • Stairways and Ladders • Diving • Toxic and Hazardous Substances
1926.58	Asbestos	Provides for protection of workers who encounter asbestos or are exposed to asbestos during construction. Existing facilities in which asbestos materials may be present generally have an industrial hygienist conduct an inventory to identify the location of such materials. During subsequent renovations, materials that might be encountered or disturbed by the renovations are usually removed by firms specializing in removal of these materials. Workers don entire suits to protect them, showers are provided to wash off fibers, and the rooms are subject to negative pressure with HEPA filters to the outside to protect surrounding spaces. Friable asbestos, such as on pipe insulation, plaster, or boilers, is considered the most dangerous. Asbestos siding and vinyl asbestos tile have the fibers encapsulated and the procedures for the removal of these materials is less stringent. These can often be removed without special notification and with wetting procedures.
1926.62	Lead	Lead was banned in paints after 1978. Construction work in pre-1978 facilities will involve protection of workers, protection of adjoining spaces, and proper collection and disposal of the materials.

Figure 6.2 (cont.)

industrial applications where cold liquids (below the freezing point of water) are moved from place to place to avoid the consequences of freezing, but their use is confined to absorption chillers and for heat transfer. Ammonia is also used in refrigeration applications.

The three most common CFC refrigerants used today are CFC-11 (R-11), which is mainly used in negative pressure centrifugal chillers of about 250 to 1000 ton capacity, and CFC-12 (R-12) and CFC-500 (R-500), which are mainly used in all positive pressure chillers. Negative pressure chillers operate below atmospheric pressure and leak inward; positive pressure chillers have the fluid above atmospheric pressure, and they leak to the atmosphere. The most common HCFC refrigerant is HCFC-22 (R-22), used primarily in positive pressure reciprocating chillers and most small commercial and residential air conditioners.

Environmental Effects

The ozone layer surrounding the earth protects the earth from harmful and carcinogenic ultraviolet radiation. Both CFCs and HCFCs contribute to the depletion of the ozone layer when they are released into the atmosphere. The chlorine from the released CFC or HCFC chemically reacts with the ozone layer and forms a gap in the layer. CFCs, which contain more chlorine than HCFCs and consequently are more hazardous to the ozone layer, are scheduled for a faster phaseout. CFCs and HCFCs have also been linked to global warming.

CFCs and HCFCs currently leak into the environment as a result of repair and maintenance on cooling systems that use these substances as a refrigerant, and by other industries that have used these chemicals as cleaning agents, principally in the cleaning of computer boards.

In keeping with worldwide environmental concerns, most industrialized nations have adopted policies that reflect concern for the environment. In 1990, the Clean Air Amendment (CAA), a sweeping piece of legislation that calls for the complete phaseout of the production of all such chemicals, was adopted as legislation in the United States. The Environmental Protection Agency (EPA) is the enforcement agency for the planned phaseout of CFCs and HCFCs in the environment. According to the timetable for phasing out the use of such chemicals, production is to be halted on CFCs initially, and the production of HCFCs is to be gradually phased out. Stockpiled materials can be used for repairs and installation. New or existing equipment that utilizes CFCs can remain and be maintained utilizing the existing stock of chemicals.

CFC and HCFC Phaseouts

Because of the ozone depletion potential (ODP) and global warming potential (GWP) of CFCs and HCFCs, governments throughout the world have agreed to phase out the release, use, and production of CFC and HCFC refrigerants.

In September 1987, the United Nations Environmental Program (UNEP) signed the Montreal Protocol, which established the production phaseout of CFC and HCFC refrigerants. This phaseout schedule has been accelerated twice: first, by the London Amendment in June 1990; and second and most recently, by the Copenhagen Amendment in November 1992. The U.S. Environmental Protection Agency (EPA) 1990 Clean Air Act has been amended to adopt and further accelerate the UNEP production phaseout regulations.

As of the date of this publication, the EPA phaseout schedule is summarized in Figure 6.3.

Recycling and Disposing of CFC & HCFC Refrigerants

Although production of CFC and HCFC refrigerants will be phased out according to the EPA Clean Air Act, they may not be disposed of or released into the atmosphere. All such materials must be recovered and recycled. There is currently no legislation prohibiting the recycling of these refrigerants. As long as CFC and HCFC refrigerants are recovered and reused without any venting or leakage to the atmosphere, all machines can use these refrigerants for their entire useful life. The refrigerants from equipment that have been converted to non-CFC or -HCFC products must be reclaimed and stored for recycling at CFC "banks." The rapidly increasing cost of CFC and HCFC refrigerants will greatly affect their use.

Losses and Venting of CFCs and HCFCs

Until July 1, 1992, venting of CFC and HCFC refrigerants to the atmosphere during service or disposal was permitted. Losses of refrigerants to the atmosphere are principally the result of leaks, improper purging, mishandling, and contamination from oil, water acid, or motor burn. Since July 1, 1992, the Clean Air Act has prohibited the venting of ozone-depleting compounds used as refrigerants into the atmosphere while maintaining, servicing, repairing, or disposing air conditioning or refrigeration equipment. Other regulations of the Clean Air Act require that ozone-depleting refrigerants be recovered and that a permitting and certifying procedure be implemented for equipment used for recovering refrigerants and for the technicians handling refrigerants.

Alternatives to CFCs

Since the environmental effects of CFCs and HCFCs are now better understood, and as the requirements for phaseouts and the strict regulations regarding emissions begin to take effect, alternatives for refrigerants are necessary. Again, CFCs will be phased out first, then HCFCs.

CFC and HCFC Phaseout Schedule		
Date	**CFC Schedule**	**HCFC Schedule**
1/1/1996	Ban on production of CFCs including R-11, R-12, & R-500*	Cap at 3.1% of 1989 CFC production plus 1989 HCFC production
1/1/2004		65% of 1996 levels
1/1/2010		35% of 1996 levels
1/1/2013		10% of 1996 levels
1/1/2020		0.5% of 1996 levels Ban on HCFC-22 and HCFC-123 production
1/1/2030		Total ban on all HCFC production
*There is a limited extension for medical users and sterilizers utilizing ETO (Ethylene Oxide) that demonstrate need.		

Figure 6.3

There is a class of refrigerant, hydrofluorocarbons (HFCs), that do not contain chlorine and are thus not an immediate threat to the ozone layer. They can be a suitable replacement in certain circumstances. In principle, there are three options to consider in the phaseout of hydrofluorocarbons: containment, replacement, or conversion of CFCs.

Containment:
Continue using the existing CFC refrigerant, but modify existing equipment so emissions are minimized and the refrigerant can be recovered and reused. This option is the least expensive. It involves an investment in containment and recycling, and leak monitoring equipment, as well as providing adequate ventilation to the space. All of these measures can be undertaken at relatively low cost.

Replacement:
Remove the existing equipment and replace it with new equipment using HFC or HCFC refrigerant, or even an absorption chiller or alternate cycle unit. Replacement is potentially the most expensive option from a first cost standpoint, and may cost approximately four times as much as containment. However, it is by far the most reliable solution. There are advantages to a new, warranted, reliable, high-efficiency unit with an expected long life span.

The issue of using HCFCs does involve some risk that the timetables for use of HCFCs will not be further shortened. For institutional users, it means another possible replacement after the year 2030, when HCFC consumption is prohibited. Figure 6.4 indicates existing and replacement refrigerants as well as phaseout dates.

Conversion:
Remove the CFC refrigerant and replace it with an HFC or HCFC alternative. Conversion costs may range between the previous two options, although in some cases, they could actually be more expensive than replacement. The range in cost for conversions of chillers is dependent on:
- The type of chiller
- The refrigerant in use
- The alternative refrigerant
- The compatibility of the chiller with the alternative refrigerant

Refrigerant Replacements		
Existing Refrigerant	**Replacement Refrigerant**	**Replacement Refrigerant Phaseout Date**
CFC-11	HCFC-123	2,030
CFC-12	HFC-134a	NONE
CFC-500	HFC-134a	NONE
CFC-22	HCFC-22	2,020

Figure 6.4

- The costs to access and remove the existing, and replace it with the new equipment
- The costs for refrigerant monitors and new ventilation systems

Compatibility of the chiller with the alternate refrigerant is the governing factor for the extent of work to be carried out on the existing chiller.

Conversion may be possible by replacing only the refrigerant, seals, gaskets, o-rings, seats, and sealants. This would be relatively inexpensive. On the other hand, the conversion could also require replacing heat exchanger tubes and impellers, changing the compressor speed, providing larger passages for increased flow to achieve compatible efficiency, flushing out and replacing the oil, and adding economizers to the compressors. This would be much more expensive and, depending on other factors such as chiller age, life cycle cost analysis, phaseout of the replacement refrigerant, and additional monitoring and maintenance required, replacement of the chiller may be a better choice.

When considering conversion of an existing chiller, it is important to contact the chiller manufacturer for advice and information regarding the correct route to take.

Refrigerant Alternatives: As outlined previously, the most common refrigerants used in air conditioning and refrigeration equipment are CFC-11 (R-11), CFC-12 (R-12), CFC-500 (R-500), and HCFC-22 (R-22).

The replacement for CFC-11 is an HCFC refrigerant, and HCFC-22 has as yet no replacement. Both of these refrigerants will be phased out by 2030, and it is expected that a replacement will be produced by that time. When replacing a refrigerant, the following factors must be considered:
- Efficiencies of the replacement refrigerant and payback for conversion costs
- Costs of applying the new refrigerants
- Ozone depletion potential of the replacement refrigerant
- Global warming potential of the replacement refrigerant
- Toxicity of the replacement refrigerant

Consequences of Nonconformance

Environmental: The environmental consequences of nonconformance to the Clean Air Act can result in continuing depletion of the ozone layer and an increase in global warming, both of which cause severe environmental and health damage.

Legal: Under the Clean Air Act, penalties for nonconformance are very severe. Civil penalties can range as high as $25,000 per violation per day, to a maximum of $200,000. Certain violations (e.g., when responsible persons had "prior knowledge" of the condition, and the event posed a potential for significant injury or death to workers or the public) also carry criminal penalties that can include fines of up to $250,000 per day ($500,000 for corporations) and, as felonies, up to 5 years imprisonment.

Oil Storage & Containment Piping

Oil storage and containment is a critical environmental topic because of the potential of soil and ground water contamination that can occur. Several states have strict codes and regulations regarding the installation and construction of new, and the testing of existing, oil storage tanks and their installation and construction. Underground oil piping is also under strict regulations. Oil storage tanks are generally classified as either above grade or underground (buried) tanks. Some of the regulations and requirements that should be expected when planning the installation of new above-ground and below-ground storage tanks and associated piping are outlined in the next section.

Before proceeding with any other part of the project, the local fire department and governing bodies should be contacted, as they usually control the storage of fuel-burning material. They typically review the location, flammability, and quantity of stored materials and issue permits for their use.

Typical Design and Code Requirements
General for all Storage Tanks:
- Permits will be required for tanks in excess of 10,000 gallons capacity.
- A dike may be required to surround the above-ground tank. The capacity of the dike should be at least 125% of the capacity of the tank.
- Manholes should be provided in the tank for access.
- All tanks should be a minimum of 10 feet away from all buildings.
- Maximum diameters and lengths of tanks for different tank capacities, whether horizontal or vertical, should be within code limits. For example, a 2,500 gallon vertical tank should be no greater than 72″ in diameter and no greater than 12′0″ high.
- Exceptions for tank sizes are permitted, when approved by local governing bodies, for tanks of very large capacities (25,000 gallons and greater).
- All tanks should have filler and vent pipes.
- Contractors installing tanks and associated piping should be fully certified.
- All tanks should be tested and inspected regularly.

Underground Storage Tanks:
- All tanks should be designed and constructed to minimize the risk of corrosion and leakage.
- All tanks should be double-walled fiberglas or steel, or a combination thereof.
- Provide leak detection monitoring of interstitial space between walls.
- Tanks should be equipped with a splash plate at the bottom of the tank at each opening.
- Steel tanks should have cathodic protection and electrical isolation.
- Tanks should have a spill containment manhole and an overfill protection device.
- Some local regulations may require that leak detection equipment and cathodic protection monitoring equipment be installed on a new underground tank.

Underground Piping:
- All piping should have secondary containment such as impervious liners or double-walled piping.
- All new piping should be constructed of noncorrodable material.
- Piping should be installed in a trench on a 6″ bed of well-compacted noncorrosive material such as clean washed sand or gravel.

- Pipes at the tank should be installed in a manner to prevent a siphoning condition.
- Where practical, piping should slope back toward the tank.

Figure 6.5 shows a detail of an above-ground fuel oil tank located in a bunker. Even though the tank is below grade, since it is housed in a concrete room, it is considered above-ground. Tanks in building basements are also normally considered to be above-ground tanks because they can be visually inspected easily.

Figure 6.6 shows a detail of a buried (underground) fuel oil tank. It requires a double wall, monitoring equipment, and tie-downs to prevent heaving when the tank is empty and the ground is wet. The detail also shows a heating element that would be used for low-grade oil, which needs to be heated when cold in order to make it flow through the piping to the burners.

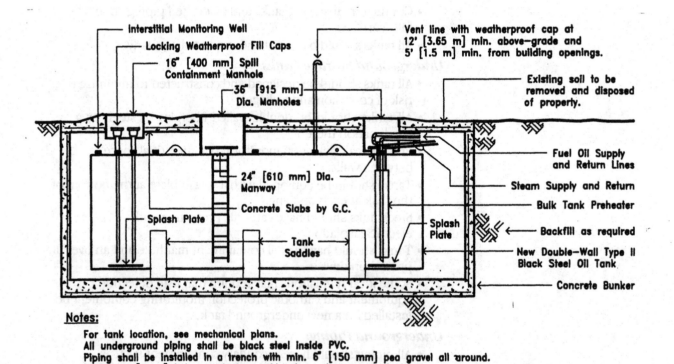

Notes:
For tank location, see mechanical plans.
All underground piping shall be black steel inside PVC.
Piping shall be installed in a trench with min. 6" [150 mm] pea gravel all around.
Tank in a "room" below grade is considered as above-grade tank.

Section of an Above-Grade Fuel Oil Tank

Figure 6.5

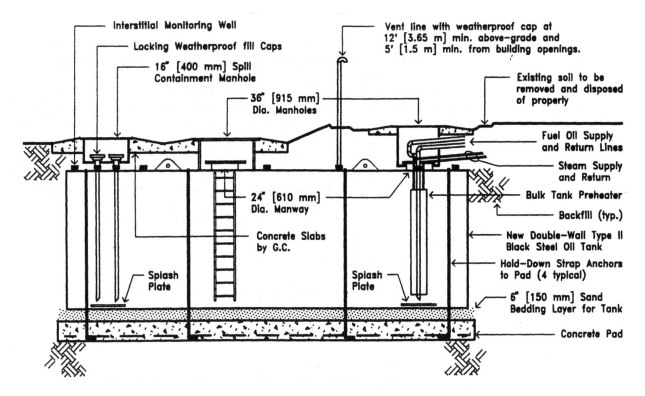

Longitudinal Section

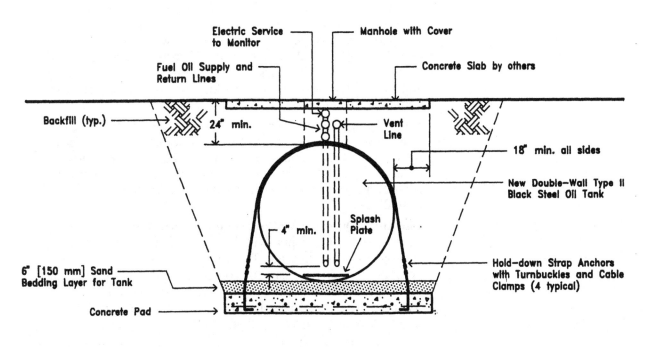

Cross Section

Sections of Buried Fuel Oil Tank

Figure 6.6

Radon Gas

Radon is an odorless, colorless, radioactive gas with a short half life (less than four days). Radon seeps into spaces, primarily basements, from the earth below. Radon attaches itself to airborne particles, which can be inhaled. Once inhaled, the particles cling to the mucus lining of the throat area, where they emit damaging radiation as they decay, causing lung cancer. There are several accepted means of drawing Radon out of basements. First, a pipe can be placed from below the basement slab up to the roof. A fan is installed in this pipe and runs continuously, drawing the gas through the gravel under the slab and out the roof vent. This system can also be installed without the fan, relying on stack effect to draw the gas out. Second, an air-to-air heat exchanger can be placed in the basement, drawing in fresh air and exhausting the radon-laden air. Third, the basement can be positively pressurized with a fan. Negative pressurization is not acceptable, as it can draw more Radon up through the slab and potentially distribute it around the building. The systems of venting Radon gas are shown in Figure 6.7.

Indoor Air Quality

A topic of increasing interest and awareness to building users and regulators alike is indoor air quality (IAQ). Modern materials and technology have both increased the potential for building contamination and provided ways to alleviate it. Following are some common factors that affect building air quality in existing facilities.

Maintaining Positive Pressure. With rare exceptions (such as laboratories and certain infectious disease rooms), buildings should be slightly positively pressured in relationship to the outside. In basic terms, buildings should be "pumped up" with more fresh air from the outside than they exhaust from toilets or general building exhaust. This is most difficult to achieve with variable air volume (VAV) systems, particularly in spring and fall when large quantities of both supply and exhaust air are being handled. Having controls in place that will properly track the air quantities of both the supply and return air while maintaining building pressurization will result in better overall air quality.

Maintaining Chillers During the Winter. Chillers that are shut down in winter (sometimes as a result of cooling tower shutdown) often result in the inability to cool air on warm "swing" winter days. This causes building temperatures to rise, resulting in greater perception of air quality problems and a general feeling of stuffiness.

Closing of Outdoor Air Damper on Freeze Warning. In winter, fresh outdoor air dampers will close to protect heating and/or cooling coils from freezing. This results in less fresh air to the occupants. While necessary to protect the coils, other designs such as "run around loops," or at least a quick response time for maintenance staff to manually open dampers helps to maintain a feeling of wellness in the building.

With the advent of sick building syndrome, IAQ has become a major topic in the building industry. Building owners, architects, engineers, building managers, and occupants are now actively seeking to ensure that indoor environments are safe and productive. While concerns over indoor air quality have resurfaced recently, proper ventilation and air filtration has evolved in an attempt to keep buildings habitable and to vent out the fumes and stenches that made people sick. The primary reason for poor IAQ

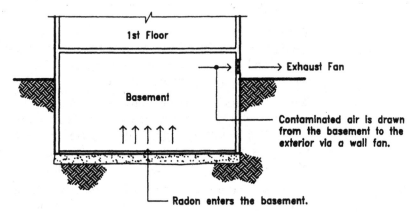

Space Depressurization

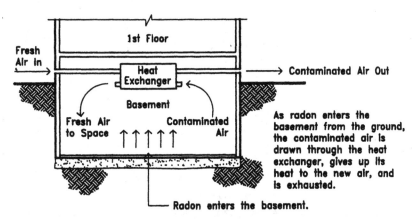

Air to Air Heat Exchanger

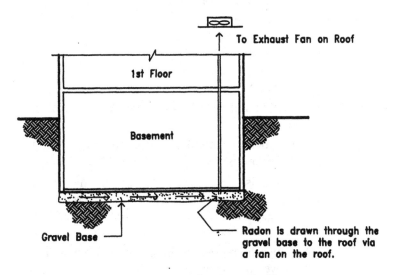

Sub-Slab Depressurization

Methods of Venting Radon Gas

Figure 6.7

recently is that modern buildings are more airtight and some modern building materials are more prone to pollute the indoor environment. Safer material alternatives should be investigated, and they need to be vented properly.

The effects of poor IAQ are extensive. Respiratory illness, allergic reactions, headaches, and drowsiness are some of the ill effects it can cause. The result can be poor health and, consequently, lack of production, lost work days, and low morale among workers.

While ventilation standards have been increased in response to IAQ, it is important to realize that ventilation is not the cause, and often is a weak cure to the problem. Realizing that if there were no indoor contaminating pollutants, there would be no indoor air quality problem, goes to the heart of the issue. Poor IAQ is caused by any combination of the following:

Excessive emissions of Volatile Organic Compounds (VOC). These originate in large measure from "wet" building materials, such as:

- Plywoods and particleboards containing urea-formaldehyde resins
- Paints
- Adhesives
- Caulking

And other sources such as:

- Cleaning, waxing, and polishing agents
- Equipment
- People
- Office products

Products in the environment, such as:

- Tobacco smoke
- Nitrogen dioxide
- Carbon dioxide
- Carbon monoxide
- Radon
- Formaldehyde
- Sulphur dioxide
- Ozone
- Asbestos

Lack of adequate fresh air ventilation to the building or space in question.

Products of combustion from heating plants and vehicles entering the building through the ventilation system or through windows.

Poor IAQ is often improved by increasing the fresh air supply and exhaust and by controlling emissions from materials. However, extensive investigations and testing are often required to find the actual sources of the poor air quality.

Although increased fresh air ventilation will dilute the air and consequently improve the quality of the air, it will not stop the emissions of the pollutants; therefore, indoor air quality is as much an architectural issue as it is a mechanical or ventilation issue.

Architectural Design and Emissions from Building Materials
While traditional construction materials, such as masonry and plaster, are relatively benign once installed, many newer materials contain volatile

organic compounds (VOCs). Most of the VOCs emitted from building materials are mucus membrane irritants and consequently there is a high rate of related symptoms for people in new, remodeled, retrofitted, or refurnished buildings where VOCs are present.

"Natural Materials"

Selecting building construction materials requires some consideration. While natural materials and sustainable design are desirable, there are disadvantages. Many natural materials will decay. Wood studs rot, and wool carpet attracts moths. To combat this decay, chemicals and preservatives are used that are deadlier than some of the artificial substitutes like metal studs or nylon carpet. There are also some natural materials such as lead, arsenic, asbestos, and formaldehyde that we all would rather not use in any case. Also, while there are newer materials that are less polluting, they are often less durable. Therefore, they will be reapplied more often, while a building is occupied, creating a higher frequency of emissions.

"Toxin-Free Products"

In the early stages of project design, the architect must be aware of the building products, their content, and the emission rates of VOCs. Specified products should have low pollutant emission characteristics. As noted earlier, many adhesives, paints, and sealants have very high VOC emission characteristics. Carpeting, too, can have a high emission of compounds such as formaldehyde and fibers. Manufacturers are producing equipment and materials with lower emission rates and many claim that their products are "low polluting," "non-toxic," and "environmentally safe." The American Institute of Architects is one source of information on environmentally friendly building materials.

Currently there are few regulatory guidelines or limits for indoor air pollutants. The number of chemicals utilized in building construction is vast, and the sensitivity of individuals to any one chemical is widely variant.

Baking VOC During Construction

During construction there are procedures that help to considerably reduce VOC emissions. It is typically necessary to run the HVAC (particularly the heating) system during construction to provide warmth for plasterers and other finish trades. The heat often will "bake" the VOCs, vaporizing them so that the exhaust air system can remove them from the building before occupancy. A building may be deliberately baked for a week at warm temperatures with a high ventilation rate to help alleviate pollutants.

Architectural Layout

Good internal building layout contributes to good IAQ. It is important to separate pollutant-generating activities such as food preparation, graphic arts, physical exercise, smoking rooms, and photography from other areas of the building. These areas should be equipped with 100% exhaust and ventilation air systems to convey contaminants directly outdoors. These areas should be kept under negative pressure with respect to abutting occupied spaces. It is important to avoid installing "fleecy" products in which microbiological organisms can reside.

Office layout and structure have a considerable affect on IAQ. Most office partitions absorb VOCs when the building or space is initially completed and release a constant rate of VOCs into the environment. In this way, many of the VOCs are not removed by the ventilation system at an early stage of occupancy. Office partitions also interfere with good air distribution and circulation, causing air pockets or voids with stale and contaminated air.

Open shelving with large amounts of paper products are another great source and home for VOCs and microbiological organisms.

Operable Windows

Operable windows in offices and new buildings were phased out in the 1960s in order to rely on mechanical ventilation systems and to control energy consumption. However, studies have suggested that incidences of poor IAQ, building-related illnesses (BRI), and Sick Building Syndrome (SBS) have been more frequent in mechanically ventilated buildings than in naturally ventilated buildings. Including operable windows in buildings could increase the IAQ and give occupants more control over their environment.

Site Layout

Site layout is also an important factor in controlling IAQ. As discussed previously in the ventilation section, locating loading areas, dropoff areas, parking areas, and high vehicular traffic areas away from air intake openings and any building openings can greatly reduce contamination of the indoor air.

Effect of Mechanical Systems and Ventilation on Indoor Air Quality

Mechanical ventilation systems can be either an effective method of controlling IAQ or a source of poor IAQ. Cleanliness, system components, filtering, air distribution, humidity control, recirculation of air, transfer of air, system layout, and ventilation rates could all have positive or negative effects on IAQ.

Cleanliness: The cleanliness of the intake louvers, ductwork, intake and recirculation filters, diffusers, and registers is important to maintain clean air circulation of ventilation air. Diffusers and registers are likely locations for dust and particulate matter to build up. A proper housekeeping program can ensure adequate cleaning. (See Chapter 11 for maintenance guidelines, and *Means Facilities Maintenance and Repair Cost Data* for specific recommendations on frequencies of maintenance tasks, with associated costs.)

Systems components such as cooling towers, cooling coils, drain pans, humidifiers, mixing boxes, and man-made fibrous insulations and liners are areas where many microbiological organisms can flourish and where organisms like *Legionella pneumophillia*, the bacteria that causes Legionnaires Disease, may be found. These organisms can be picked up in the airstream and circulated throughout the building. Cooling towers must be kept clean and treated with chemicals to avoid contamination. Water introduced into a duct system is a major cause of mold. Leakage of coils, backed-up condensate line, and steam leaks all warrant priority status on the repair list. Cooling coils and their drain pans should be cleaned regularly, and drain pans should be checked to ensure that they are sloped to drain. Humidifiers should be dry steam-type and not water-spray-type. Unitary portable humidifiers should be maintained and cleaned regularly. Mixing boxes should be inspected and cleaned regularly. Fibrous insulations should not be used as duct liners and should be avoided where possible. Chapter 23, "Health Care Facilities", reviews the impact and design criteria of these items for hospitals.

Filtration: System filters should be designed and selected for recirculation air and for outside air. Low-efficiency filters are not recommended. Recirculation filters should be capable of removing all VOCs from the airstream. Outdoor air filters should be selected based on their capacity to sufficiently remove from the air the contaminants that are common in the outdoor environment of that area. City projects require filters adequate for vehicular fumes; specialized or HEPA filters may be required for industrial applications or for special applications in hospitals.

Air Distribution and Exhaust: Proper air distribution is critical to overall air quality. Not only should the proper quantity of air be delivered to a space, but the distribution of the air to match the space layout and the density of occupants is an essential component of overall IAQ design. Moveable office partitions can cause problems by being too close to the ceiling and blocking distribution of the air to the occupants. Sometimes moveable partitions that worked well in an initial layout do not work well with a new layout.

Air movement and transfer, local exhaust, and dedicated supplies can prevent bad air from circulating through the wrong spaces. Kitchens and laboratories should be equipped with hoods and kept under negative pressure; graphic arts rooms and smoking lounges should have dedicated exhaust fans and should be kept under negative pressure. Office areas should be kept under relatively positive pressure. Air should not be recirculated from any areas such as kitchens, bathrooms, cafeterias, laboratories, print rooms, and similar spaces where there is a buildup of air pollutants.

Basements often have a buildup of humidity where microbiological organisms and mold will thrive. As these contaminants build up, they propagate throughout a building in time. Basements should be dehumidified and should have a source of ventilation.

Ventilation Standards: When deciding on ventilation rates, an analysis should be made of the building structure and materials. It is important to provide fresh air to compensate for VOC emissions and to provide adequate fresh air for the occupants.

Specific Code Requirements: Currently the only codes and regulations set to improve IAQ are those codes that set minimum values to the volumes of fresh air and air change rate that a space must receive. In the early 1970s the required fresh air rate per person was considered high (15 – 20 cfm per person). Following the oil crises in the mid-seventies, this level was reduced considerably to 5 cfm per person. However, because of the resulting poor IAQ, ASHRAE revised their standard (62-1989 Ventilation for Acceptable Indoor Air Quality) to increase the recommended fresh air supply to 15 – 35 cfm per person for most occupancies. It is expected that most state and local codes will adopt the ventilation rates ASHRAE recommends.

One consequence of increased ventilation standards is increased energy costs for the additional fresh air. Studies have indicated that while there is indeed a cost increase, the total energy impact for most buildings is under 10%. Schools, hotels, and retail buildings will be particularly affected by increased ventilation standards as a result of their high number of occupants. As these ventilation standards are implemented, the volume of fresh air and the energy consumed to condition that fresh air will be evaluated with new technology to improve the IAQ of buildings.

Summary of Procedures for Meeting Code Requirements

With the host of regulatory issues and code requirements, effective facilities management means initiating a system to deal with meeting the standards. The basic steps are:

1. Identify the applicable standards. Obtain a copy of the local, state, and federal laws and related professional accrediting societies that affect your facility's operations.

2. Determine the specific standards that affect your maintenance program.

3. Determine how well current systems meet the applicable standards.

4. Identify changes necessary to "meet or beat" the standards.

5. Determine costs, manpower, and equipment necessitated by the changes and obtain funding for them.

6. Prioritize the needed changes according to health effects, time, and available funding.

7. Prepare an action plan showing which items will be done and when.

8. Implement the program and achieve compliance.

Refer to the back of the book for the Bibliography/Recommended Resources.

Part 2

Maintenance Requirements by Building Component

Part 2

Introduction

Part 2 is a review of major building components that most commonly require maintenance and repair. It is organized roughly according to the UniFormat, a system of recording construction elements in the sequence in which they are actually installed. Part 2 begins with "Foundations, Substructures and Superstructures" — and ends with, "Landscaping" which would be among the last items completed in a new construction project. (See Chapter 14 for more on the UniFormat.)

Specialists in each of the listed building elements have presented guidelines on the items most likely to require attention in the form of inspections, maintenance, and repair. This information serves as a useful overview for facility or maintenance managers organizing a maintenance program or building audit, whether using in-house staff or outside contractors. The guidelines provide an idea of the relative labor, code, and inspection requirements of various systems. Managers can apply this information to planning the work of in-house staffs, and to overseeing outsourced work. Facility owners or prospective owners should find these guidelines helpful in assessing future building operating requirements and costs.

The chapters in this section do not follow an identical format from one to the next, although each is intended to address the most pertinent issues for that particular building element. For HVAC, for example, inspections and testing are a significant requirement, whereas landscaping, other than issues like ADA access requirements, is not tied to codes in the same way, and therefore, does not involve the same type or frequency of inspections for safety. The landscaping chapter also takes a different approach in that it focuses substantial coverage on the enormous financial benefits of incorporating maintenance programming considerations into new landscape construction at the outset.

Part 2, used in conjuction with Part 4, "Special Maintenance Considerations by Building Type," should provide facility managers with a good basic knowledge of the building systems of greatest concern in their type of building, together with what must be done to keep those systems in safe and efficient operating order.

The Facility Audit

The maintenance planning and budgeting process must begin with an inventory of the facility's deficiencies. The facilities audit is a vehicle for producing such an inventory by assessing the existing physical condition and functional performance of buildings, grounds, utilities, and equipment. The facilities or maintenance manager uses the audit results to plan major and minor, urgent and long-term maintenance and repair, for short- and long-term financial planning. The basic principles outlined below can be applied to any scale operation, from a single structure to a facility consisting of multiple building complexes in dispersed locations.

Other terms for the facilities audit include *inspection, survey, review, analysis*, and *site assessment*. All have in common a physical visit to the various parts of a facility, with a corresponding record of conditions observed.

Basic Structure

The audit process has three basic phases: designing the audit, collecting the data, and presenting the findings. In this approach, the audit's scope must first be determined. Next, the audit team is selected, and a set of forms designed or obtained from a source like *The Facility Manager's Reference*, by Harvey H. Kaiser. (See Bibliography/Recommended Resources.) A survey team then records field observations on these forms for individual building components. Finally, information from the forms is summarized, and priorities set, presented, and used to:

- develop cost estimates for capital repair and replacement projects
- define capital repair and replacement projects to address deferred maintenance items
- prioritize remaining deferred maintenance items
- eliminate conditions that are either potential safety hazards or could damage property
- define regular maintenance requirements
- restore functionally obsolete facilities to a usable condition
- identify energy conservation measures
- compare a building or facility's condition and performance to other buildings or facilities

Detailed coverage of the audit process is beyond the scope of this book and can be found elsewhere, such as in *The Facility Manager's Reference*. The audit is a starting point for many of the topics covered in this book, including capital planning, prioritizing deferred maintenance items, establishing reserve funds, benchmarking, analysis of the benefits of outsourcing and CMMS and, of course, evaluating existing building systems in order to prepare a maintenance program and budget.

Chapter 7

Foundations, Substructures and Superstructures

Stephen C. Plotner

The best way to evaluate and analyze an existing structure or facility is from the bottom up. By beginning an inspection of existing conditions at the bottom, the inspector, whether the building owner, facility manager or plant engineer, can observe the structure as it was built. Structural elements can then be identified and followed up through the building. This chapter will identify, following the UniFormat divisions of Foundations, Substructures and Superstructures, the major structural components of a building, deficiencies that can be observed during a facility audit, and their possible causes. (See Part II, "Introduction," for more on the UniFormat.)

Foundations

The foundations of an existing building may consist of stone rubble, concrete, wood or steel piles, or other material. The foundation material and, if possible, its thickness and other characteristics, should be noted on a facility audit, because this information will become essential when analyzing causes of deficiencies, and planning and estimating any subsequent repairs. Foundations can take the form of strip footings, step footings, spread footings, equipment footings, grade beams, piles, pile caps, caissons or other special types. Analysis of the condition of foundation elements and the cost to repair any deficiencies requires very specialized knowledge and may require the expertise of a structural and/or geotechnical engineer. The following are some common deficiencies that have been noted in foundations, and their possible causes.

Settlement, Alignment Changes, or Cracks

Most causes of foundation settlement and misalignment are related to the condition of the underlying soils. Soil load-bearing capacity could be reduced due to shrinkage, erosion or improper compaction of soils. Adjacent construction could disturb or even undermine existing building foundations. A change in the depth of protective backfilled soil could subject foundation footings to frost upheave. Foundation deficiencies may also relate to design considerations. For example, static building equipment loads and other live loads could now be exceeding the design loads, and vibration from heavy equipment could be exacerbating this problem. Structural changes or modifications to the building could be another cause of settlement or misalignment. Also, forces from past seismic activity could have produced foundation deficiencies.

Surface Material Deterioration

Concrete can spall due to the corrosion of reinforcing steel, penetration of water into aggregate pores or cement voids, chemical reaction between cement and soil, and/or chemical reaction between cement and concrete aggregate. Steel piles and other ferrous metals can corrode if they come into contact with acid-bearing soils. Wood piles can decay if subjected to insect infestation or if they come into contact with air due to a change in the groundwater table.

Substructures

Substructure elements include slabs on grade, as well as basement foundation walls and their exterior moisture protection, and interior insulation. Deficiencies in substructure elements are easier to recognize than those in foundations because they are readily visible. The following paragraphs describe some common deficiencies in substructures and their possible causes:

Slabs on Grade

Deficiencies in concrete floors and other slabs on grade usually manifest themselves first at the joints. Joint-related distress is influenced or exacerbated by moving loads that cross the joints, but it is usually caused by other factors such as underlying soil conditions, changing moisture conditions, improper construction materials and/or methods, and temperature. Underlying soils can settle unevenly due to inadequate compaction. A change in the groundwater table or groundwater flow can cause the erosion of underlying soils or heaving of the slab due to hydraulic pressure. Heaving or buckling of a slab could be caused by the freezing of moisture within poorly drained underlying soils. Curling of a concrete slab is caused when the top of the slab dries faster than the underside during original construction and is evidenced by the edges of the slab having "curled" up to a slightly higher elevation than the rest of the slab. If the edges of two adjacent sections of slab have curled, spalling of the concrete at the edges could be caused by differential vertical movement as loads cross the joint and depress the curled edges back down to the underlying base.

It should be noted that all concrete will do two things—harden and crack. Cracks should be expected in any concrete structure whether it is a slab, beam, column, wall, or other element. The building owner or facility manager should be alert to the development of new cracks, the noticeable widening or a lengthening of existing cracks, and cracks that are exuding any foreign material, especially rust-colored substances. Such conditions indicate a more serious deficiency, and should be investigated by a qualified engineer.

Foundation Walls

Frequently, foundation wall problems are linked to foundation deficiencies. Settlement or misalignment of foundation elements causes forces to work on foundation walls, resulting in cracks, moisture penetration, and buckling. The results could lead to structural failure. As with cracks in floor slabs, the building owner or facility manager should be alert to the development of new cracks in foundation walls, the noticeable opening of old cracks, and cracks that are exuding any foreign material, especially rust-colored substances. Again, such conditions should be investigated by a qualified engineer.

Moisture Penetration

A change in groundwater table conditions could be the cause of moisture penetration through the slab-on-grade or foundation wall if perimeter foundation drains and sump pits and pumps are nonexistent, inadequate, or ineffective. Excess underground moisture could also be caused by leaking utilities, such as the water service, storm drainage, roof drain collectors, and sewer pipes. Surface water runoff could be a problem if the exterior grades are not sufficiently sloped away from buildings. Inadequate or defective waterproofing will allow this excess moisture to penetrate the foundation walls and slabs via cracks, construction joints, and utility penetrations. Interior condensation could be caused by an inadequate vapor barrier, insulation, ventilation, and/or dehumidification.

Superstructures

The most basic elements comprising a building's superstructure are columns, floor systems, and roof systems. Stairs, ramps, balconies, and canopies could also be included. The primary materials encountered in a superstructure inspection are concrete, steel, and wood. Problems with various components of the superstructure may not be recognizable or apparent at first, because attention will be focused on readily observable deficiencies such as cracks, the movement of materials, moisture penetration and discoloration of elements comprising the exterior closure, roof and/or interior finishes. However, the observation of outlying deficiencies will require further investigation to determine whether they are limited to these surface materials and building elements, or if the cause goes deeper to a deficiency with one or more superstructure components. After the cause of a superstructure deficiency has been identified, immediate action may be required to prevent further damage to the building. Such action might include the removal of excess loads, the repair of spalled concrete, or the enhancement and/or replacement of a structural member. Estimates for the repairs can be completed in-house by following a formal planning and estimating process, and with the informed use of published cost data such as *Means Facilities Maintenance & Repair Cost Data* and *Means Facilities Construction Cost Data*, but may require the services of a qualified engineer.

Concrete

Concrete is a composite material and subject to more types of deterioration than steel or wood. Failure of a concrete structural member may have been caused by incorrect design or improper construction methods, rather than by a more recognizable outside force or influence, and may require the analysis of the original design criteria and the laboratory testing of actual material samples to determine the cause. Some of the other outside forces that can cause deficiencies in concrete members are as follows.

Misalignment of concrete members could be caused by the settlement of underlying foundation elements, changing soil and/or groundwater conditions, or changing load conditions. Excessive deflection in concrete members could be caused by changes in design loads. Cracks could be caused by changes in design loads, stress concentrations, extreme temperature changes, excessive expansion and/or contraction, and secondary effects of the freeze-thaw cycle. Scaling, spalls and pop-outs could be caused by extreme temperature changes, moisture penetration, reinforcement corrosion, environmental conditions and mechanical damage. Stains could be caused by a chemical reaction between reinforcing steel and materials in the concrete mixture, or by other environmental conditions. Exposed reinforcing could be caused by moisture penetration, corrosion of steel, insufficient concrete cover, and mechanical damage.

Concrete Repair and Maintenance Illustrated, published by R.S. Means Co., Inc., contains a thorough treatment of the subject of concrete repair and maintenance, including factors that influence the behavior of concrete, test methods, and survey procedures for evaluating existing concrete structures, illustrated repair techniques, and maintenance strategies.

Steel

Steel is one of the most versatile structural materials because it is strong, has stable chemical properties, is permanent if maintained, and is ductile, adaptable to prefabrication, easily erected, weldable and reusable. Despite its resilience, however, steel must be regularly inspected and repaired in order to preserve its structural integrity. As with any structural-type members, not only should the members themselves be inspected, but also their connection to other members. When performing an inspection, notes should be made on the facility audit as to the present condition of the structural steel, the observable surface deteriorations, such as defects and rusting, and the dates on which any particular members or elements were repaired. Movable structures and structural components subjected to moving loads should be inspected every three months to one year, and stationary structures should be inspected every one to five years.

Deficiencies in structural steel members can be caused by various factors and influences. Misalignment could be caused by the settlement of underlying foundation elements, changing soil and/or groundwater conditions, or improper fabrication. Excessive deflection could be caused by inordinate expansion and/or contraction, changes in design loads, or fatigue due to vibration or impact. Corrosion could be caused by electro-chemical reactions, failure of protective coatings, and excessive moisture exposure. Surface deterioration could be caused by excessive wear or mechanical damage.

Methods of protecting structural steel members include painting, the application of other protective coatings, encasement in concrete or similar materials, and armoring or covering with other protective materials.

Wood

The use of wood as a structural building element in commercial buildings has been greatly reduced due to stringent building and fire codes. This trend is reversing itself, however, in the small office/condo buildings being built in various areas of the country. Wood has been used for centuries in the construction of homes, light and heavy commercial buildings, and many other types of facilities. It has been used extensively because it is readily available, resilient, durable, easily manufactured into useful sizes and shapes, easily fabricated and fastened, and it has a high strength-to-weight ratio, low heat conductivity, high resistance to electric current, and good weathering qualities when treated. Some types of wood are more durable than others and, in fact, some parts of the same species are more durable than others (heartwood is more durable than sapwood). Inadequately maintained wood structures will usually result in extensive deterioration, which will require replacement instead of repair. In fact, most of the repair of wood or timber structures is by replacement.

Deficiencies in wood members can be caused by various factors and influences. Misalignment could be caused by the settlement of underlying foundation elements, changing soil and/or groundwater conditions, improper design, or improper construction and fastening methods. Excessive deflection and cracking could be caused by inordinate expansion and/or contraction, changes in design loads, fatigue due to vibration or impact, failure of compression members, poor construction techniques, or general material failure. Rot and decay could be caused by direct contact with moisture and/or condensation, omission or deterioration of a vapor barrier, poor construction techniques, and damage by rodents or insects.

Preservative treatments will protect wood against decay and insect attack. Research has shown that available preservative treatments will extend the life of wood many times over its natural, unprotected life span. The cost of the initial treatment to prevent wood decay is many times lower than the cost of replacing the member in the future. Therefore, any lumber that will be used on the exterior of buildings, and in the interior, but in contact with or in close proximity to the ground, and in below-grade installations, should be treated with a wood preservative treatment.

Earthquake Protection

It is possible to design and retrofit earthquake protection measures, such as confinement strengthening or the use of seismic isolators, for components and elements of existing foundations, substructures and/or superstructures. However, this subject is beyond the scope of this book. Also, due to the fact that each building is unique and each location may behave differently when subjected to seismic forces, the services of a qualified structural and/or seismic engineer will be required. Before any cosmetic or remedial repairs are performed to earthquake-damaged structures, a qualified engineer must be consulted to determine if there is any underlying structural damage which would create an unsafe condition.

Conclusion

This chapter has identified the major structural components of building foundations, substructures and superstructures, and has pointed out commonly observed deficiencies of each, along with their possible causes. With the structural integrity of a building at stake, it cannot be stressed emphatically enough that these primary structural elements of any facility, large or small, must be thoroughly inspected on a periodic basis. The ongoing operations within a facility are dependent on a building with strong stable foundations, substructures and superstructures.

Chapter 8

Roofing

Ken Brzozowski, ACS, AIPE, ASTM, RCI

While regularly scheduled maintenance has become an integral part of many building and facilities operations, most facilities completely ignore one of their most critical and costly assets: the roof. For example, the U.S. Air Force has recently indicated that its new and replacement roofing costs range from $3 to $9 per square foot[1]. Using these figures, even a modestly-sized facility involves a major expenditure when a roof must be installed. The economic justification of roof maintenance is demonstrated in the facility's ability to spread costs over a number of years, thus avoiding a large, unexpected outflow of dollars. There are, of course, other good reasons for avoiding roof problems through a program of inspections and preventive actions. Low worker morale, operation downtime, poor public image, and damage to goods are just a few of the problems that can result from roof leaks.

Roofing maintenance is complicated by the fact that the roof is undoubtedly the most abused building component. It is subjected to all the weather factors, such as high and low temperatures, wind, hail, rain, and snow, plus the harmful rays of the sun. Flat or low-slope roofs are forced to carry the weight of snow build-up and ponding water. The building's occupants may use the roof as recreational and storage areas; equipment platforms; and exhaust areas for chemicals, cooking fats, and a variety of other harmful agents. This list of abuses is by no means complete, as building owners and nature are continually creating new and unexpected challenges for the roof membrane.

Program for Preventive Action

There are four key areas of focus in a preventive maintenance program (PMP) for roofs:

- **Determine the roof condition:** At the outset of the program, the roof should be thoroughly inspected. The visual examination should be complemented by some type of nondestructive analysis survey. The three most common of these nondestructive methods are based on *infrared, nuclear,* or *capacitance* technologies. Infrared technology is actually based on thermal differences between wet and dry areas of a roofing system. For this reason, it is useful only on insulated assemblies. It is also complicated by heat discharges onto the roof, heat sources under the roof deck, etc. Nuclear detection devices count hydrogen atoms and identify moisture by greater concentrations of the hydrogen found in water. They work equally well on insulated and uninsulated membranes and are not complicated by thermal effects. Capacitance methods function by detecting changes in a weak electrical field imposed on the membrane. These changes are created by the presence of moisture and, it is claimed, can be used to pinpoint the specific component(s) of the roof that is (are) wet. Nuclear and capacitance methods involve the use of a grid pattern on the roof with measurements taken at the intersection points, while infrared gives an overall scan of the roof field. These surveys are available through many independent contractors as well as consultants and material manufacturers. The inspection and survey establish a starting point on which to build a program for immediate needs as well as future actions. It is critical that this inspection be performed by someone knowledgeable in roofing.

- **Develop comprehensive files and records:** A background file should be developed for all roofs and roof sections. This file should include all warranties associated with the roof, past inspections, repairs that have been made, as-built drawings, and any other pertinent information. This file should be updated whenever new information is developed.

- **Schedule regular inspections:** Visual inspections should be made at least twice a year, preferably in the spring and fall. If weather conditions are severe, the frequency should be increased. Again, the person examining the roof should be well-versed in both inspection techniques and in the potential problems that can occur with roofing systems.

- **Develop a program for preventive action, restoration, and repair:** Several factors must be included in this phase of the program. Interior operations, condition of the roof system, available budget dollars, and type of roofing system are just a few of the many considerations in deciding the steps to be taken to correct roof problems. If the building owner is not expert in these areas, assistance is available from a variety of sources. The roof should be brought up to a condition that will allow regular maintenance to preserve its watertight performance.

The Roof Inspection

Actual roof inspection techniques will vary depending on the type of system in place. The three most common types of roof membranes are built-up systems, modified bitumen membranes, and single-ply sheets. The most common types of metal roof systems, whether architectural or structural, are through-fastened systems and standing seam systems. For an actual "on roof" inspection, the best approach begins with an understanding of the problem areas of each of these major roof systems.

By far the most reliable source of this information is *Project Pinpoint*. *Project Pinpoint* is a program conducted by the National Roofing Contractors Association (NRCA) that involves an annual survey of its member contractors regarding the types and numbers of membranes being installed and the problems being observed. Figure 8.1 shows the main problem areas listed by the NRCA from *Project Pinpoint* data in the period 1983-1992, in decreasing order of occurrence.

A review of the problem frequency information, such as that in Figure 8.1, provides the roof inspector with a good starting point when examining the various types of roofing systems.

A comprehensive roof inspection is multi-faceted and involves the membrane as well as all the building components that may affect the integrity of the roofing system. This discussion can only touch on the elements involved in a thorough inspection. It is important to realize that many leaks attributed to the roof are actually the result of water penetration through other building components. Examples are parapet walls that are not properly waterproofed, and window frames that are not sealed correctly, or have deteriorated from moisture entering through deteriorated wall flashings.

The following is a very general checklist of areas that should be part of a roof inspection:

- **Supporting Structure:** Examine exterior walls for cracks, deterioration, moisture stains, damage, failing caulk, and any other problems. Check the interior walls and roof deck.
- **Roof Condition:** Look for clogged drains, physical damage, new equipment/alterations, and debris. Check the surface for bare spots, slippage, degradation, and excessive ponding. Examine the membrane for blistering, splitting, etc.
- **Flashing Condition:** Flashings occur where the roofing membrane is interrupted or terminated; they should be inspected for any defects, as well as proper height. A low flashing

Most Common Roofing Problems			
Built-Up Roof Systems	**APP & SBS Modified Bituminous Systems***	**EPDM Membranes****	**PVC Membranes*****
Blistering	Lap/Seam	Lap/Seam	Shrinkage
Splitting	Flashing	Flashing	Embrittlement
Ridging/Buckling	Shrinkage	Puncture/Tear	Flashing
Flashing	Puncture/Tear	Shrinkage	Puncture/Tear
Slippage	Embrittlement	Wind Uplift	Lap/Seam
Wind Damage	Wind Uplift	Fasteners	Wind Uplift
Embrittlement			

* APP and SBS are two polymer modifiers used to improve asphalt properties.

** EPDM (Ethylene-Propylene-Diene Monomer) is a rubber membrane that greatly dominates the single-ply portion of the low-slope roofing market.

*** PVC (Polyvinyl Chloride) was the most common plastic roofing sheet used from 1983 to 1992.

Figure 8.1

allows water to come in over its top during heavy rains. Examine all perimeter walls and components for problems.

- **Roof Edging:** Check all edge metal for rusting, deterioration, loose fasteners, and any other potential point of failure or water entry.
- **Roof Penetrations:** Inspect equipment base flashings and equipment housings for water entry points; look for discharges that may damage the membrane. Check all vents for attachment problems and possible chemical release.
- **Expansion Joints:** Look for open joints, punctures/splits, rusting, securement, proper fastening, etc.
- **Pitch Pans:** These are metal collars placed around roof penetrations that are filled with hot bitumen or roofing cement. Although not recommended, they are in wide use. Inspect these for fill material shrinkage, attachment, the presence of bonnet covers, and any other problem.
- **Drains/Scuppers:** Examine for debris, proper hardware, membrane seal, etc.

The above is by no means a complete list of the steps involved in a roof inspection. It should, however, illustrate the complexity involved in a proper roof examination.

Metal Roofs

Metal roofs represent a substantial inventory of square footage and may have reached a point at which maintenance is critical to extend their service lives. In conducting a roof inspection, and in planning maintenance requirements, consideration must be given to the following concerns:

- Very low slopes or deflection of panels.
- Exposed fasteners.
- Excessive movement of components and failure to accommodate for this condition in the system's design.
- System components that are less substantial than required for the application.
- Penetrations and rooftop equipment.
- Heavy traffic on the roof because of equipment maintenance or other needs.
- Surface coatings that are not performing adequately.

Repair and Maintenance Options

The alternatives available for maintaining a roofing system depend on the type of membrane that is in place. Built-up roofing membranes (BUR), modified bitumens, and elastic or plastic single-ply sheets are each handled differently and require different materials for repair and maintenance. This chapter is not meant to provide detailed how-to instructions, but it will cover some of the options available for each of the major roofing types.

Built-up Roofing and Restoration

Built-up roofing (BUR) membranes have the greatest number of treatment and repair alternatives. BURs can be based either on coal tar (a derivative of coal) or asphalt (a derivative of petroleum). It is very important to know what bitumen type was used in the membrane construction as, in general, problems will result if coal tar and asphalt are mixed indiscriminately.

Asphalt roofs may be surfaced with gravel, granules, or any of a great variety of coatings. A roof finished with some type of coating (and no gravel or granules) is referred to as a *smooth-surfaced membrane*. Coal tar roofs are always surfaced with some type of gravel.

One of the main advantages of BURs is their multi-ply construction. Because they are applied in a minimum of three layers, BURs can sustain damage or weather erosion to the upper plies without causing a leak into the building. This is a key point regarding the maintenance advantages of multi-ply roofs. A good inspection program would be expected to locate problem areas and result in corrective action prior to the development of serious roof deterioration.

With gravel roofs, a general upgrading typically begins with a process known as "hydrovacuuming." In this step the gravel is removed and, at the same time, the roof surface is washed. The roof is now in an ideal state for a thorough inspection, and any defects can be repaired with a variety of cold mastics and reinforcing mats or scrims. Alternatively, hot coal tar pitch or asphalt and compatible roofing felts can be used to reinforce questionable areas. Additionally, if sections of the roof were found to require replacement, this work could be accomplished at this time.

The entire roof is recoated after all repairs and replacement work is completed. The owner is now in a position to choose any one of several hot- or cold-applied surfacing materials. Some of the choices include:

- A "resaturant" material, a solvent-based coating formulated with oils to soften the aged or brittle bitumen that it contacts.
- A "cutback" coating, a solvent-thinned bitumen containing some amount of fibers for reinforcement and viscosity control.
- A "rubberized," cold-applied coating that is formulated to accept greater roof movement with improved fatigue resistance.
- An application of molten bitumen, either coal tar or asphalt, whichever matches the bitumen type of the roof.

Into each of the above coating materials, a sufficient amount of gravel is applied to provide protection from weather, hail, traffic, and other abuses.

Granule and Smooth-Surfaced Built-up Roofs

The repair and maintenance of granule and smooth-surfaced roofs are more straightforward than with gravelled systems. The protection and longevity advantages of gravel do make inspection and remedial actions more difficult. The non-gravelled roofs require repair of any defective areas and replacement of sections in which leaks have led to water-saturated insulation.

After all the necessary work is performed to ensure membrane watertightness, a coating is applied for weather protection. These coatings are different from the products used on gravelled roofs and are generally one of the following types:

- *Asphalt emulsions* are water-based coatings known for their excellent weather and ultraviolet light resistance. They are often modified with some type of polymer to improve their performance properties in areas such as elongation, fatigue resistance, flexibility, etc. The materials are also available with various types of fibers added for increased strength and abrasion resistance.
- A great variety of *reflective coatings* are used as membrane surfacings to protect the underlying membrane from UV rays, to lower the roof surface temperature, and to reduce air conditioning costs. These materials can be solvent- or water-based and are formulated using a number of polymer types. Some examples are acrylics, urethanes, and synthetic rubbers. Asphalt-based aluminum coatings are also popular as a

surface finish layer. Reflective coatings can be used directly over the membrane or as a final protective material on emulsions.

Flashings on Built-up Roofs

Vents, equipment stands, edges, and penthouses represent just a few of the junctions in the roof membrane that must be sealed. These areas are critical to the watertightness of the roof—some say as high as 90% of roofing leaks are associated with flashing details. Any inspection should emphasize these points of the roof where repairs are often made. In general, the mechanics of flashing repair will be similar to those made in the field of the roof. It is always good practice to surface the flashings with some type of protective coating. If a protective coating (such as granules or an aluminum foil surfacing) is not already built into the topmost flashing sheet, it is essential that such an application be made in consideration of the critical nature of this part of the roof membrane.

Modified Bitumen Repair and Restoration

Modified bitumen systems combine the advantages of single-ply and built-up membranes. They offer the waterproofing of multi-ply, built-up systems, yet possess the flexibility of single-ply products. Modified products offering a great deal of specification latitude are installed using techniques familiar to most contractors.

There are two major types of modified bitumen products; both use asphalt as the starting material, but their polymer modifiers provide quite different characteristics. SBS (styrene-butadiene-styrene) produces rubber-like characteristics in asphalt. SBS products are softer and elongate more than APP (atactic polypropylene) materials, which use the second major type of modifier. APP produces improvements in the plastic properties of asphalt. This feature results in good high-temperature characteristics, excellent resistance to ultraviolet light rays (leading to good weathering properties), and better low-temperature properties than standard asphalt.

APP and SBS Repair

Because APP products are almost always installed using torching methods, this technique is also preferred for repair situations. Once the area to be repaired is identified, it should be thoroughly cleaned and primed with an asphalt primer. A patch sheet is then torched over the membrane break and the edges are sealed with a torch and trowel to form a tight bond. Standard asphalt mastics do not perform well on APP membranes.

There are several repair options available for SBS modified asphalt roofs. SBS materials can be applied by torch, hot asphalt or cold adhesive, and repairs can be made by any of these methods. Torch repair methods are similar to the techniques used for APP products. Hot asphalt is typically used only when problems cover a large area, as set-up of equipment is costly. Cold repairs can be made with special mastics or with sheet patches using appropriate adhesives. The sheet used for patching must be completely compatible with the membrane on which it is being placed.

Coatings for APP and SBS Membrane

An SBS membrane must always be protected from the sun's rays by some type of covering. This covering can be factory-installed (granules or metal foil) or field-installed (coatings or gravel). An important part of SBS membrane maintenance is ensuring that this coating is not disturbed or worn away. If repairs are made or the coating is eroded, it should be replaced as quickly as possible. At one time, there was some disagreement in the industry as to the necessity of coating APP membranes. Most manufacturers and specifiers, however, have taken the position that APP sheets should be protected in the same way as the SBS products.

The coating products used for SBS and APP membranes are generally the same as those used for built-up roofs. Emulsions (standard and polymer-modified) and reflective coatings are very often the surfacing materials. Since SBS and APP sheets are more flexible than standard asphalts, coatings with higher flexibility and elongation properties are preferred.

Elastic and Plastic Single-Ply Maintenance and Repair

The single-ply membranes offer the least number of choices with regard to maintenance options. Single-ply materials are specific in terms of the adhesives and patching materials that can be used for repair. Only the materials recommended or supplied by the manufacturer should be employed when correcting problems.

When seams are found to be defective, the best course of action is to place a 6–8″ wide strip of the same sheet over the suspect area. This strip should be applied using the same procedures and the same adhesive as those used when the seam was originally made.

Several suppliers are working to develop alternative maintenance and repair products for the single-ply market. This, of course, represents a significant need in the industry and offers great potential for the material manufacturers. Self-stick tapes are being explored for correcting seam problems and applying patches. In addition, a number of new coatings are in the field trial stage. Some of the challenges in identifying maintenance products are the difficulties in bonding to membranes such as EPDM (Ethylene-Propylene-Diene Monomer) and the greater movement and flexibility the single plies exhibit.

Metal Roofing Maintenance and Repair

It is of considerable importance that a thorough evaluation be made to determine the requirements or degree of maintenance necessary for a metal roof. Metal roof construction varies significantly, but the areas that almost universally require remedial treatment are:

- Fasteners
- Seams
- Projections or penetrations
- Rooftop equipment
- Rust
- Failed coatings

It is important to identify existing problems and select the appropriate level of maintenance. The available options include:

- Paints or rust-preventive materials.
- Coatings that also offer a degree of waterproofing.
- Systems containing several products to address a variety of existing problems/conditions.

For example, if seams are a problem, a single coating system should be selected that uses tapes or reinforcing membranes with mastics. If rust is the major problem, the expense of a full system employing sealants, mastics, and tapes would be unnecessary.

In actuality, maintenance needs are seldom as clear-cut as indicated above. A combination of treatments will most likely be necessary, especially if a large roof area is involved.

Following are some guidelines on the materials used in maintaining and repairing metal roofs.

Paints

For the purposes of this chapter, paints are considered to have little or no waterproofing capabilities. Their three primary purposes are to:

- Treat existing rust and inhibit the formation of new rust.
- Improve the aesthetics of the roof.
- Produce a more reflective surface and reduce temperatures in the building.

The products used are as numerous as the types of paints and treatments available. Alkyd rust paints, zinc oxide primers, one- and two-part epoxies, aluminum-pigmented materials, and rust conversion products are all employed for this application.

Sometimes primers are used in conjunction with top-coat product. Some manufacturers are incorporating rust-inhibiting chemicals into their surfacing products and suggesting or requiring two coats. This eliminates the need for having two separate products on the job.

Paints offer more weatherability, reflectivity and hiding than primers because of their higher pigment content. Paints generally are higher in viscosity than primers and therfore do not have the same ability to penetrate rust.

It is critical that the surface be prepared properly before painting. All loose rust must be removed by some means of abrasion and the substrate must be performed to determine compatibility, as well as to check for bleed-through.

Painting must not be considered a means of correcting leaks. Paints can improve reflectivity and aesthetics, but leaks must be repaired by other means.

Primers

These materials may be solvent- or water-based and are low in viscosity to ensure penetration and sealing of rust. They typically have rust-inhibiting additives that act to seal the rust and prevent access to oxygen in the air.

Other additives also can be formulated into the primers that chemically react with the rust to inhibit its spread.

Coating Options

Coatings materials are more heavily-bodied than paints. Consequently, they can be applied in greater thickness or build. Coatings generally have somewhat better mechanical properties than paints. Specifically, they can be formulated to provide good elongation and recovery characteristics, as well as good fatigue resistance. Due to these attributes, coatings supply a greater degree of waterproofing properties than paints and will seal pinhole leaks. They should not, however, be viewed as a long-term solution to leaks at bolt heads, joints or other areas of the metal roof that are prone to problems.

Included in the coatings category are asphalt-based aluminums, rubberized asphalt cutbacks, modified asphalt emulsions, and latex-based products. These materials serve the functions of paints (aesthetics, reflectivity, and rust treatment), but they also provide an additional degree of waterproofing not offered by lower viscosity materials.

Sealants/Mastics

These products are much heavier in viscosity than coatings and can be used at much heavier rates of application. They can be packaged in cartridges or cans and are supplied in trowel, brush or even spray grades.

Sealants and mastics are used to stop leaks. Their heavy body allows the placement of enough volume of material to create a substantial water barrier. They are used primarily in seam areas, over bolt heads or other types of fasteners, and in general where different components of a system come together.

These products many times are used in combination with some type of reinforcement mat or satin. The reinforcement is embedded between two layers of the sealant or mastic to give greater integrity and strength to the application. Many manufacturers are using polyesters as the reinforcing mat because of their ability to elongate and recover through many cycles (fatigue resistance).

Sealant and mastic materials can be made from many different polymer and resin materials. Acrylics, urethanes, butyl and silicone rubbers are just a few of the chemical types that can be used. Many times, the sealant simply is a higher viscosity version of the coating. This ensures compatibility between these two materials on the roof.

Tapes

To avoid the necessity of the mastic-scrim application, tapes now are being offered. These materials are self-sticking and have the scrim already incorporated into the tape. Polyester and polyethylene are two of the more common reinforcements being used in tapes. The advantage of tapes, besides easier installation and less dependence on the installation skills of the mechanic, is that there is no waste of time while curing is taking place. This allows the rest of the application to proceed without delay.

Surfacing

A number of optional surface finishes are available. The dark-colored or black coatings can be left as is or finished with aluminum or white paints. Colored granules can be embedded in the surface coating if desired. Specifically matched colors are being offered by many suppliers as well.

Planning for Maintenance

A roof designed without consideration of maintenance issues is destined for early failure. Specifically, this means that particular maintenance situations should be anticipated and provisions should be made in the design and selection of the roof to accommodate them. Designing a maintainable roof should be the first priority of every specifier. Everyone, however, from the owner, manufacturer and installer to the building maintenance manager, needs to play a role in achieving this objective.

The purpose of proper design, installation, and maintenance is to keep the roof in a leak-free condition and thus reduce its overall life cycle costs. Once water enters a roofing system, the roof's natural deterioration is greatly accelerated. Moisture in a membrane-type roof causes delamination of the various component layers in the assembly. It also causes degradation of the insulation (if any is present) and serious damage to the decking system. Additionally, wet insulation loses its resistance to heat flow, resulting in greater heating and air conditioning costs.

As indicated earlier, built-up roofing provides the greatest number of maintenance options. Modified bitumen materials offer good maintenance characteristics as well, because in most cases they are applied as two- or three-ply assemblies. At this point in time, the single-ply roofs are the least maintainable of the major roof types. In any case, steps can be taken to extend the service life of any roof membrane. Some roofs, of course, will necessarily have to be replaced. Theoretically, when part of a roof has deteriorated, it should have received the greatest amount of stress or abuse.

Reroofing allows the specifier to know the special factors to be considered when designing the replacement membrane for these areas.

The following are significant points that should be part of a design or specification when installing a roof:

Slope to drain: If all roofs had proper slope to allow for efficient and effective removal of water, there would be a significant reduction in problems. The combined action of ponded water and the sun's ultraviolet rays is devastating on any roof membrane. Coatings are also subject to this deterioration, which is why they need to be reapplied after several years.

The slope-to-drain issue is compounded in several ways. First, there is a natural tendency for roof decks to deflect and settle over a period of time. This is especially true in the case of lighter gauge metal decks. Drains are usually installed at building support beams, where a roof low point occurs. But, as time passes, roof deck deflection or sagging takes place. If the slope is minimal to begin with, ponding will likely occur. Further aggravating the problem is the fact that leaks will most likely occur in areas of the roof that are subject to the greatest degrading forces. Thus ponding will probably take place over the points of leakage into the building, making conditions even more intolerable.

Solutions to the ponding problem are not easy or inexpensive. Ideally, proper slope should be built into the deck when the building is constructed. However, because of the extra cost, as well as the need to heat and cool greater interior space, this is not typically done. There are several options available. One is to lower the drains, if possible. This will help the situation, but so-called "birdbaths," or random low spots, will still exist. A second choice is to remove the existing membrane and install tapered insulation to obtain proper slope. This can also be accomplished with lightweight concrete, which is cast-in-place and formed to achieve proper slope. Alternatively, the insulation or concrete may be placed on the existing roof and another membrane installed on the tapered layer.

Slope-to-drain is probably the most significant design criterion that can be employed to extend roof life. It eliminates ponding, reduces deterioration of the membrane surface, and minimizes water leaks into the building. Most roofing slope problems are, however, not faced until after the fact when the deck and membrane have been in place for a number of years.

Proper flashings: The sensitivity and importance of flashings have been discussed previously. Flashings should be installed using the best materials available; there should be no attempt to reduce costs in this area of the roof. Because flashings are typically subjected to a great deal of localized movement, only materials and adhesives capable of withstanding this movement should be used. For this reason, single-ply sheets are often used as flashings even if the membrane is a built-up system. The modified bitumen products are available with polyester reinforcements. Polyester is highly flexible and has excellent elongation, recovery, and fatigue-resistance properties. Consequently, these products are also gaining favor for flashing details.

If the roof under consideration is on a new building, every attempt should be made to minimize the number of flashings needed. This is not always possible, but a reduction in flashings is a reduction in potential leaks. If vents can be routed to side walls or if equipment can be located beneath the roof level, this should be done. Otherwise, the best advice is to make sure flashings are well installed with high quality materials. In a roof inspection, close attention should be paid to these critical parts of the roof.

Accessibility to potential trouble areas: Gravelled or ballasted roofs are more difficult to maintain than smooth-surfaced membranes, in terms of locating and repairing leaks. Ballasted or free-floating single plies are especially difficult in this area, as membrane leaks may be located some distance from where the water enters the interior. Since there is no attachment between the roofing components, water can run between layers before it finds entrance to the building.

The location of equipment on the roof can also pose problems. HVAC units, piping, and any other equipment placed near or on the roof surface make it difficult to perform repairs if leaks occur beneath them. Everything should be done to make provisions for the potential of roof repair under equipment. One option is installing all equipment at a height sufficient to allow work underneath.

Protection in high use areas: As stated earlier, it is best not to place equipment on the roof surface. If this cannot be avoided, some type of protection should be installed to handle worker traffic, dropped tools, burning cigarettes, etc. Roof pads, pavers, walktreads, bridges, and any other protective measures should be used to avoid damage to the roof membrane.

Budgeting for Maintenance

The financial advantages of maintaining a roof in first-rate, watertight condition have been mentioned. Protection of valuable office and factory equipment, along with finished goods and raw materials, is essential for sound corporate performance. Delaying a $100 roof repair over the part of the building that houses a million-dollar computer mainframe is foolhardy, and can be extremely costly. A long-term, regular roof maintenance plan will have long-term financial advantages.

A typical preventive maintenance program costs about 10 cents per square foot per year. This allows for good maintenance procedures and funds any restoration programs that are needed. On the other hand, a tear-off and replacement will cost, at minimum, about $3 per square foot. If factors such as proper disposal and landfilling of asbestos-containing roofing are considered, the cost is much higher. See Figure 8.2 for a comparison of typical maintenance costs vs. typical replacement costs.

It is easy to see how a thorough roof inspection might detect a small flashing break or an open seam that requires repair. A single repair cost is nominal at this stage of the problem. This "small" break could lead to disastrous consequences, however, if the repair is not performed promptly. If the roof were allowed to continue to deteriorate, the final result could be a very expensive roof replacement and serious damage to the building interior.

See the Yearly Roofing Maintenance Budget graph in Figure 8.3. In general, a roof in good to excellent condition will cost about 8 to 12 cents per foot to maintain on a yearly basis. A budget below this figure may mean that necessary work is being avoided. If a maintenance budget is much above the 12-cent figure, this may mean that there are major problems or that the roof is nearing the end of its useful service life. It may be time to take a

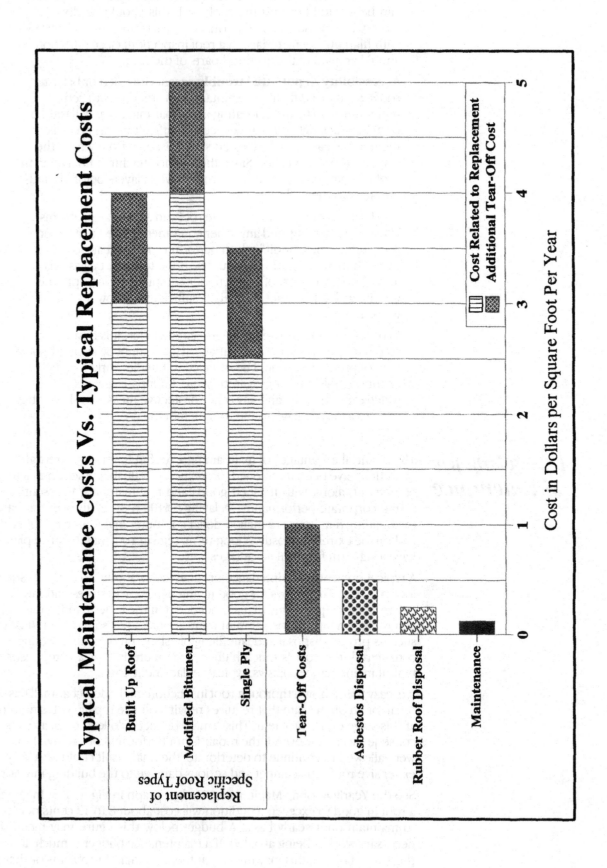

Figure 8.2

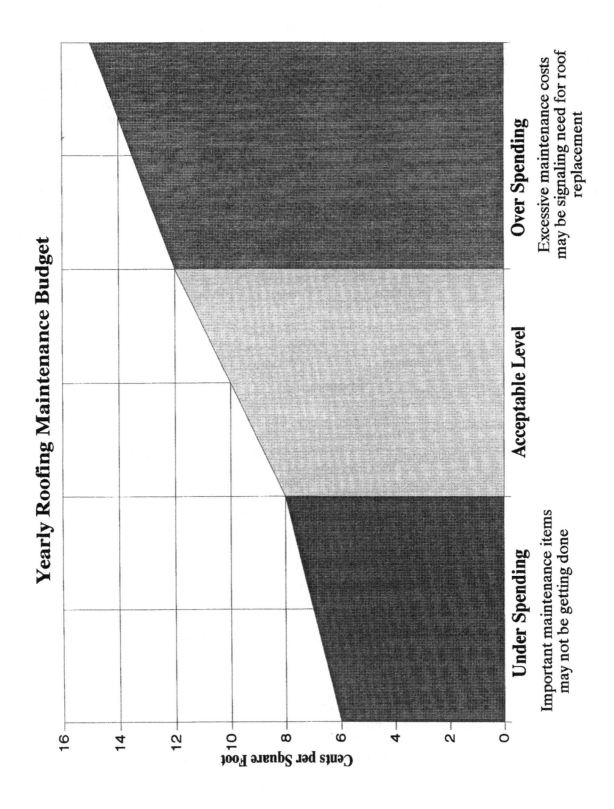

Figure 8.3

hard look at the roof's condition and consider some replacement work. Remember, roofs have a finite life expectancy. Good maintenance practices will produce maximum benefit from a roofing system, but every roof eventually fails.

Roofing System Guarantees

The matter of roof guarantees has recently become a favorite subject of discussion in industry groups, meetings, and periodicals. The National Roofing Contractors Association, for example, issued a Consumer Advisory Bulletin in August 1992 emphasizing the need to focus on the quality of the roofing system rather than the length of the guarantee. This, of course, is good advice when it comes to the purchase of any goods or service and has been largely ignored in roofing until the last few years.

The use of guarantees in marketing roofing products and systems accelerated in the 1970s when the single-ply membranes were introduced. Companies new to the industry needed credibility for themselves and their membranes, which were unconventional at the time. Suppliers continued to "up the ante" in terms of guarantee length, which probably resulted in the many bail-out clauses found in guarantee documents today.

It must be understood that manufacturers' guarantees are written by their legal departments and, as a result, attempt to protect the material supplier as much as possible.

A number of examples can be cited to demonstrate the various limitations that are found in guarantees. Only a few will be discussed here, as a comprehensive review of this subject is beyond the scope of this work.

Length of Guarantee

The terms of guarantee documents typically range from 5 years to as long as 31-1/2 years. As the lengths of guarantees increase, they begin to be modified with conditions and stipulations that are required to keep them in effect. For instance, some will require inspections by the material supplier every five years. The guarantee is continued only when the procedures recommended are performed using the supplier's products. Many guarantees are issued on a prorated dollar basis; thus, ten-year coverage would mean that the supplier's liability would decrease by 10% each year. While this is occurring, of course, the cost of replacing the roof is increasing each year because of inflation and other cost factors, placing an ever-larger burden on the building owner.

Other Limitations

There are many conditions found in guarantees that can negate a manufacturer's liability. These include:

- Lack of inspection at time of application.
- Repairs, alterations, or additions without approval of manufacturer.
- Failure to pay all bills.
- Failure to notify manufacturer of transfer of building ownership.
- Failure to have roof inspected within 60 days of installation anniversary.
- Failure to notify manufacturer of leak or defect within prescribed time limit.
- Change in building usage.
- Repair work by contractor other than approved contractor or use of unapproved repair materials.
- Lack of positive drainage.

- Exposure to gale force winds. (The Beaufort Scale of Wind Speeds identifies a moderate gale as winds of 32 − 38 mph.)

The above discussion appears to be an indictment of all roof guarantees. It is not meant to be so, but frequently an owner views the guarantee as a "security blanket" and fails to even read the document. If it was read, the responsibilities placed on the owner would be understood, and more emphasis would be placed on the quality of the membrane system. A surprisingly large number of roofing material suppliers have gone out of business in the last few years. Although not all of these actions have been the direct result of the guarantee liability, the owner is still left with basically a useless piece of paper. Parent companies are not liable for the responsibilities of subsidiaries, so the building owner can rely only on the good faith of the parent if a subsidiary cannot meet its obligations.

The message should be clear: Do not substitute a roof guarantee for a quality roof system. Even if the manufacturer lives up to all the conditions of the guarantee, the consequences of roof problems or leaks are not worth the money saved by installing a marginal system.

Despite the negatives cited above, there are real advantages to the roof guarantee. Some of these are:

- A pre-roofing conference is required to review installation techniques, material, and potential problems.
- An approved and properly trained applicator is required.
- Regular inspections are usually an integral part of the installation process.
- The building owner is alerted to the need for regular inspections after the roof is installed.
- The vast majority of manufacturers honor their commitments and resolve any problems associated with their materials.

Conclusion

This chapter has covered only a small part of a large and complex subject. Roofing is full of surprises: First, it is much more complicated than most believe, and second, it is much more costly than ever expected. This second point can be managed with a responsible maintenance program. It will certainly reduce expensive financial surprises. Large unplanned expenses are something no facility manager desires.

A good roofing program must involve individuals with training and experience. If not available within the building owner's organization, this expertise must be obtained through the services of a competent contractor, consultant, or material supplier.

[1] Firman, D. M., "Maintenance needs: An owner's perspective," *Professional Roofing*, August 1990, p. 23.

Chapter 9

Interior Finishes

Robert W. Mewis and Mary P. Greene

Introduction

Interior finishes are a primary focus of maintenance in most facilities. The extent of the maintenance program is naturally related to the function, and consequently, image requirements of the facility. For example, a corporate headquarters will likely have more costly finish materials, more frequent replacement of finishes (for reasons of design enhancements, if not higher traffic), and more stringent standards for general maintenance than that same corporation's satellite facility. A manufacturing facility will undoubtedly concentrate the substantial part of its maintenance effort on mechanical, electrical and equipment condition, rather than the aesthetic appearance of its visitor or office space. Likewise, the various parts of a building may have different maintenance standards and frequencies. The front entrance/lobby of a building is not only subject to the greatest amount of daily traffic, but it is also the most visible, and the first element visitors experience in the facility. Consequently, it may receive more frequent and careful cleaning and immediate attention to repairs, as compared, for example, to office space on the third floor.

Maintenance of interior finishes, over the long run, represents a substantial portion of the overall maintenance budget, along with electrical and mechanical systems. In addition to cleaning and repairs, carpeting, flooring, acoustical tiles, and wallpaper are periodically removed and replaced, and walls are repainted, not only to address wear and tear problems, but to enhance or respond to changes in the organization's mission and/or image. Interior finishes may also change in response to churning within a facility, where downsizing or reorganization leads to reconfiguring the space. In some instances, the maintenance manager may not be responsible for all interior space improvements, which may be the responsibility of tenants.

Selection of Materials

The interior finish materials selected for new construction or remodeling projects has a significant impact on the life cycle, and therefore long-term costs, of maintaining and repairing walls, floors, windows and ceilings. For example, vinyl wallpaper has a longer life and stands up better to cleaning than paper. Epoxy paint is more durable than regular paint, and certain grades of floor tile have much better resistance to dents, stains, and chemical exposure. These attributes, together with longer warranties, are reflected in the initial cost of materials. A facility area that is planned for only temporary use might not warrant the quality level selected for an established department for which no changes are currently planned.

Maintaining a Data File and Inventory of Materials

It is the responsibility of the maintenance department to keep a record of the finish materials—from paint types, colors and brands to carpeting, furnishings, baseboard materials and window treatments—used in new construction or remodeling projects. If you inherit this responsibility from a predecessor who did not maintain these kinds of records, it will be necessary to gather information from plans and specifications and contractors' submittals that may be in the possession of another department. If this information cannot be gleaned from existing files, it is a good idea to establish a file and begin identifying and recording existing finish materials. Thereafter, all finish materials used in future maintenance, repair, and improvement projects should be recorded. Paint colors, bases (oil or water), finishes (e.g., eggshell), and manufacturers should be standardized as much as possible to simplify the selection and procurement of the paint.

The Role of Building Users

The people who use a facility have a major impact on the condition of interior finishes. There are many ways of guiding building users, including physical features, such as bulletin boards and handrail/bumper guards, in order to protect finishes. Several of these elements are discussed in this chapter in conjunction with the element they are designed to protect. In addition to these barrier or alternate-use-type solutions, guidelines can be created and distributed or posted to help employees and the maintenance staff maintain a safe, pleasant and effective environment. Issues addressed might include storage recommendations to prevent code violations and unsightly obstacles, guidelines for hanging pictures, and instructions to contact the maintenance department to conduct any moves of furniture, fixtures, or boxes. Appropriate signage may be helpful in some situations. The guidelines for both maintenance staff and building users should be based on manufacturers' recommendations and the overall maintenance program.

Interior Materials Maintenance and Repairs

The following sections cover the basic cleaning, maintenance and repair issues for interior finishes. Recommended frequencies will depend on how the area is used, and the image or standard the facility owner wishes to maintain. (Frequencies and crew guidelines can be obtained from *Means Facilities Maintenance & Repair Cost Data*.)

Coverage in this chapter includes permanent partitions (plaster, drywall, ceramic and industrial tile, wood and masonry) and demountable partitions. Also included are windows and doors; wall, floor and ceiling finishes; and signage. To address repair issues, each section begins by identifying the most common defects, or damage, that affects that particular item.

The approach taken in this chapter differs somewhat from some others in Part 2. Interior finishes, unlike, for example, mechanical, electrical and conveying systems, are often maintained and repaired by a facility's own in-house maintenance staff. Therefore, this chapter includes some specific guidance on maintenance and repair procedures that might be directly applied to define maintenance standards.

Permanent Partitions

Plaster

Defects: The most common defects found in plastered interior partitions are cracks, holes, loose sections and water stains. Cracks and loose sections could be signs of improper installation, uneven structure settlement, or underlying problems. Wood structural components are subject to rot and decay, steel can corrode, masonry and concrete will show signs of aging. Each of these problems may become apparent only when the plaster shows signs of deterioration. Water stains can be the result of a leaky roof or water pipe, condensation problems, or simply spillage from an upper floor. Any water-related problems should be corrected immediately, as they will only get worse. Water problems must be fixed before beginning repairs to the plaster surfaces.

Repairs: Many types of plaster are available and used on interior partitions. Compatibility of original to repair material is important. If there is no available information on what type of plaster was used as a base material, trial and error may be the only method for finding the proper repair material. Most cracks and small holes can be repaired by first removing all loose material. The edges of the crack or hole should be chamfered back toward the outer surface, forming a canyon shape for a crack, and a crater shape for a hole. The edges of the area to be repaired should be wetted prior to applying the patching plaster. This is done to prevent the existing plaster from wicking the moisture out of the patch material.

Larger repairs require that material be removed down to the lath in order to provide a bonding surface for the patching material. Once this has been accomplished, the repair procedure is the same as for small cracks or holes. Occasionally, additional reinforcement in the form of metal lath or heavy-gauge screening material may be needed for large patches. Again, all edges of existing plaster material must be kept wet to prevent rapid drying of the patch material. Light sanding to remove ridges and valleys completes the task.

Very large damaged areas or many repairs in one area may justify removal of the plaster entirely to a natural break such as a door opening or corner. An entirely new section can then be installed in its place. Though this may seem disruptive and messy, it can be the best financial as well as aesthetic decision.

Drywall

Defects: Defects in drywall are very similar to those found in plaster walls. Cracks, holes, nail or screw pops, water stains, and improperly installed joint tape are the most common problems. As in plaster walls, the causes of these problems could be improper installation, uneven structure settlement, or underlying problems, such as structure decay. Any water stains are probably the result of similar problems found in plaster walls.

Repairs: Small drywall defects, such as cracks or holes, are easily repaired with drywall joint compound. This is the same material used to seal joints between drywall sheets and cover nail or screw heads during new installations. Several coats may be required depending on the severity of the repair, because drywall joint compound shrinks when it dries. Light sanding between coats is necessary for proper adhesion. A very fine dust is created when sanding joint compound. Steps should be taken to contain this dust and to protect furnishings, fabrics, and other finishes.

Larger repairs can be accomplished by cutting out the damaged area with a sharp utility knife and inserting a patch of similar drywall material. Using joint compound as "glue" and joint reinforcing tape at the edges of the repair, a large patch can be blended into the wall surface.

Very large repairs can be accomplished in much the same way. Entire sheets of drywall can be removed and replaced if necessary. Standard-sized sheets are 4' x 8', 4' x 10', and 4' x 12'. Once the damaged sheet has been demolished, the new sheet is attached to the framing members, all joints are reinforced with joint compound and joint tape, and all screw or nail heads are also covered with joint compound. After several coats have been applied and sanded, the replaced sheet will blend in with the existing wall.

Ceramic Tile

Defects: Some common defects in ceramic tile-covered interior partitions are cracked or broken tiles, loose or deteriorated grout, missing tiles due to failure of the mortar adhesive or deterioration of the substrate, and tiles that have been damaged by fasteners for towel bars, toilet paper holders, paper towel holders, soap dispensers, or other accessories.

Cleaning: Ceramic tile requires periodic cleaning with nonabrasive cleaners. If any grout is cracked or missing, be careful to keep the cleaning solution out of the area. Moisture will deteriorate the mortar adhesive behind the tile and cause further problems. Damaged tiles found during cleaning should be repaired immediately.

Repairs: Cracked, broken or damaged tiles should be completely removed. The tile must be carefully chipped away, along with the mortar adhesive behind it. Do not damage the substrate. Any remaining grout attached to the surrounding tiles must be removed as well. Prior to installing a replacement tile, the area should be wetted. Spread the mortar on the substrate using a trowel that is notched to the proper depth for the tile and mortar. By starting from one edge and snapping the tile into place, air is prevented from becoming trapped behind the tile. Twisting the new tile aligns it. Allow the mortar to set before grouting. Setting tile with an organic adhesive requires a different procedure. Follow the manufacturer's instructions.

Occasionally, old ceramic tiles need to be re-grouted, or new, replacement tiles need to be grouted. Grouting serves two purposes, the first being to complete the waterproofing of the ceramic tile installation, and the second to enhance the aesthetic value of the tiles. Before re-grouting a large area, all loose grout must first be removed and the ceramic tile cleaned and made secure. Grout is mixed according to the manufacturer's specifications. A soft rubber trowel is used to spread the grout mixture over the entire surface. The trowel should then be used to force the grout mixture into the joints between tiles. After all the joints have been filled, the mixture should set for the time recommended by the manufacturer. Then, using a clean sponge that is rinsed often, the remaining grout is cleaned from the surface of the ceramic tile. The new grout should dry completely before the surface is exposed to moisture.

Acoustical Tile

Defects: Common defects for acoustical tile walls include broken or loose tile, tiles damaged by moisture or physical abuse, and poor maintenance.

Cleaning: Acoustical tile is best cleaned by vacuum. Hand dusting or light washing with a mild detergent are other acceptable methods. Soft by nature, acoustical tile is easily damaged; therefore, care should be taken when performing any of the above operations.

Repairs: Repairs to acoustical tiles are performed by simply replacing the individual tiles as required. Mounting clips or adhesives are used for fastening to interior partitions. Clips can be salvaged when damaged tiles are removed, and re-used for installation of replacements. Adhesives are readily available in trowel-on or tube form. Adhesive manufacturer's instructions should be followed for best results.

It is a good idea when purchasing acoustical tile to buy extra quantities for future replacements. In this way, you can be sure of matching the original pattern.

Wood

Defects: Dents, cracks caused by structural failure or uneven drying, shrinkage or swelling, pest infiltration and contact damage are all common defects to wood interior partition finishes. Wood changes with changes in moisture conditions. Heating and air-conditioning cause wood to shrink, swell, cup, bow or twist. Proper care reduces the scope of each of these defects, but nothing will eliminate them entirely.

Cleaning: Frequent dusting with a soft cloth or vacuum cleaner will keep wood clean. Depending on whether the wood is painted, stained or sealed, other cleaners can be used. Painted finishes can be washed with water and a mild detergent, though care should be taken to prevent the wood from becoming too wet. Sealing finishes such as varnish, shellac, polyurethane or lacquer can be waxed and polished with a furniture polish. Stained finishes should only be dusted or vacuumed.

Repairs: Some "defects" in wood are considered part of its inherent beauty and are best left alone. Dents, chips, nail holes, knots and non-structural cracks may be acceptable or desirable depending on the surrounding architecture. Small imperfections in wood surfaces can be filled with a wood filler and sanded to the desired smoothness. Patching wood is difficult, as the patch will seldom match the texture, color or grain pattern of the surrounding wood. It is best to remove and replace the entire piece or panel. The steps necessary to perform this task are as different as each type of installation, but some general tips include being careful not to damage surrounding finishes, removing all nails or screws in both the damaged piece and the substrate, and removing material to a natural break, such as a door opening or corner of the partition.

Masonry

Defects: Common defects in masonry interior partitions are cracks — both horizontal and vertical, stains, chips, holes and marks. Cracks can be a sign of structural failure and should be thoroughly examined before any attempt at repair is made. Staining could be from abuse or from efflorescence — the light-colored powder that shows up on masonry walls as a result of salt in the mortar mix water. Chips, holes and marks on masonry walls are usually the result of abuse or fasteners used for attachment of plaques, paintings, etc.

Cleaning: For interior applications, masonry is best cleaned with soap, water, and a stiff brush. Methods such as chemical washing, high pressure washing, and sandblasting are more applicable to exterior applications, but they could be employed on interior masonry as long as effective containment measures for dust, dirty water and fumes are employed, and all other furnishings and finishes can be adequately protected. Cleaning of efflorescence requires the use of a solution of water and muriatic acid.

Repairs: Masonry walls not exposed to the weather will remain in good condition for long periods of time. Only occasional repairs will need to be made. Repointing, the replacement of loose or soft mortar joints, is done by first chipping or grinding out the old mortar to a depth of 1/2" – 2". The area must be cleaned thoroughly, and wet down to prevent wicking of the moisture into the surrounding brickwork. A tuck-pointing tool and trowel is used to pack new mortar into the joints as necessary. Each new joint will need to be "tooled" for consistency in shape and depth.

Cracks in the mortar could be the result of structural stress. The source of the stress should be eliminated before repairing the crack; otherwise it will reappear. Once the structural stress is eliminated, cracks can be repaired in a manner similar to repointing.

Broken or chipped masonry units will need to be replaced. First, the damaged unit is chipped out with a chisel and mason's hammer. When all the surrounding mortar has been removed, mortar is spread on the bottom and two ends of the new masonry unit, which is then slid into place. A tuck-pointing tool and trowel are used to finish off the mortar joints. The new joints are tooled to match the surrounding joints.

Demountable, Portable or Folding Partitions
Operational Maintenance
The only operational maintenance required is cleaning and lubrication of any moving parts or tracks. The frequency of this maintenance depends on the amount of use to which the partition tracks and hinges are subjected. Lightweight lubricating oil or spray silicone should be used. Tracks can be cleaned by sweeping, vacuuming, or brushing with a stiff brush.

Finish Maintenance

Defects: Demountable or folding partitions are used for a variety of applications. The materials used for their finishes is also wide-ranging. Typical defects are the same as those that occur on other interior partitions, as they are subject to the same daily rigors. Please refer to those materials for defect information.

Cleaning: Cleaning is generally the same as for similar materials used for interior partitions.

Repairs: One advantage to demountable, portable and folding partitions is that if damage occurs, the entire panel can be replaced. While this could involve some time and effort, it may be the best solution to the problem. Small defects may be repaired in the same manner as for similar materials used on permanent interior partitions.

Windows
Metal & Wood Frame

Frames: The most common problems with the frames of interior glazed openings are surface defects. Please see the next section on finish maintenance. Operational hardware, if any, is not exposed to the same conditions as exterior glazed openings; therefore, only cleaning and lubrication may be necessary.

Finish Maintenance

Defects: Defects such as rust spots on steel frames, chipping paint, scratches, dents and loose fasteners all show up on interior glazed openings. Causes of the above defects are typically excessive moisture, abuse, and everyday wear and tear.

Cleaning: Identifying small surface defects should be part of a regular cleaning program. During cleaning, the frame should be checked for structural integrity, damage, and operation. A solution of soap and water applied with soft brushes or cloths should be sufficient for keeping the frames clean. The frame cleaning can be done most efficiently with the glass cleaning to save time and effort. The frame should be cleaned after the glass.

Repairs: Loose, chipped or peeling paint and scratches should be scraped to remove all material that is not properly adhered. Sanding to feather out the edges of old paint will make the new coat look better and smoother. The frame must be thoroughly cleaned before painting to remove all dirt, grease, and foreign matter. Use a good quality latex or oil-based paint, and apply a smooth, even, thin coat over all surfaces.

Window frames that are damaged, have dents, or show other signs of abuse may require replacement. Chances are that if the frame has been damaged enough to require replacement, then the glazing also must be replaced.

Glass

Types: Glass is usually used within the interior of a structure to provide borrowed light. Interior windows are used to separate rooms within a building while borrowing the natural or man-made light for a room with no outside windows from an adjoining room. Many types of glass are available, but not all of them are used for interior applications. Float, insulating, laminated, wire, and specialty glass, such as beveled, sandblasted, etched and patterned, are all available. Banks, secure facilities, jails and prisons may require high-strength or bulletproof glass. Hospitals, medical clinics, and treatment facilities may require lead glass in x-ray rooms.

Mounting Methods and Maintenance

Rubber Gasket: Some interior frames require rubber gaskets to hold the glass in place. The rubber is subject to drying out if not occasionally treated with a silicone protectant. If the rubber has deteriorated enough to justify replacement, first remove the old gasket by cutting, pulling, and prying it out. The replacement gasket should be of the same size and shape. Once the proper gasket has been located, a special rolling tool is necessary to push the gasket into place between the frame and the glass. A small amount of silicone makes the installation go a little easier.

Putty: Dried-out, chipped, cracked, or broken putty on wood or metal window frames is a common problem. Even with a good coat of paint, putty deteriorates and needs occasional replacement. First, all the old putty is removed by scraping, chiseling, or prying it off. The glass is then removed and the putty bed behind it cleaned out. For wood frame windows, a coat of linseed oil is recommended to condition the wood prior to installing the new putty. A thin bead of putty is placed around the window frame, and the glass inserted. Glazier's points, small diamond-shaped pieces of flat metal, hold the glass in place. The triangular cross-sectional area between the glass and the window frame is filled with more putty. A flexible putty knife is used to press the putty into place and smooth the visible surface at the same time. This process takes a little practice. After the putty has dried sufficiently, the window is painted and the glass cleaned.

Mechanical: Quite possibly the only problems found with the components used for mechanically holding glass in place are loose, corroded, or faulty fasteners. Loose fasteners require tightening; corroded or faulty fasteners require replacement.

Interior Doors, Frames, and Hardware
Door and Frame Maintenance

Defects: Common defects found on interior wood doors include: poor fit due to improper installation; deterioration of the door shape; loose door parts such as rails, stiles, or panels; failure of the finish material; cracked or broken glass; and split or cracked wood on the door or frame. Improper door frame alignment can also occur due to excessive shrinkage/swelling.

There are very few "common" defects in metal doors. Most will retain their shape and size and will fit into the frame with accuracy unless the hinges fail. Physical damage from collision or structural failure may occur and probably will require professional repair or replacement.

Larger metal doors such as those used in warehouses, airplane hangars, factories, and similar structures can be rolling, sliding, folding, or coiling. These types of doors are usually powered by electric motors and are operationally sensitive. Most manufacturers require a program of preventive and scheduled maintenance in order to honor a guarantee. Lubrication, track cleaning, and minor adjustments are typical maintenance items.

Cleaning: Dusting and cleaning frequency is based on the particular conditions surrounding the door. At least twice a year, doors should be washed with a mild detergent appropriate to the type of door. Larger doors, such as industrial metal doors, may require washing with a pressure washer. Potential cost implications for larger door maintenance may be a requirement for scaffolding or hydraulic lifts for access, heated or air conditioned air loss, security or safety concerns.

Repairs: Repairs to the larger doors in a facility, such as warehouse coiling or sliding doors, hangar doors, etc., should be left to professionals. Most manufacturers have a local representative who will perform any repairs efficiently and safely.

Metal doors may also require professionals to perform necessary repairs. As previously stated, any metal door with damage could possibly be replaced instead of repaired. The facility manager would need to make the financial decision of repair vs. replacement. Part of this decision would be the potential aesthetic effect of a repaired door compared to a new door.

Wood doors, depending on the severity of damage, can often be repaired by in-house personnel. Alignment problems can often be repaired with the door in place, or removed and repaired close by. Other repairs, such as sagging, cracks, broken panels, or loose hinge connections, may require the door to be removed and taken to a shop. This creates a number of problems if the door is required to maintain separation. A temporary replacement may be necessary. Again the cost implications of repair vs. replacement will need to be weighed.

Note: Most door operational components that require maintenance fall under the "Hardware" category. Please refer to that section for more information.

Finish Maintenance: Defects such as rust spots on steel frames, chipping paint, scratches, dents, and loose fasteners all show up on interior doors. Causes of these defects include exposure to excessive moisture, abuse, and everyday wear and tear.

A regular cleaning program should include identifying small surface defects so that they can be repaired before they become larger. During this cleaning the door and frame should be checked for structural integrity, damage, and operation. A solution of soap and water applied with soft brushes or cloths will be sufficient for cleaning.

Loose, chipped, or peeling paint and scratches should be scraped to remove all material that is not properly adhered. Sanding to feather out the edges of old paint makes the new coat look better and smoother. Thoroughly clean the door and frame to remove all dirt, grease and foreign matter before painting. Use a good quality latex or oil-based paint, and apply a smooth, even, thin coat over all surfaces.

Hardware

The only possible way for the facility manager to effectively maintain all the hardware in a facility is to first inventory each component. The inventory should include the make, model, date installed, installer, manufacturers' recommended maintenance and repair instructions, and a record of any repairs or maintenance performed.

Fasteners: Each time the hardware is scheduled for preventive maintenance, the fasteners should be checked and tightened as necessary. Many problems associated with loose hardware could be alleviated with this one simple step. For example, a loose screw that holds a door hinge in place is not tightened. The extra stress on the other screws causes two of them to break. The hinge can no longer work properly; therefore, the door does not close completely. Consequently, the door is damaged by a forklift attempting to fit through a space that was previously accessible. Now, instead of a 5-minute tightening of the screws, there is a $1,200 bill for door replacement.

Lubrication: As with any item that requires lubrication, the more often it is done the better the component will work, and the longer it will last. A good preventive maintenance program, based on the manufacturer's recommendations, will keep the facility hardware in good condition. At least twice a year, all nonvisible parts should be cleaned and lubricated. Visible parts should be done weekly. Such a pattern should keep the hardware replacement costs down.

Adjustments: Adjustments to moving parts should be carried out as recommended. Compliance issues, such as use of lever handles to meet ADA requirements, are part of hardware maintenance. Other codes regarding access and egress apply as well. A good hardware maintenance program, together with an accurate inventory, will provide the facility maintenance manager with the information needed to stay ahead of these issues.

Wall Finishes

Wall finishes include paint and wallcoverings. Common problems and cleaning and repair requirements are outlined below.

Paint

Defects: Since interior paint is not subject to the same harsh environment as exterior paint, it does not commonly show the same types of defects. Usually, interiors are redecorated well before the interior paint begins to deteriorate. The most common reasons for redecorating are for a change in aesthetics, cleanliness, and general appearance.

Cleaning: Interior paint can be washed with a mild detergent and water fairly frequently without damaging the finish. (When repainting, an appropriate paint should be selected that will stand up to cleaning requirements.) For heavier cleaning, a solution of tri-sodium phosphate and water can be used. Frequent dusting and vacuuming will keep interior paint finishes in good condition for many years.

Repairs: Over time interior paint fades or discolors. As a result, spot painting can show up like a sore thumb. Unless a natural break, such as a door opening or a corner, is available to hide the perimeters of spot painting, the entire surface will require repainting. This is relatively inexpensive and is commonly done. The psychological effect of new paint on employees, customers, or facility users is usually a major benefit. Any changes made show concern for the interior environment.

Wall Coverings

Defects: Wallpaper and wall coverings generally experience the same kinds of defects as the surface to which they are applied. Maintenance problems are usually linked to the wall itself. Some signs of adhesive degradation may be shown when wall coverings begin curling at the edges. Seam sealer can be applied to correct minor problems.

Cleaning: Cleaning of wallpaper and wall coverings is the same as painted walls. Please see that section for more information.

Repairs: While repairs can be made to wallpaper and wall coverings, the time it consumes often makes this work uneconomical. Only if matching paper left over from the original installation is available and in good condition, and the host paper has not faded so much that a good match is impossible, will a wallpaper or wall covering repair look reasonably good. A better solution would be to re-paper the entire wall, which is one of the reasons that paint is considered.

Floor Finishes

General

The most used part of any facility is its floors. A high percentage of labor hours, equipment, and money goes into the maintenance and repair of floors and floor finishes. Careful selection of the proper materials, use of high quality installers, and a comprehensive maintenance program will stretch the facility or maintenance manager's floor budget. Common defects in flooring materials are often caused by poor quality materials and installation, structural instability, environmental conditions, inappropriate use, or abuse.

Cleaning

Care in the choice of equipment and products for floor cleaning is key to extending the life of these systems. The materials must adequately perform their function without causing undue damage or wear. Many manufacturers offer full lines of products for a variety of flooring materials and should be used as a valuable information resource on the care and maintenance of flooring materials in conjunction with the use of their products.

Repairs

There are as many repair methods as types of floors. These methods will be discussed further within each individual type of flooring.

Resilient Flooring

Defects: A generally softer texture and porous surface make resilient flooring susceptible to defects such as tears or punctures and small scratches in the surface that attract dirt and dust. Improper installation can result in corners and seams becoming loose, increasing the opportunity for damage. A good maintenance program helps to head off some of these common problems.

Cleaning: There are many types of resilient flooring. Linoleum, rubber, and vinyl are just a few examples. Each type has its own cleaning frequency, method and solutions. It is not safe to assume that one method will be good for different types of flooring. Again, check with the flooring manufacturer for information.

Repairs: Resilient flooring comes in sheet and tile form. Availability of original tiles, being small in size and relatively easy to replace, lend themselves to use in areas of high traffic and wear. Tile color and pattern is the one major obstacle to repairs. Stockpiling surplus materials during the initial installation is a good idea.

Sheet goods can be repaired, but the same question arises as to whether to repair or replace. Replacement is often the best answer.

Masonry

Defects: Masonry floors provide favorable qualities such as high resistance to wear, staining, and environmental deterioration. These same qualities can also lead to certain defects. The hard, brittle surfaces and physical mass of masonry floors lend themselves to chipping, dusting, and temperature absorption, although each of these problems can be overcome.

Cleaning: Vacuuming or sweeping rids masonry floors of unwanted dust and dirt. Regularly scheduled cleaning minimizes any staining problems that may arise. Ceramic tile, marble, and terrazzo floors can be cleaned with water and a mild detergent. Slate and stone, on the other hand, become very slippery when washed, and unsealed brick absorbs the moisture. Many of the above types of masonry floors benefit from a periodic sealing with a penetrating sealer. Refer to manufacturers' instructions for sealing applications.

Repairs: Masonry stock repairs are best left to a trained tradesperson. Installing each type is usually a specialty, and repair methods vary. In general, the damaged flooring material will need to be chipped or hammered out, old dry mortar removed, and a new section replaced with new mortar. For terrazzo floors, sections in between expansion joints may require replacement. In this case, a demolition hammer may be needed to demolish the section before new terrazzo can be cast in place.

Wood

Defects: Common defects in wood flooring are wear, dents, cracks, stains, and loose strips or planks. Wood shrinks and swells with rising and falling moisture conditions, causing problems with whatever finish is applied to it. In addition to manmade defects, wood has several natural defects such as cupping, twisting, knots that become loose or fall out, pitch pockets, worm holes, and fungus. Some of these, such as knots or worm holes, may be desired, as they give the finished product a certain aesthetic effect. Others, such as pitch pockets or fungus, must be removed or repaired.

Cleaning: Wood floors can be swept or vacuumed daily. Occasionally, a solution of water and special wood soap can be mopped on to clean thoroughly. Waxing is recommended, with the frequency dictated by the amount of use the floor receives.

Repairs: Refinishing a wood floor is the best way to restore its beauty. The old finish must be removed by power sanding, and a new finish applied in multiple coats. The floor cannot be used during this process. The resulting down time can present scheduling problems, particularly in a facility that is refinishing a space in heavy demand for large groups, such as a gymnasium. Moisture penetration from above or below could ruin a wood floor. If a tightly laid wood strip floor were to be exposed to water for even a short period of time, the resulting buckles in the floor would need to be cut out and replaced. If decay or rot is a problem, replacement of the floor is probably the best decision.

Ceilings

Finish Maintenance

Acoustical Tile

Defects: Common defects for suspended acoustical tile ceilings include broken or loose tile, tiles damaged by moisture or physical abuse, and poor maintenance.

Cleaning: Acoustical tile is best cleaned by vacuum. Hand dusting or light washing with a mild detergent are other acceptable methods. Soft by nature, acoustical tile is easily damaged; therefore, care should be taken when performing any of the above operations.

Repairs: Repairs to acoustical tiles are performed by replacing the tile. Most acoustical tile ceilings are laid into a grid of metal supports. The replacement of a damaged tile can be performed by just about anyone who can climb a step ladder. Other mounting methods, such as a concealed grid or adhesive mounting, require more work. The damaged tile in this case will need to be destroyed in order to remove it. Many of these types of ceiling tiles interlock with each other. The replacement is best left to a professional. The choice of ceiling tile may depend on this factor.

Ceramic Tile, Drywall & Plaster

Defects, repairs, and cleaning of ceramic tile, drywall, and plaster ceilings are the same as for ceramic tile, drywall, and plaster walls. Please refer to that section for more information.

Psychological and Compliance Issues

Psychological

The care and maintenance of a facility has a major effect on those who use it. A beat-up, run-down school building will only get worse, since its users will associate the condition of the structure with a lack of concern by its owners. The message this sends is that the building is expendable, so further damage to it is not a concern. On the other hand, the school department staff that removes graffiti immediately from the bathroom walls is sending the message that damage to the structure is not acceptable.

In business, a clean, well-maintained building promotes the image of the company. Employees will see the respect that the organization affords to the structure, and feel that a part of that respect is directed to them. Visitors to the company will be impressed by the condition of the offices and have a positive perception. Money invested in maintaining interior finishes is well spent on many levels.

Compliance

Compliance with local and state building codes, as well as with the Americans with Disabilities Act, falls under the facility or maintenance manager's umbrella. Many required modifications involve interior finishes and components. Compliance takes the form not only of upgrades and modifications, but also maintenance and upkeep of those improvements. Part of the yearly building audit should address compliance-related conditions that need work, and identification of any requirements that have changed in the past year.

Protecting Interior Finishes During Remodeling & Major Repairs

The maintenance manager needs to establish procedures for protecting finishes and furnishings during major repair or construction projects in or adjacent to space that is currently in use for a facility function. In some cases, it may be the contractor's responsibility to provide such protection; in others, it must be done by the maintenance department. Either way, the maintenance manager will likely be involved in ensuring that this important part of the job is done correctly. Depending on the space and the functions it houses, there may be a host of specific and unique protective measures required. Following are some examples:

- Removing doors and even door frames, or fabricating protective devices to cover frames.

- Using kraft paper to mask areas. (When taping, do not use duct tape, which leaves a sticky residue.)
- Removing or replacing existing carpet.
- Storing furnishings. This may be done off-site at a professional storage facility if an appropriate location cannot be found on-site. In some cases, furnishings are sealed in protective wrapping and placed in the center of the affected room. (An air-tight wrapping protects furnishings from fine dust particles such as are generated by sanding finished drywall.)
- Assigning one of the available elevators (where there is no freight elevator) for transport of construction materials only, for the duration of the project.
- Establishing specific procedures for relocating furnishings so that finishes are not damaged and the continuing functions in the affected space are not disrupted (e.g., computers are not unplugged, etc.).
- Changing filters in heating systems when projects are complete in order to dispose of accumulated dust.
- Re-keying doors after the project is complete to eliminate the possibility of inappropriate access.

In addition, the maintenance manager might conduct before and after inspections to document the original condition of the space prior to construction, in order to be able to identify any damage done in the course of the work.

Conclusion

Interior maintenance and repair is affected by many variables, such as building type and use, and the image the facility owner wants to convey. Depending on the type of facility, general maintenance of interior finishes and components is very often the most discretionary expense in the maintenance budget. This is because failure to perform a full maintenance program does not usually compromise the integrity of the building — at least not in the short term. Furthermore, estimates for this type of work, especially painting, can vary considerably among contractors and in-house staff, due to different methods of measuring surface areas, and evaluating the complexity of a project.

To deal with these challenges, it is necessary to establish an appropriate maintenance and repair program, grounded in historical data, and linked to the organization's mission and operations budget. Components in such a program include a database of existing finishes and materials (to simplify repairs); historical labor hours and costs; and use of resources, such as published productivity standards. Such a program enables management to make informed decisions about the level of maintenance they wish to provide.

Chapter 10

Conveying Systems

John Corcoran and the Schindler Elevator Corporation Staff

Conveying systems generally include elevators, escalators, dumbwaiters, material handling conveyers, moving ramps and walks, parcel lifts, and motorized car distribution systems. This chapter focuses on elevators and escalators, since these are the systems most facility managers frequently encounter.

Maintenance of elevators and escalator components is typically performed by outside service contractors because of the specialized skills, tools, and parts required. However, in-house facility maintenance plays a major role in keeping these systems clean and safe, thereby protecting the organization's investment in these costly and crucial building components.

Elevators — Operating and Safety Features

The following section is an overview of elevator components and their functions.[1]

Hall Controls and Displays

Hall Position Indicator: This fixture, usually used in the main lobby of multiple elevator installations, alerts waiting passengers to the location and direction of travel of the elevator cards in the hoistways. Hall position indicators are typically mounted above the elevator entrance; illuminated lenses indicate the car's floor location and direction of travel. If specified, hall position indicators may be combined with hall lanterns.

Hall Lanterns or Car Lanterns: These fixtures indicate the direction of travel of the arriving car. The lenses illuminate to show the car's direction of travel. An audible tone announces the arrival of the car by sounding once for an Up car and twice for a Down car. Hall lanterns are mounted in the corridor above the elevator entrance or adjacent to the entrance and jamb. With this feature, passengers can see the direction of travel before the car arrives. Car lanterns are mounted on the entrance column of the elevator car. They allow waiting passengers to see the direction of travel once the elevator doors have opened. Most elevator installations contain either hall lanterns at each floor or car lanterns on each car; both fixtures are not usually used together.

Hall Pushbutton Station: This fixture, located in the wall beside an elevator entrance, calls an elevator to the floor when the button indicating the desired direction of travel is pressed. At intermediate floors the fixture contains two buttons, Up and Down. At the top and bottom terminals, the station reflects the only direction of travel. On most fixtures, a light illuminates when the button is pressed and it remains lighted until the call is answered. Never press both the Up and Down buttons when registering a call; this may delay services for all passengers.

Car Controls and Displays

Car Position Indicator: This fixture advises elevator passengers of the car's position, its direction of travel, and its arrival at a selected floor. Many car position indicators are mounted in the transom above the car door; illuminated numbers indicate the floor location of the car. Other types of car position indicators may be located above the car station.

Car Station: All elevators are equipped with a main car station. As with other elevator fixtures, car stations vary in appearance and design depending on the system furnished, but most contain these basic features:

- Call register pushbuttons
- Door open button
- Alarm button
- Communication system intercom or telephone
- Restricted access switches

When specified, elevator cars are equipped with an auxiliary car station to enhance passenger convenience and to speed the registering of calls. The auxiliary car station contains only the buttons required for passenger use and can be mounted on the front of the return panel opposite the main car station or adjacent to a rear door.

Call Register Pushbuttons: The Call Register pushbuttons are used to select a floor destination. They illuminate when pushed to indicate that a call has been registered for that floor. As required by code, raised numerals and Braille markings indicating the number of each floor are installed adjacent to the pushbuttons to aid people with disabilities.

Door Open Button: When there is a need to keep the car doors open longer than the normal programmed time, such as during passenger loading, pressing and holding the Door Open button extends normal door-opening time or re-opens a closing door.

Alarm Button: When pressed, this button activates an audible signal to notify building personnel of a possible emergency situation in an elevator. The signal is audible both inside the car and outside the hoistway.

Emergency Stop Switch: When activated, this switch stops the elevator and sounds its emergency signal. This switch may be replaced with a key switch in most jurisdictions to prevent nuisance use and unnecessary sudden stops. To restore normal operation after activating the stop switch, return the switch to its normal position. The switch may be located in the car station or the locking service panel, depending on local codes.

Communication from the Car: A telephone (provided by the owner and located in a telephone compartment) is used for emergency communication from inside an elevator. In some cases, an intercom speaker is provided to allow communication between building personnel and elevator passengers.

Locking Service Panel: Your system may include controls that are either key operated or located in a locked compartment. These switches are not for passenger use and should be used only by authorized personnel.

- *Power or MG (Motor Generator) Switch*: Normally left On to permit constant use of the elevator. When turned Off, the elevator is not operable.
- *Stop Switch*: When activated, this switch stops the elevator and sounds its emergency signal. To restore normal operation after activating the stop switch, return the switch to its normal position. In some jurisdictions this switch is located where it is accessible to passengers.
- *Hand Switch*: When On, this switch removes a car from normal operation to allow servicing by a qualified service technician. This switch should be used only by trained elevator service technicians.
- *Fan Switch*: Controls the operation of the ventilation fan and should be On when the car is in use.
- *Light Switch*: Controls car lighting and must be On whenever the car is in use.
- *Service Switch*: This switch is marked HE or Service. When On, this switch cancels all previously registered car calls and corridor calls assigned to the car and reassigns the calls to other cars. Pressing a car button after the switch is in the On position will send the car to the requested floor, where it will remain with its doors open until another car button is pressed. The switch must be turned Off to restore routine use of the car.

Elevator systems may have other switches to provide special functions tailored to a particular building. Your manufacturer's representative can provide instructions for their use.

Elevator Doors

Elevator doors open automatically when the car arrives at a floor. The "door open" interval may be adjusted by trained elevator technicians to increase the door open interval or to reduce the door speed. The following safety features are included in the door system for the protection of elevator passengers.

- A sensitive, mechanically operated switch that extends the full height of the door to automatically re-open a closing door when it comes in contact with an obstruction.
- An electronic sensing device that controls the door closing to conform to the traffic movement across the threshold. If the photoelectric light beams across the door are broken by a passenger or an object, this device causes doors to remain open or to re-open if they have started to close.
- In lieu of the above, many newer elevators are equipped with electronic door sensors using infrared technology. They are also available as an upgrade to older systems. These devices cause doors to remain open or to re-open if the sensor detects a passenger or object in its path.
- If doors are prevented from closing after an extended period of time, the doors start to close at a reduced speed ("nudging"), and the door warning buzzer sounds until the doors are fully closed. If the safety switch described above is held during nudging, the doors will stall but will not re-open.
- Emergency Exit: Emergency exits are not for passenger use. *Warning: These exits, located at the top or on the side of a car, must remain locked at all times unless under the direct supervision of trained elevator technicians or emergency*

personnel. Passengers should not attempt to use emergency exits except under the direct supervision of trained elevator technicians or emergency personnel. Failure to heed this warning can result in severe personal injury or death.

Monitoring and Control Systems
Elevator Monitor Center

The elevator monitoring center is a part of most multiple-elevator, high-rise installations. It may be located in the main lobby or in a remote location. The station provides a centralized point at which to observe the elevator system operations and activated features. The elevator monitoring center is the "master control" of most multiple-elevator installations. It usually includes a screen that displays the elevator system status. It may contain a variety of features depending on the design of your system but, in every case, access should be limited to authorized facility management personnel. These systems are designed to meet the unique requirements of the facility in which they are installed. Your manufacturer's representative should provide your staff with complete operating instructions for your particular system.

Firefighter's Service and Emergency Operation Feature

Every building's disaster or fire drill procedure should include information regarding the proper operation of the firefighter's service feature. This is an elevator control system designed to protect passengers and aid firefighters in combating fires. It is activated by a key switch or by the building's smoke detectors on each floor. The switch may be located in the elevator monitoring center or separately at a designated floor.

Firefighter's service provides for the automatic and immediate return of all elevator cars to a designated floor where they will park, with doors open, and will not go to other floors. All call buttons, emergency stop switches, and signal lights, except the car position indicator, will be inoperative. In some systems, if the sensor is activated at the designated floor, elevators will be recalled to an alternative emergency exit floor. During a fire or emergency, the elevator can be operated only from the car using the in-car key-operated emergency switch. Only trained emergency service personnel should operate elevators with this key-operated switch. The keys to this switch should be kept where they are immediately accessible to authorized personnel, but where they are not available to the public or any unauthorized personnel.

Warning: In case of fire do not use elevators; use stairs or other emergency exits. Use of elevators during a fire can impede fire fighting efforts and can result in severe personal injury or death.

Emergency Power Control

Many buildings have an emergency (standby) power generator system or other emergency power source. Where emergency power is available, elevators may be capable of operating. In the event of a power outage, the emergency power system controls power distribution to the elevators. The emergency power control panel may be located in the elevator monitoring center or in a separate locked station.

The emergency power system varies among elevator systems. However, all accomplish the basic functions of allowing operation of each car to permit passengers to exit, establishing reduced service for the duration of the emergency, and returning the elevator system to normal service when building power has been restored. In most emergency power systems, building personnel can communicate with passengers by telephone or intercom.

In most systems the changeover from regular power to emergency power will occur automatically. In some systems the changeover must be activated manually, usually by means of a keyed or controlled-access switch.

Car Selection: For multiple elevator high-rise buildings, the most common system provides automatic selection. In the event of a loss of normal power, all elevators in motion make an emergency stop. When standby power becomes available, a signal is sent to the controller that automatically starts returning the cars (one at a time) to the designated (usually main) floor, opens their doors, and takes them out of service. This is done in a predetermined sequence. When all cars have returned to the designated floor, the system automatically places one or more preselected cars in service for normal operation on standby power. If the preselected car is not in operating condition, a second choice is made.

In some automatic systems, an override feature permits manual selection of the car to be put into normal service. This is done by activating two switches in the emergency power control panel: the override switch plus the switch for the car selected to run.

Some systems require manual selection of cars operated under emergency power. When there is a loss of normal power, all elevator cars in motion make an emergency stop. When standby power is available, an indicator light at the control station signals that the cars can be operated. Qualified building personnel at the station can then return the cars one at a time to the main floor and take them out of service. This is done by manual operation of individual car switches on the emergency power control panel. Typically, all cars are returned to the main floor and a car is manually placed in service to operate under emergency power.

Restoring Normal Power: For those systems where emergency power operation is activated automatically, the restoration of normal building power will cause a signal from the ATS (automatic transfer switch) to notify the elevator system that a transfer from emergency to normal power is impending. This signal directs the elevator to stop at the next available floor until the transfer to normal power occurs. This avoids the possibility of the car making an emergency stop during the transfer sequence. Individual car override switches should be returned to Off position when normal building power has been restored.

For systems that require manual transfer between normal and emergency power, care should be exercised to ensure that any cars in use are stopped at a floor before transferring back to normal power. Otherwise, an emergency stop may occur during the transfer, which may be unsettling to the passengers.

Emergency Lighting
In the event of a power loss, a battery-powered light will automatically provide low-level illumination sufficient to view car station controls in each elevator.

Promoting Safe Ridership
Improving safety requires commitment by building owners and operators. Maintaining a facility provides an opportunity for the all-important day-to-day contact with the equipment and the people who ride on it. That contact provides a unique opportunity to promote safe ridership. Be sure that people in your organization make it their job to keep an eye on usage of

the elevators and escalators, and to intervene when unsafe actions are observed. Maintaining good housekeeping at the entrances to elevators and escalators also helps prevent slipping and tripping situations.

Because elevators and escalators are among the safest forms of transportation, passengers have a tendency to take them for granted and, sometimes, to relax their safety awareness. Make sure passengers remain aware of safe ridership practices and pay attention while using elevators or escalators.

Communicate to Your Passengers

It helps to remind people of basic ridership rules. For escalators, a sign can remind passengers to face forward, hold the handrail, and avoid the sides of the steps. Elevator passengers can be reminded of the importance of using stairways, not elevators, in the event of a fire. These messages are short enough that they can be posted conveniently at the elevators and escalators themselves.

Posting signs is probably not enough to implement an effective safety program. You might consider verbal announcements, bulletin board notices, safety meetings, or articles in employee newsletters. The Schindler Elevator Corporation offers customers "Quick Tips," brief suggestions that can be included in newsletters or on bulletin boards. Schindler also offers a safety video and brochure.

Preventive Maintenance

Preserving elevators' smooth and safe operation requires regular preventive maintenance performed by a qualified service organization. Follow the guidelines listed below as part of a conscientious accident prevention program. Failure to adhere to these guidelines could risk serious injury or death.

Operating Guidelines:

- Report any unusual equipment operation to trained maintenance personnel. Shut down equipment if potentially dangerous conditions exist.
- Adopt a rigid preventive maintenance program in accordance with the manufacturer's recommendations. Do not allow untrained or unskilled individuals to perform any maintenance or repairs on the equipment.
- Do not use elevators during fires, blackouts, or power shortages except under the special Firefighter's Service and Emergency Power conditions described earlier in this chapter. Under no circumstances should passengers or untrained individuals be permitted or encouraged to use elevators during such emergency situations.
- In certain unusual circumstances, elevator cars may level slightly too high or too low at a floor. Always verify proper leveling before entering or leaving the car. If misleveling occurs, shut down the car and summon trained maintenance personnel.
- Immediately report all accidents to the maintenance company and your insurance carrier.

Maintaining the Environment:

- Keep elevator interiors clean.
- Clean the door sill area in both cars and corridors.
- Ensure proper lighting at all times in corridors and in the elevator car.

- Secure handrails, ceiling panels, and ceiling light panels to prevent them from falling on passengers.
- Follow code requirements for the machine room, and limit its access to trained elevator technicians. The elevator machine room houses the machinery that powers the elevators, electronic panels that control the operation of the elevators, plus peripheral devices designed to provide safe and dependable operation. Machine rooms contain high voltages and rotating equipment, which pose severe danger to those not experienced with elevator operations. Local elevator codes specifically prohibit the storage of all non-elevator related material in machine room areas and special care must be taken to keep all flammable material away from this area.
- Maintain machine room temperature for hydraulic elevators near normal room temperature. Failure to do so can result in elevator malfunction.
- For elevator control systems, manufacturers have normally requested that a machine room have properly regulated humidity and a temperature range of 55 to 90 degrees Fahrenheit. This type of environment will serve to promote reliability and to increase the effective lift of the elevator motors, coils, and other electrical devices.
- When a hydraulic elevator has not run for a number of hours (such as overnight), it may be necessary for designated personnel to operate the cars for a few trips. This circulates warm oil into the jack and minimizes leveling difficulties that may be encountered during normal passenger use after initial start-up.

Enforcing Proper Use:
- Do not permit overloading of passengers or freight.
- Do not permit individuals to ride or play on top of the elevator cars.
- Do not move furniture or other materials above or beneath elevator cars or allow material moved to extend through the car top emergency exit.
- Do not permit unauthorized people to have access to emergency door keys, so as to enable them to open hoistway doors when the car is not present. Such keys should be used only by trained elevator or emergency service personnel.
- Only qualified elevator technicians or trained emergency personnel should attempt to remove passengers from stalled elevators. Unauthorized or untrained personnel attempting to rescue trapped passengers may place themselves and the passengers in danger.
- Do not attempt to use emergency car top exits. These exits are provided for use by trained elevator technicians or emergency personnel only.
- Advise passengers on the proper use of the elevator system.

Care and Cleaning
Most surfaces of an elevator, such as walls and floors, can be maintained using normal cleaning materials and procedures. Avoid using abrasive or corrosive cleaners on or around the fixtures. Do not use ammonia-based cleaners on lacquer finishes as they may strip the finish off of metals and woods. Stainless steel and bronze fixtures require special care, as explained on the following page.

Stainless Steel and Bronze Fixtures: A quality product such as a stainless steel cleaner and polish works well on brushed stainless steel surfaces, and a quality bronze polish should be used for bronze fixtures. Apply a light coat of cleaner and polish with a soft cloth. Do not spray any cleaners directly onto pushbuttons or any other elevator fixture to avoid damaging the pushbuttons and causing them to look unsightly.

Preparing Elevators for Extreme Weather

Many parts of the country are subject to the effects of hurricanes, flooding, and other extreme weather conditions that can damage elevator equipment. The following tips can help in preparing for such events in order to avoid or minimize elevator damage.

In advance (right now): Inspect the elevator machine room ventilation openings, windows, and doors for possible rain leakage. Install metal splash guards over ventilation openings to help prevent water from reaching electrical panels. Install weatherstripping around machine room doors that open to the outdoors.

Immediately before a storm hits: Close up all vents and openings in the top of the hoistway to prevent water from getting into the elevator shaft. Barricade machine room windows as necessary. If elevators are enclosed within the building (hoistways not exposed to the elements), run each car to a stop near the center of travel. For two-stop elevators, park the car at the top floor. If elevators are exposed to the elements (observation cars on the outside of the building), run the elevator cars to the floor below the top floor. In each case, after parking the car, shut it down with its keyed switch. Park it with its doors closed. Go to the elevator machine room and turn the elevator electric disconnect switches to Off. There is one switch for each elevator, usually located adjacent to the machine room door. This will help ensure that no unauthorized person can use the elevators during the storm.

During the storm: Never use elevators during a hurricane, even if the building has an emergency power generator. Rising water or wind-driven water can cause electrical short circuits that could disable an elevator, even if emergency power is available, and lead to passenger entrapment.

After the storm has passed: When you get an official "all clear," but before you restore power to elevators, inspect the machine room for water on the floor or the control panels. Also check the pit area for water damage. (If the building does not have a "walk-in" pit, your manufacturer's service representative can help you gain access.) If you find water in these areas, call your service representative for an inspection before operating the equipment. Be sure to open any vents or openings at the top of the shaft if you sealed them just before the storm.

Keeping Up with the ADA

The Americans with Disabilities Act (ADA) is having a significant impact not only on building owners and managers, but also the elevator industry. In a variety of ways, today's elevators are significantly more accessible to people with disabilities than they were years ago. Signaling systems, button designs and locations, even the size and shape of elevator cars have been affected by ADA requirements.

Existing Buildings

Of course, there are tens of thousands of elevators in operation that were built prior to the ADA becoming effective in 1992. The ADA recognizes that immediate, total compliance with all ADA guidelines is often a practical and financial impossibility. However, the ADA does impose a duty to improve accessibility in older buildings in a variety of situations.

Of significant importance is the distinction between buildings that are "public accommodations" and those that are "commercial facilities." Generally, the requirements for public accommodations (such as stores, malls, hotels, restaurants, hospitals, and certain professional offices) are more stringent. For such buildings, "readily achievable" steps must be taken immediately to bring existing buildings into compliance. Determination of "readily achievable" hinges on the difficulty and cost of an alteration and the available financial resources. For either class of building, if alterations are made to the building, the ADA requires that the altered area be made accessible to the greatest extent feasible.

Bringing Elevators into Compliance

Existing elevators may not comply with all requirements of the ADA. When you choose to make ADA alterations, a variety of improvements can be introduced to improve accessibility. Here are a few of the most popular items, each of which can be readily installed on most existing elevators, usually at modest cost:

Braille plates: Braille and raised numbering can be easily added to the car pushbutton station, and to the elevator door jambs at each landing.

Hall lantern gongs: If you have existing hall signal lanterns, they can be modified to provide "one-ring-up, two-rings-down" operation.

Floor passing chimes: The addition of a chime inside the elevator, which sounds as the car passes each floor, alerts the visually disabled to the movement and location of the car.

Escalators/Moving Walks

Building owners and managers should know about their escalators' basic functions so they can help educate their tenants and employees about using them correctly and safely. The most efficient form of vertical transportation, escalators are power-driven, continuous-inclined stairways capable of moving 4,000–7,000 people per hour. They rise from four feet to over one hundred feet, going floor-to-floor, or sometimes even skipping floors.

Today's escalators are driven by a microprocessor-controlled motor (often mounted at the top of the escalator) that turns the main drive shaft. The drive shaft powers the step chain that moves the steps. It also powers another chain that keeps the handrail moving in sync with the steps. The steps are simply wedged-shaped risers that move along this continuous conveyer. In addition to the steps and handrails, which both travel in a continuous loop, there are the balustrades, which are the sides of the escalators. Many of today's escalators have glass balustrades for a clean, open look.

Escalators are supported by a strong steel truss, which is a frame to which the escalator components are mounted. It holds the tracks on which the steps travel, as well as the drive unit that powers the escalator steps and the handrails.

Operating and Safety Features
Operating Controls

For the owner or operator, the only controls for the operation of the escalator or moving walk involve starting and stopping it. Authorized building personnel should be trained in the starting and stopping of escalators and moving walks so they will be prepared in an emergency.

On-Direction Switch: The On-Direction switch is key-operated and conveniently located at both the top and bottom of each escalator. The exact location of the switch may vary depending on the model. The direction in which the escalator travels is controlled by the On-Direction switch. The switch is marked to indicate a clockwise turn for Up operation and a counter-clockwise turn for Down operation. Turn the key to initiate travel in the desired direction. Function and location are similar with moving walks. If you choose to change the direction of travel in which your escalator has been operating, contact your service technician for guidance.

Warning: Be sure no one is on the escalator or moving walk when it is started.

Off/Emergency Stop Button: The Stop button is a red pushbutton located on the top and bottom newels for convenient accessibility in the event of an emergency. The newel is located at both ends of the escalator. (Many older models have this feature next to the On-Direction switch under the newel. Escalators should be checked immediately and all building personnel informed as to location of this button, if they do not know it already.) The pushbutton functions both as a normal Stop and Emergency Stop button.

Warning: Never stop the escalator or moving walk when passengers are riding, except in an emergency. Doing so may cause injury to passengers who may not be prepared for the stop. If an emergency stop is necessary when riders are present, alert the riders before pushing the Emergency Stop button. Re-start the escalator or moving walk only after the emergency condition is resolved.

Safety Features

Escalators and moving walks are equipped with numerous safety features to minimize the chance of accidents or injuries. Following are some of the safety features that come with escalators or moving walks, depending on the model. Familiarize yourself with your particular equipment and the safety features included. Contact your manufacturer's representative with any questions.

The features described below apply to escalators and moving walks except where indicated.

Hold Handrail Sign: This sign was designed to promote safety and comfort, and to help educate the public in the proper ridership of escalators. The sign, which should meet ADME A17.1 standards for color and format in effect at the time of installation, is affixed to the balustrade at both sides and ends of the escalator.

Handrail Entry Protection Device: This device helps prevent hand injuries by using a safety switch, which automatically stops the escalator should an object become wedged between the handrail and balustrade.

Narrow Pitch Step Treads: Step treads are made of die-cast aluminum and feature narrow grooves for passenger comfort and safety. The narrow pitch treads help prevent objects from being entangled at the combplate, and provide a more stable platform on which to stand.

Combplate and Combfingers: Combplates are an extruded aluminum assembly located at each escalator entrance and exit. They provide a stable entrance and exit platform. The combfingers are sometimes of a contrasting high-visibility yellow color. They assist the passenger in distinguishing between the stationary combplate and the moving step. Combfingers protrude into the step tread and help prevent objects from being caught under the combplate.

Grooved Risers: Grooved risers allow a combing action between the back of a step and the adjacent riser of the next step. This helps prevent objects from being caught between the steps.

Understep Lighting: Fluorescent lamps mounted below the steps at entrances and exits clearly outline the separation between adjacent steps. This feature helps the passengers avoid stepping on the separation when stepping onto and off of the escalator.

Soffit Guards: Soffit guards may be installed between the inclined deck of the escalator and the building's ceiling soffit to prevent objects from being wedged at the junction.

Safety Switches: There are a number of safety switches designed to detect unusual or unsafe conditions. Most will stop the escalator automatically when activated. Prior to re-starting your escalator, contact trained escalator maintenance personnel for an equipment check. The particular features of your system should be discussed with your service representative.

Preventive Maintenance

Preserving the smooth and safe operation of an escalator requires regular preventive maintenance performed by a qualified service organization. Certain steps must be taken to promote the safe operation of the equipment. Follow the guidelines listed below as part of a conscientious accident prevention program. Failure to adhere to these guidelines could risk serious injury.

Operating Guidelines:

- Report any unusual equipment operation to trained maintenance personnel. Shut down equipment if dangerous conditions exist.
- Adopt a rigid preventive maintenance program in accordance with manufacturer's recommendations. Do not allow untrained or unskilled individuals to perform any maintenance or repairs on equipment.
- Your escalator maintenance company should regularly (at least monthly) apply a silicone friction reducer on skirt panels as part of general procedures. Do not remove the silicone during your routine building maintenance procedures.
- If understep lighting is burned out or inoperable, notify your escalator maintenance company.
- Report any unusual noises or vibrations to your escalator maintenance company.
- Stop the escalator or moving walk and remove any debris accumulated on the steps or combplates.
- Monitor for broken combteeth and see that repairs/replacements are undertaken immediately by qualified technicians.

- Always remove the start-up key from the On-Direction switch.
- Be sure all safety features are in place and operable.
- If an escalator or moving walk makes an automatic emergency stop, contact trained maintenance personnel for an equipment check before returning it to operation.
- Report all accidents immediately to the maintenance company and your insurance carrier.

Enforcing Proper Use:

- Instruct employees and qualified personnel as to the location and use of emergency stop buttons.
- Instruct all employees and passengers to avoid hitting escalator balustrades (side panels) with objects.
- Do not permit use of an inoperative escalator as a stairway.
- Maintain crowd control.
- Provide alternative vertical transportation for people with disabilities, the elderly, children in strollers, or persons carrying parcels or freight.
- Advise passengers regarding proper use. Schindler offers a brochure and video on escalators as well as elevators. See the Bibliography/Recommended Resources for more information.

Care and Cleaning

The balustrades of an escalator or moving walk come in various finishes. Following are guidelines for care of system components.

Glass: Glass should be cleaned with a quality glass cleaner. Do not lean over the handrail to clean outside glass surfaces while standing on the escalator or moving walk, as a loss of balance may occur.

Stainless Steel and Bronze Areas: Stainless steel portions of the balustrades should be treated with a quality stainless steel cleaner and polish, and bronze surfaces with a quality bronze polish. Apply a light spray coating and polish with a soft cloth. Avoid spraying this product on the emergency stop buttons, on keyed switches, or over traffic areas.

Skirts: Skirts, the vertical panels adjacent to the moving steps, should be carefully treated with a friction-reducing material by your escalator maintenance company. Do not apply cleaner or polish to the skirt area as they remove the friction-reducing material.

Aluminum Steps: Brush the aluminum steps with a stiff broom to dislodge most dirt or debris. Do not attempt to use liquids to clean the steps without first consulting a trained technician. Liquids could enter the area below the steps and result in an unsafe condition. Do not polish the steps.

Synthetic Yellow Step Inserts: Use a mild detergent and soap solution. Do not use any other cleaners, including steam jet blowers. Avoid contact with oil and other lubricants.

Handrails: Handrails should be cleaned periodically with a mixture of light detergent and water.

Warning: All cleaning of escalators and moving walks should be performed when the system is not in operation and passengers are not present.

Preparations for Extreme Weather Conditions

While you cannot relocate an escalator to avoid storm damage, you should always shut it down just as you would an elevator. Then, after the storm, check for the presence of water, particularly in the well area at the bottom. Always be certain that equipment has escaped damage before restoring it to service. Contact your manufacturer's service representative if there are any questions.

System Upgrades

Many new products are being developed that can be easily and economically introduced onto existing elevators and escalators. They have been designed to improve the performance, reliability, safety, security, appearance, and accessibility of elevators and escalators without a major capital investment. These upgrade items generally fall into the category of maintenance and repair. Following are some examples.

- A new security product permits lockout of selected floors, keyless access to designated floors, and monitoring of your system.
- There are many available communications systems that help achieve compliance with ADA guidelines for new buildings. Other accessibility improvements can be introduced at modest cost. These include Braille and raised numbering next to the car pushbuttons and on the door jambs, improved door protection that does not require contact with persons or objects to reverse the doors, audible and visible signals to inform passengers of car position or direction, and relocated buttons at ADA guideline height.
- Invisible light curtains, rather than mechanical safety edges, can provide reliable door edge protection while eliminating many moving parts. New landing switches, which allow the system to activate required actions as the car moves throughout the shaft, have no moving parts and outlast older electromechanical devices, giving you more reliable service.
- Fire emergency service can be added, allowing the automatic return of all cars to the designated floor, and use of the elevators only by authorized personnel during the emergency. Introducing emergency lighting, which activates automatically during a blackout, provides peace of mind during power emergencies.
- For escalators, new modification products include safety combfingers, hold handrail signs, safety brakes, and speed reduction kits. Refinished steps can be installed that let you restore a like-new appearance at a fraction of the cost of new steps.

Conclusion

Elevators and escalators are crucial components in a building's operation since they are relied on by building users for access to their work, recreational or living space. When this equipment is out of service or presents any potential safety hazard, it is the maintenance department's responsibility to take prompt corrective action, including contacting their service representative. An ongoing preventive maintenance program is essential to minimize disruption of service.

While actual servicing of equipment components is largely performed by skilled contractors, the facility's in-house maintenance staff is responsible for keeping accessible surfaces and surroundings clean, well-lit, and safe. Communicating the correct cleaning and operating procedures to staff and enforcing those procedures is a major focus of maintaining conveying systems. Facility operators must also convey information on proper safety and use to building occupants to protect their well-being and the equipment itself.

[1] These features are based on those found in elevators manufactured by the Schindler Elevator Corporation. There may be variations in other manufacturers' elevators.

Chapter 11

Mechanical Systems

William H. Rowe III, AIA, PE,
John J. Moylan, and J. Walter Coon, PE, CSP

This chapter addresses the major maintenance requirements of mechanical systems, including heating, ventilation and air-conditioning (HVAC), as well as plumbing and fire protection. A substantial portion of a facility's annual budget is typically devoted to HVAC equipment operation and maintenance. It is also a key consideration in terms of the building users' comfort, and its successful management has a substantial bearing on the image of the maintenance department. Well-maintained plumbing systems also play a key role in the building occupants' comfort and the facility's image. Persistently leaky faucets and fixtures that don't operate properly make a poor impression. The importance of maintaining fire protection systems cannot be overstated; it is an essential investment in life safety and property protection.

HVAC Systems

Maintenance and repair of HVAC systems is one of the largest components of a facility's budget. Mechanical systems experience significant wear and tear, since they have moving parts and require cleaning and upkeep to continue functioning. In addition, HVAC systems consume the greatest portion of energy for electricity, fuel, and utilities.

Additional expenses related to HVAC systems include the cost of capital improvements that are beyond ordinary repair. In some cases, HVAC systems fail as the result of poor design. In other cases, they may fail because of changed conditions such as a different use for part of a building that requires a redesign of portions of the system. For such a situation, an architect and/or engineer is engaged to redesign the system. The resulting capital expenditures are outside the normal costs of facilities maintenance and repair.

Initiating an HVAC Maintenance Program

To avoid crisis management in operating building systems, a facilities audit (see the Introduction to Part 2) should be prepared to provide an overall picture of the facility's condition, with a focus on areas where action or planning will reduce annual expenses. In the case of HVAC systems, a routine, or general, maintenance program and a preventive maintenance program are the greatest factors in reducing overall costs to the facility. It takes only one broken pipe or failed transformer to set in motion a succession of costs for emergency action, temporary equipment, replacement of damaged finish materials, overtime for workers disrupted, and repair of failed systems. These costs are staggering when compared to the cost of routine inspections, which often reveal problems that can be addressed fairly easily, early on.

The facilities manager needs to have an overall program for the facility that includes the procedures necessary to keep the mechanical operations functioning. If a building's mechanical systems are designed properly, implementing an annual maintenance program for HVAC equipment is the central task. HVAC systems are maintained by proper upkeep of the individual pieces of equipment.

Reviewing Operating/Maintenance Manuals

A survey or audit of the facility's equipment includes a review of available records on that equipment. Operating manuals from manufacturers and maintenance logs are among these resources.

Information on every item of equipment should be available in the library of information for a facility. For a new building, this information may take the form of several three-ring binders containing equipment operating manuals or "cut sheets," which provide detailed performance data, dimensions, requirements, diagrams, and installation and maintenance procedures from the manufacturer, all properly indexed. The use of computers to store drawings, details, and data is common. Manufacturers offer much of this information electronically, including in CAD format.

Information on any hazardous materials associated with the equipment, as well as emergency procedures to be followed, should also be listed in the operating, or maintenance, manual. Material Safety Data Sheets (MSDS) must be available on site for all hazardous chemicals and materials. For particular emergencies such as fire, theft, power failure, or release of hazardous materials, the manual should list instructions for how to respond. Directions for building management, security, custodial staff, tenant fire wardens, and emergency stations for building staff should be specifically called out. Home and emergency phone numbers should be listed with back-up numbers when authorizations for particular responses are necessary. (Staff should be drilled several times a year on emergency procedures. Training in response to fire, hazardous materials, OSHAct, and safety procedures is mandatory under certain circumstances.)

Properly arranged and indexed, the maintenance or operating manual is the key information on which effective building upkeep is based. In addition, the manual should include a list of contractors and vendors who have serviced the equipment, as well as a separate file of logs for repair and preventive maintenance that have been performed on each piece of equipment. The equipment is often abbreviated for convenience; e.g., "AHU-01-CS" may designate air handling unit #1, which is a part of the cooling system. The facility manager of a campus that has many buildings may use "tags" with the abbreviations of specific buildings that house particular pieces of equipment. When the equipment is older, or when no cut sheets are available, data sheets such as that shown in Figure 11.1 can be used to provide the necessary information. This chart can be modified for procedures that are already in place for a particular facility.

Keeping Logs

Logs should be kept of all maintenance procedures. Daily (or hourly) temperature readings of equipment should be recorded, along with readings for pressure, outdoor and indoor temperature, and humidity. After a short time, a building operator develops a "sixth sense" of how the building is operating, and how it should respond throughout each day and each season. With time, all equipment loses some efficiency. By keeping accurate logs

EXHIBIT D1: EQUIPMENT DATA SHEET SAMPLE

Equipment Name _____   Designation _____

Location _____   Associated System _____

Manufacturer _____   Model No. _____

Serial No. _____   Date of Mfr. _____

Vendor/Agent _____

Purchase Order No. _____   Date _____

Status: New ( ) Rebuilt ( )   Warranty Term _____

Start-up: By _____   Date _____

O&M Instruction Videotapes Available: Yes ( )   No ( )

   Name and Number _____
   Source of Tapes _____

Spare Parts

   Complete List: Yes ( )   No ( ); Name and Date _____
   Inventory List: Yes ( )   No ( ); Name and Date _____

Preventive Maintenance Actions and Time Required:

   Chemical Treatment _____   Coil and Pan Cleaning _____
   Filter Changing _____   Lubrication _____
   Motor Starter _____   Power Contactor _____
   Pump Seal _____   V-Belt Drive _____

Scheduled Routine Operation and Maintenance Actions and Time Required:

   Daily
   Semiannually
   Pre-season
   Overhaul: Minor _____   Major _____

Routine O&M Action Description, Skill Level, Tools and Consumables:

   Action Name: _____
   Technician(s) Skill Level _____
   Special Tools/Appliances Required _____
   Consumables Required _____
   Service Cart Type _____

Maintenance History—List for Each Maintenance Action

   Date _____   Work Order No. _____
   Description of Maintenance Action Performed _____
   _____
   _____
   Cost of Maintenance Action Performed _____
   Technician Name(s) _____
   Comments _____

   Technician Report (Comments and Recommendations) _____
   _____
   _____
   _____
   _____
   _____
   _____

ASHRAE GUIDELINE 4-1993

Figure 11.1

of standard readings and notes of unusual developments such as excessive vibration, it can be determined when a piece of equipment begins to make the transition from a slow, incremental loss of efficiency to a more accelerated rate of failure. Unusual readings entered into the logs are often early warning signs that a piece of equipment is beginning to malfunction; the ability to predict this and to adjust the equipment to maximize performance and minimize costs from failures is obviously valuable.

Predictive maintenance techniques, such as infrared photos of electric switchgear, vibration reading of motor mounts, ultrasound, ammeter readings, and other non-destructive testing is used to identify equipment problems that could lead to failure.

Common Repairs and Maintenance of HVAC Equipment

HVAC systems can be classified into five basic system types: *Generation Equipment, Distribution Equipment, Terminal Equipment, Controls,* and *Accessories.* (These are discussed in detail in *HVAC: Design Criteria, Options, Selection,* 2nd Ed., 1994, R.S. Means Co., Inc.) The following is a general description of the basic systems and their common problems. It is followed by a more detailed checklist of items that should be inspected and repaired for all systems. The problems cited are listed for the purpose of general information; actual conditions will require more extensive analysis. Refer to the reference material in the Bibliography/Recommended Resources at the end of this book for detailed information on resources.

Generation Equipment

Generation equipment includes all the major pieces of equipment that generate heating or cooling and their associated equipment: boilers; furnaces; heat exchangers; chillers and cooling towers; and pumps, fans, and accessories. Major pieces of generation equipment, which produce the bulk of heating, cooling and ventilation for the building, are usually housed in a mechanical room or on the roof.

Generation systems contain the most equipment and typically require the most attention of all HVAC systems. Some common repairs and maintenance procedures follow:

Heating Systems:
- Repair fuel pump or gas valve.
- Adjust and clean burners.
- Adjust electrical system; check for blown fuses or failed low-voltage transformers.
- Adjust air quantity to achieve full burning and lower stack temperature.
- Replace steam trap interiors where appropriate.

Cooling Systems:
- Repair failed compressors.
- Lubricate equipment.
- Drain system in winter.
- Repair frozen coils.

Ventilation System:
- Adjust air quantities.
- Check for proper operation of fans and dampers.

Distribution Equipment

Distribution equipment includes devices that connect generating equipment to the terminal units. The distribution system includes pipes and ducts that transport steam, air and water, and any pumps, fans and accessories that are not already considered part of the generation system.

Distribution systems are the least troublesome of the HVAC systems. Nevertheless, maintenance problems do occur. The most common repairs and maintenance procedures are listed below.

Piping:

- Repair leaking pipes.
- Flush system to clean scaling and dirt buildup.

Ductwork:

- Repair air leakage from loosened joints or faulty connections at flexible ductwork.
- Replace ductwork corroded by chemicals in industrial environments.
- Clean ductwork of dust or mold buildup. (This task is sometimes performed in response to complaints or IAQ diagnosis.)
- Adjust dampers; check for improperly closed fire dampers and high-leakage motorized dampers.

Fans:

- Adjust isolators, belts, motors or drives to limit excessive noise or vibration.
- Check wear at bearings.
- Tighten connectors that have experienced gradual loosening of parts over time.
- Replace worn belts and pulleys. Check alignment.

Pumps:

- Align motors; check bearings; lubricate.
- Adjust for noise and vibrations.

Terminal Equipment

Terminal equipment includes the devices at the end of duct or pipe runs that transfer the heating or cooling from the distribution system to the space to be conditioned. Examples of terminal devices are convectors, radiators, fan coils, diffusers, grilles, and registers.

Terminal equipment tends to be relatively trouble-free. Except for minor adjustments and repair of leaks or steam traps, these devices have few operating parts and tend to last a long time without extensive repairs.

Controls

Controls are the devices used to regulate HVAC systems. They include thermostats, temperature and humidity controls, manual and automatic valves and dampers, DDC systems, timers, outdoor reset devices, and any other device used to regulate the setting or operation of the HVAC system. Controls may be manual, electric, electronic, or pneumatic.

Control systems must be constantly maintained. They require active supervision and are the first systems adjusted when a system malfunctions or fails. Some typical repairs and maintenance procedures are:

- Adjust space or equipment temperature settings.
- Repair linkages.
- Replace control valves.
- Repair leaky dampers.
- Repack leaking valves.

Accessories

Accessories include equipment such as flues, chimneys, draft inducers, motor starters, noise and vibration equipment, air cleaning and filtration, hangers and supports, insulation, and related materials that are necessary for the complete operation of an HVAC system.

Repairs and maintenance of accessory equipment are usually incorporated into the preventive maintenance program. Cleaning of filters and visual inspection for proper hanger supports are common maintenance activities that are performed outside of a crisis atmosphere.

Preventive Maintenance and Repair of HVAC Equipment

The most cost effective procedure to implement for any building is a preventive maintenance (PM) program. By spending a small amount of time in minor maintenance and in assessment of performance during the year, costly replacement and repair can be avoided. A preventive maintenance program for a facility consists of maintenance for each piece of equipment at least once a year. For equipment that is stationary such as pipe and ductwork, this maintenance requires only that the equipment be looked at to see that it is functioning, that there are no leaks, and that it is not vibrating excessively. Minor maintenance such as lubricating, changing filters, or adjusting belts also takes place in the PM program. (See the current edition of *Means Facilities Maintenance & Repair Cost Data* for recommended preventive maintenance tasks, crews, costs, and frequencies.)

Because the PM schedule is fixed based on the needs of the equipment to be serviced, it can be arranged in a logical order for overall efficiency. In buildings where the activities are generally routine, such as office buildings, the PM program can proceed predictably. Many of the tasks can be performed by staff. Activities that take considerable time or special tools, such as the takedown of a chiller for annual cleaning, can be performed by outside specialists.

Some PM programs are performed off-season or during off-hours. Boilers are serviced in summer, chillers and cooling towers in winter. In some cases, work after hours or on weekends is necessary for equipment that must be used during regular business hours.

Repair activities are not predictable—they almost always come at the wrong time. Repairs usually involve some troubleshooting, both to diagnose problems and to determine when and how to fix them. When equipment breaks down, an assessment is made as to how quickly the repair must be done. If water is gushing into a tenant space or air conditioning is not being supplied to a manufacturing space with strict environmental controls, an emergency exists—and immediate action must be taken. When one of three modular boilers fails in the spring, or a standby pump is getting too noisy, the repair is generally scheduled in a more routine manner. With some luck or foresight, the building operator will have anticipated some of these failures based on the history and age of the equipment. Spare parts or backup systems may have been put in place to allow for a transition as older equipment becomes obsolete or dysfunctional from ordinary wear and tear.

When equipment is too dysfunctional to maintain or repair, it is replaced as a capital expense. Capital expenses are discussed in Chapters 1, 2, and 18.

Figure 11.2 lists some of the common areas requiring maintenance and repair for the five basic types of HVAC systems. A more complete set of checklists can be found in *Means Facilities Maintenance & Repair Cost Data.*

Plumbing Systems

Maintenance and repair of a facility's numerous plumbing system components is especially important because, according to health and safety codes, buildings must not be occupied without functional plumbing. Plumbing systems are constantly in operation, maintaining pressure, temperature, and water seals between the occupied spaces and the septic or sewage disposal system.

Types of Plumbing Systems

Plumbing systems found in most facilities include:

- Potable water
- Hot water
- Sanitary (drainage waste and vent)
- Storm (roof and various area drains)
- Natural, manufactured, mixed, and various medical gas systems

Certain facilities contain additional specialized plumbing systems and fixtures. These include:

- Hospitals, which have medical gas systems carrying oxygen, vacuum, nitrogen, nitrous oxygen, and compressed air. Hospitals and laboratories also have acid waste systems, bed pan washers, autopsy tables, and therapeutic baths.
- Schools, gymnasiums, athletic facilities, and hotel/motels, which have specialized swimming pool and filtration systems.
- Restaurants and other food preparation facilities, which utilize various plumbing systems to serve dish, utensil, and tray washers; pot washers; coffee urns; cooking units; ice makers; and refrigerators.
- Dental offices, which require cup sinks and vacuum and compressed air systems.

Initiating a Plumbing Maintenance Program

The first step in beginning a maintenance program is a complete audit of the various systems and a count of all of the fixtures and plumbing apparatus such as pumps, water heaters, pressure regulators, and filters.

A set of the building's plumbing plans and specifications is helpful to have, as are operating and maintenance manuals, valve tag charts, parts lists, and manufacturers' cuts. All of these materials are usually readily available from the architect or the plumbing contractor for a recently constructed building.

Planning for Emergencies

Part of initiating a maintenance program is to make decisions pertaining to service—emergency or otherwise—such as how much will be performed in-house and how much will be contracted out to a plumbing contractor or professional service organization. It is imperative that these decisions be made before an emergency arises.

Emergencies such as broken water pipes, backed-up drains, gas leaks, and lack of hot water have to be addressed immediately. As with HVAC equipment, when plumbing emergencies occur during non-working hours, telephone numbers should be available so that custodians or security personnel can contact authorized maintenance people. The maintenance

Maintenance Schedules for HVAC Equipment		
Equipment Type	**Description of Maintenance Required**	**Frequency***
Generation Equipment		
Boiler	Check temperature and pressure control, safety protective devices.	M
	Check furnace, heating elements, blower, motors; fill out maintenance checklist and report deficiencies.	Q
Warm Air Furnace	Check burner and blower, controls, lubrication, filter, flue pipe, damper and stack; fill out maintenance checklist and report deficiencies.	Q
Heat Exchanger	Check steam modulation valve and condensate trap, temperature gauges, controls; fill out maintenance checklist and report deficiencies.	S
Fuel Oil tank	Check storage tank pressure, fuel oil pressure, controls, fire protection devices, liquid level, tank corrosion and pump operation; fill out maintenance checklist.	A
Deaerator Tank	Check steam pressure regulating valve operation, controls; blow deaerator tank; check tank and piping for leaks; test sulfite on deaerator water sample; fill out maintenance checklist and report deficiencies.	S
Split System	Clean condenser and evaporative coils; blower, drain pan, fans, motor and drain piping, lubrication; replace filter; check fan belts, controls, compressor oil level, refrigerant pressure; fill out maintenance checklist and report deficiencies.	Q
Compressor	Check lubrication, controls.	S
	Operation check of air compressor, oil level, oil filter, oil pressure, temperature, piping leaks; check intercoolers and aftercoolers for high cooling water temperatures; check excessive vibration, noise, and overheating; fill out maintenance checklist and report deficiencies.	M
Air Conditioning Package Unit	Check belts, controls, refrigerant pressure, lubrication; replace filters, perform operational check of unit; fill out maintenance checklist and report deficiencies.	Q
	Clean coils, evaporator drain pan, drain piping; check compressor oil level, electrical wiring.	A
Heat Pumps	Check unit for proper operation, noise and vibration; clean evaporator coil; check fan belt; inspect piping and valve for leaks; replace air filter; check refrigerant pressure; check the reverse cycle valve; check defrost cycle.	Q
	Check electrical wiring; clean condenser coil, drain pan, drain piping, lubrication.	A
Air Handling Unit	Clean coils, drain pan, blower, motor; check controls, noise, vibration, belts; replace filters; clean dampers and louvers; check lubrication; fill out maintenance checklist and report deficiencies.	Q
Air Washers	Clean the strainer.	W
	Remove lint, dirt from spray nozzle, piping, damper blades, linkage, drain sump, float ball on water level float switch.	M
	Check entire spray section, drain, fan, lubrication; fill out maintenance checklist and report deficiencies.	Q
Chiller	Check unit for proper operation, oil level, temperature, dehydrator.	W
	Run system diagnostics test; check controls, safety limits, refrigerant system, noise/vibration, evaporator and condenser for corrosion.	M
	Check wiring, connection, oil filter, piping and valve leaks, blower, condenser coils, intake screen, lubrication, superheat and subcooling temperature; clean the economizer gas line; fill out maintenance checklist and report deficiencies.	A

* W—weekly; M—monthly; Q—quarterly; S—semi-annually; A—annually.

Figure 11.2

Maintenance Schedules for HVAC Equipment (cont.)		
Equipment Type	**Description of Maintenance Required**	**Frequency***
Condenser (Air cooled, water cooled)	Check unit for operation, noise/vibration, controls and wiring, lubrication; check fan/blowers; inspect piping and valves for leaks; fill out maintenance checklist and report deficiencies; test motor amperage.	Q
Cooling Tower	Check operation of unit for water leaks, noise/vibration; clean and inspect hot water basin; check controls and electrical wiring, lubrication, fan/blower, belts, drain and flush water sump; check make-up water for leakage, chemical water treatment system, freeze protection, heat tracing, float valve; fill out maintenance checklist and report deficiencies.	S
	Clean strainers	W
	Check nozzles	M
Evaporative Cooler	Check unit for proper operation, noise and vibration, belts, clean the evaporative louver panel; check lubrication, controls; inspect wiring; clean sump; fill out maintenance checklist.	A
Absorption Chiller	Check unit for proper operation, excessive noise and vibration; check the purge pump vacuum oil level, lubrication; clean strainers in all lines; check controls, wiring, connections, piping leakage; fill out maintenance checklist.	Q
Burner	Check safety devices, flame failure protection, controls.	M
	Check and lubricate burner.	Q
	Check and clean burner gun and ignition; fill out maintenance checklist and report deficiencies.	A
Dehumidifier	Check the outlet air temperature, controls, filters; fill out maintenance checklist and report deficiencies.	M
	Check desiccant wheel, motor vibration and noise, blower, electrical wiring, lubrication.	S
Humidifier	Operate humidistat through its throttling range to verify the activation and deactivation; inspect steam trap for proper operation; check controls, electrical wiring; clean and replace the steam/water nozzles; check steam lines for leakage; fill out maintenance checklist and report deficiencies.	S
Distribution Equipment		
Pump	Check for proper operation, motor-to-pump alignment, leaks on piping, seals, packing gland, controls, lubrication, vibration, noise, overheating; fill out maintenance checklist and report deficiencies.	S
	Clean out trash from pump intake; check for pump corrosion; check float switch (sump pump only).	S
Fan	Start and stop the fan with local switch; check fan noise, sheave alignment, belt tension, vibration, overheating, fan belts, electrical wiring, controls, lubrication; fill out maintenance checklist and report deficiencies.	S
Ductwork	Check for leakage; report deficiencies. Check insulation for mildew or moisture.	S
	Clean grease filters at kitchen hoods.	W
Piping	Inspect all joints, leakage; report deficiencies.	Q
Condensate Drains	Inspect all joints, leakage; report deficiencies.	Q
Air Filters	Clean and replace the filters; check the indicators for defective tubes and broken ionizing wires; secure filter and fan unit; fill out maintenance checklist and report deficiencies.	Q
Duct Coil	Inspect all joints, leakage; clean the coils; fill out maintenance checklist and report deficiencies.	Q

* W—weekly; M—monthly; Q—quarterly; S—semi-annually; A—annually.

Figure 11.2 (cont.)

Maintenance Schedules for HVAC Equipment (cont.)		
Equipment Type	**Description of Maintenance Required**	**Frequency***
Terminal Equipment		
Fan Coil	Check coils and piping for leaks, damage, and corrosion; lubricate blower shaft and fan motor bearing; clean the coil, drip pan, drain line; fill out maintenance checklist and report deficiencies.	Q
Fire Dampers	Remove and replace access door; clean out debris/dirt against damper; check lubrication; replace fusible link and check that the blades operate freely; fill out maintenance checklist and report deficiencies.	A
Grilles, Registers and Diffusers	Check operation; clean out dirt; report deficiencies.	A
VAV Boxes	Check the VAV controls, electrical/pneumatic tubing connections, lubrication, cycle actuator; verify that blades fully open and close; fill out maintenance checklist and report deficiencies.	S
Unit Heaters	Inspect coils, connections, piping, check fan and motor for vibration, noise, fuel system, lubrication, controls, electrical wiring; fill out maintenance checklist and report deficiencies.	A
Steam Traps	Clean and inspect the trap; fill out maintenance checklist and report deficiencies.	Q
Baseboard Unit, Convectors	Check coils and piping for leaks, damage and corrosion; clean the coil; fill out maintenance checklist and report deficiencies.	Q
Expansion Tank	Check piping for leakage, air vent, pressure; fill out maintenance checklist and report deficiencies.	Q
Steam and Hydronic Air Vent	Check for operation; report deficiencies.	Q
Controls		
Valves, Manual	Open and close valve to check operation, lubrication, leaks, valve exterior; fill out maintenance checklist and report deficiencies.	A
Valves, Motorized	Check valve for proper operation, controls, lubrication, leaks, valve exterior; fill out maintenance checklist and report deficiencies.	A
Automatic Dampers	Check for freedom of movement of the damper and lubricate at bearing points, controls; fill out maintenance checklist and report deficiencies; Check linkage and actuators.	Q
Manual and Splitter Dampers	Check movement of the damper manually, lubrication; report deficiencies.	Q
Temperature Sensor	Clean with care; report deficiencies.	S
Gauges	Clean with care; report deficiencies.	S
Thermometer	Clean with care; report deficiencies.	S
Pressure & Temperature Switch	Clean with care; report deficiencies.	S
Aquastat	Clean with care; report deficiencies.	S
Humidistat	Clean with care; report deficiencies.	S
Accessories		
Water Treatment	Check safety devices, controls and wiring, piping leakage; check and clean strainer, grease fitting; fill out maintenance checklist and report deficiencies.	Q

* W—weekly; M—monthly; Q—quarterly; S—semi-annually; A—annually.

Figure 11.2 (cont.)

department will decide how best to respond. The focus of any response will be to have all systems functioning as soon as possible without disrupting the facility's operations.

Preparing a Maintenance Log

A log should be kept of all the plumbing fixtures, appliances, and ancillary equipment. The log should indicate the age and condition of each item, as well as its history of operation and repair. It should also list any spare parts along with their location.

Rather than log each individual plumbing fixture in larger facilities, a log page can suffice for each toilet room, or in the case of a hotel/motel or nursing home, individual guest or patient rooms. Whatever the function of the facility, a workable unit of measure can be determined.

If the plumbing systems contain pumps or shell and tube heat exchangers, records should be maintained showing inlet and outlet pressure and temperature readings. Changes observed in these readings might indicate air accumulation in the lines or the need to punch the tubes in the exchanger coil. The electrical current drawn by the pump motor should also be recorded.

The log can be used as a checklist as well as a repair record for the various items, noting those that require excessive repair. From this a proper maintenance schedule can be determined.

Inspecting the Plumbing Equipment for Proper Operation

The simplest form of inspection of the facility's plumbing is observing the daily use of the equipment and the reliability of its function.

Questions to be answered during inspection include:

- Are adequate water pressure and volume being maintained?
- Is the design or set water temperature being maintained?
- Do all faucets or flush valves close completely or do they continue to drip or dribble after closure?
- Do any of the plumbing fixtures show discoloration, indicating rust or other waterborne minerals, which would indicate the need for a water treatment or filtration program?
- Are any of the fixtures' drains noisy or slower to empty than what is desirable? This might indicate that the fixture traps need to be emptied or cleared out, or perhaps a more severe action is necessary. A chemical drain cleaner can break up clumps of hair or other forms of buildup inside the traps or on the interior walls of the drainage piping system itself. Note that the use of any chemical cleaner can be hazardous; the manufacturer's directions should be strictly followed.
- Are any odors present? Check for dry trap seals in the floor drains or plumbing fixtures.
- Does exposed pipe covering or insulation appear moist, indicating a possible leak or condensation?

In addition to the daily "inspection by use" of these systems and fixtures the janitorial crew should be alerted to notify maintenance personnel of any perceived malfunctions.

During a heavy rainstorm, area drains and roof drains should be observed for slow operation, indicated by a backup of standing water or water cascading from the roof down the exterior walls, roof hatches, or into ducts serving exhaust fans or other rooftop HVAC equipment. The screens or strainers on roof and area drains should be checked periodically for blockage. While on the roof, also check the plumbing vents for stoppages such as birds' nests, hornets' nests, and the like.

Continual occurrences of any of the above should alert the maintenance department to the possibility of a defect in the system design or a malfunction of a mechanical device within the system.

Common Repair Items

The most frequently needed plumbing repairs typically involve flushometers on water closets and urinals, and washers and/or seals on faucets or other water outlets. The need for repairs is identified by routine inspections, or in the course of daily use of plumbing fixtures by the building's occupants. If the cleaning crew is part of the facility's in-house maintenance department, they are a good source of information on any problems and needed repairs.

Some of the modifications required by the Americans with Disabilities Act may also be included in the repair/upgrade category. Major remodeling of restrooms, hallways, etc. may be categorized as capital improvements, but smaller adaptations, such as pipe covering insulation, may be considered maintenance and repair items.

Unlike HVAC and fire protection, plumbing maintenance generally does not involve regularly scheduled maintenance *procedures*. Instead, it relies chiefly on regular inspections to reveal problems or potential problems. The observations of building users and the cleaning crew are also a primary source of information about fixture condition. An effective reporting and response system prevents minor maintenance issues from developing into long-term nuisances or costly, major repairs, thereby protecting the maintenance department's budget and its reputation.

Fire Protection Systems

"For want of a nail a shoe was lost, for want of a shoe a horse was lost, for want of a horse a rider was lost. For want of a rider, a battle was lost—all for the want of a nail."

This time-worn saying is very applicable to maintenance and testing of fire suppression and detection systems. Once a fire protection system is installed, it can remain dormant and unnoticed for years because it does not enter into the day-to-day operation or creature comfort of the facility. But that one day when a fire occurs, the fire protection system is expected to operate at 100% of the design and installation efficiency it had the day it was commissioned to protect life safety and property.

Fire protection suppression systems are mechanical. Detection alarm systems are both mechanical and electrical, and both have components that can fail, for a variety of reasons. Yet the owner of a multi-million dollar manufacturing plant may have a contract to have his home furnace and air-conditioning unit inspected and maintained every fall and spring, while neglecting the fire suppression system at his plant, a system that protects millions of dollars of equipment and hundreds of employees. All too often, systems are not tested for years following their installation, and no maintenance program is put in place.

The National Fire Protection Association (NFPA) considers the inspection, testing, and maintenance of suppression systems important enough to have developed a one-hundred-page standard—not just a guide or a recommended practice—that mandates an inspection, testing, and maintenance program for water-based fire protection systems. (See the Bibliography/Recommended Resources at the end of the book for helpful references.)

Even the basic wet pipe sprinkler system demands a maintenance and testing program, depending on the size and complexity of the system. An appropriate maintenance program should be established, scheduled and performed by in-house maintenance personnel. If they are not qualified or available, or if there are multiple or complex systems or a massive protected area, a testing and maintenance contract should be established with a reliable sprinkler company.

Many problems can occur over time that can compromise fire protection systems. For example, the main control valve from the water supply to the wet pipe system (or, for that matter, to any other water-based suppression system) might have been temporarily closed to drain the system during a power failure when the building heat malfunctioned in the winter, or when a damaged sprinkler head was replaced, and it might not have been re-opened. Records show this scenario occurs all too often. Sprinklered buildings have burned to the ground because the water supply valve was closed.

The water supply valves should be inspected weekly — they are required to be indicating-type valves. All it takes is a glance to determine if the valve is open or shut. (See "Main Control Valve" in *Fire Protection: Design Criteria, Options, Selection*, R.S. Means Co., Inc.)

Dry Pipe Systems
Buildings with dry pipe systems require without question an in-house inspection maintenance schedule, or the services of a sprinkler contractor under a maintenance contract. Water and air pressure gauges should be checked weekly to ensure that the air pressure is adequate to keep water from entering the dry piping in a freezing environment. At least annually, gauges should be checked to determine that they are working properly and accurately.

Of equal or greater importance is an inspection of the system in the fall prior to freezing weather. Drum drip drains must be emptied of condensation, or they will surely freeze, crack, and trip the system. Accelerators, exhausters, air compressors, and/or shop air supplies with pressure maintenance devices must be inspected, tested, and maintained in perfect working order to maintain a dry pipe system in effective and efficient order. (See "Dry Pipe Sprinkler Systems" in *Fire Protection: Design Criteria, Options, Selection.)*

An hour of inspection of the dry pipe system along with minor routine maintenance can save untold hours of cleanup maintenance and thousands of dollars in water damage, not to mention eliminating fearful hours or days while the dry pipe system is shut down for repairs and recommissioning, leaving the property and personnel unprotected.

No matter what type of suppression system is installed for protection of property and life safety, uninterrupted continuity of operation is critical. To achieve it, a scheduled inspection and maintenance program must be instituted when the system is put in service.

Fire Pumps
Fire pumps are the heart of any water-based suppression system. The owner would not have spent thousands of dollars for fire pumps if the required level of protection did not mandate increased volume and/or pressure to operate as designed.

Fire pumps are a perfect example of the "For want of a nail..." story. The failure of a simple solenoid or pressure switch can completely eliminate the effectiveness of a million dollar investment in the suppression system protection supplied by the fire pumps.

NFPA Standard 20 specifies the frequency of inspection, testing, and maintenance required for fire pumps, and NFPA Standard 25 indicates the frequency of inspection, testing, and maintenance for all types of water-based fire protection systems and devices. The American Fire Sprinkler Association (AFSA) and the National Fire Sprinkler Association produce detailed forms to be used to dictate and record inspection procedures and frequency, and maintenance required. A sample form from NFPA Standard 25 is shown in Figure 11.3.

Special Systems

Foam water systems are installed to protect special hazards. "Special" indicates that these conditions require more than water to control and suppress hazardous combustion scenarios. For example, foam water deluge and monitor nozzle systems for aircraft hangars are tremendously costly systems, but the values protected warrant this monetary outlay. Alas, these hangar systems are easily, and often, forgotten a few years after installation.

The proportioning system, the deluge valves, and the pressure maintenance pumps are all sophisticated systems with every device and piece of equipment extremely complex, and with hundreds of small, but operationally critical, devices that must all work perfectly and in sequence if the total system is to function effectively and efficiently. A periodic inspecting, testing, and maintenance program for these special hazard systems is essential. If not instituted, and carried out efficiently by experienced personnel, there is no point in installing the system because the pipe, valves, foam concentrate tanks, complex proportioning system, and other elements provide nothing more than a dangerous false sense of security.

The shelf life of aqueous film forming foam (AFFF) properly stored is indefinite *if* it is not contaminated with water or a foreign substance. Therefore, to ensure its quality—its ability to perform efficiently in suppression of a jet fuel spill fire—a sample must be submitted to the manufacturer every year for tests. If it is contaminated, the entire volume of AFFF must be replaced. The cost of this replacement warrants proper maintenance and inspection to prevent the contamination of the foam concentrate.

NFPA Standard 11 offers information on the testing, inspection, and maintenance of foam water systems. NFPA Standard 12 specifies critical inspection and maintenance of both high pressure and low pressure carbon dioxide (CO_2) systems. If these inspection and maintenance schedules are not followed for CO_2 systems, these dry agent systems will not provide the protection for which they were designed. The program should be conducted only by personnel experienced with CO_2 systems. CO_2 systems use unique storage units and operating devices and equipment, and the maintenance layperson should not perform anything more complex than the weekly or monthly inspections and maintenance.

Maintenance and Testing

Simple systems like the standpipe and hose systems also require periodic inspections and maintenance, especially small (1-1/2") hose systems. Unused hose that has been sitting on a rack for years is not reliable unless periodically inspected and tested.

Many inspection, testing, and maintenance programs can be performed by in-house maintenance personnel who have had some training in the

Report of Inspection & Testing
of Wet Standpipe System
ALL QUESTIONS ARE TO BE FULLY ANSWERED AND ALL BLANKS TO BE FILLED

Inspecting Firm: (Contractor) _____ Inspection Contract # _____

Name of Property: _____

Inspector Name: _____ Date: _____

Page ____ of ____ **Date of Previous Internal Pipe Inspection** _____

Inspection Frequency: ❏ Monthly   ❏ Quarterly   ❏ Annually   ❏ Other: _____

			Y	N/A	N
A-1.1	Supply Water Gauge	____psi			
A-1.2	System Water Gauge	____psi			
A-1.3	Top Floor Gauge:	____psi			
A-1.6	Class of Service: I II III				
A-2.1	Hose Valve Size: ____ inch				
A-2.2	Hose Valve With Adapter Size: ____ X ____ inch				
A-2.3	Hose valve with ____ inch hose:				
A-2.6	Type and size of Nozzle:				
	ADJUSTABLE	____ inch			
	STRAIGHT STREAM	____ inch			
	FOG	____ inch			
	NON- ADJUSTABLE	____ inch			
A-3.1	Indicate the Type and Record the information for the TOP FLOOR Hose Valve:				
	Pressure Reducing Valves Inlet Pressure Set	____psi			
	Pressure Reducing Valves Outlet Pressure Set	____psi			
	Pressure Restricting Valve Inlet Pressure Set	____psi			
	Pressure Restricting Valve Outlet Pressure Set	____psi			
	Pressure Regulating Valve Inlet Pressure Set	____psi			
	Pressure Regulating Valve Outlet Pressure Set	____psi			
	(Attach supplemental sheet recording the GPM and Pressure Setting for EACH FLOOR Hose Valve.)				

		Y	N/A	N
A-4.1	System in Service on Inspection			
A-4.2	System Equipped with Flow Switch:			
A-4.3	System Equipped with Alarm Check Valve:			
A-4.4	Trim Piping leak tight:			
A-5.1	Control Valves Sealed Open:			
A-5.2	Control Valves Locked/Tamper Open:			
A-5.6	Backflow Asmb. Valves Sealed Open:			
A-5.7	Backflow Asmb. Valves Locked/Tamper Open			
A-5.8	Backflow Assembly Operating OK:			
A-6.1	Wall Hydrant Sealed Open:			
A-6.2	Wall Hydrant Locked/Tamper Open			
A-6.6	Valve Area Clear of Obstructions:			
A-6.7	Valve Area Accessible:			
A-6.9	Wall Hydrant plainly visible:			
A-6.10	Wall Hydrant easily accessible:			
A-6.11	Wall Hydrant Identification Plate in place:			

		Y	N/A	N
A-6.12	Roof Manifold Control Valve Closed:			
A-7.1	Tamper Switches appear Operational:			
A-7.2	Alarm Devices Appear Operational:			
A-7.5	Exterior of devices in good condition:			
A-7.6	Exterior bells, gongs unobstructed:			
A-7.7	Exterior Fittings free of Water Leakage:			
	Main drain:			
	Alarm bell line:			
A-8.1	Hose Valve free of Physical Damage:			
A-8.2	Hose valve outlets with cap:			
A-8.3	Hose Valve outlet thread in good condition:			
A-8.6	System free of visible water leaks:			
A-8.8	Hose valve outlets equipped with Reducing Hose Adapter:			
A-9.1	Inspection of Cabinet per NFPA 1962:			
A-9.2	Inspection of Hose per NFPA 1962:			
A-9.3	Inspection of Hose Nozzle per NFPA 1962:			
A-9.6	Wall Penetrations Caulked / Sealed:			
A-10.1	Roof Manifold Equipped with Hose Valves:			
A-10.2	Roof Manifold Hose Valve Caps in Place:			
A-10.3	Roof Manifold Swivel rotation is non-binding			
A-10.4	Roof Manifold valves good condition:			
A-10.5	Roof Manifold Ball Drip operational:			
A-11.1	Caps or Plugs on FDC:			
A-11.2	FDC Swivel rotation nonbinding:			
A-11.3	FDC location plainly visible:			
A-11.4	FDC easily accessible:			
A-11.5	FDC Identification Plate in place:			
A-12.1	Piping Free of Physical Damage:			
A-12.2	Piping (Exterior) is free of corrosion:			
A-12.3	Piping appears to be leak tight:			
A-12.6	Ball Drip drain drip tight:			
A-12.7	Main Drain at Supply ____ (inch) : ___ psi			
A-12.9	Signage / Identification Plates in Place:			
A-15.1	**ALARM PANEL CLEAR**			
A-15.2	**ALL SYSTEMS IN SERVICE:**			
A-16.1	**COMMENTS:**			

(All "NO" answers to be fully explained.)

INSPECTOR'S INITIAL _____   OWNER/DESIGNATED REP. INITIAL _____   DATE _____

(AFSA Form 94-108A)

Page 1 of 2

Figure 11.3

QUARTERLY TESTING OF WET STANDPIPE SYSTEM

	Y	N/A	N
B-1.1 Main Drain ___ (inch) Flow at Riser: ___ psi			
B-2.1 Alarm Devices Operated:			

REFER TO NFPA 1962 FOR TESTING OF STANDPIPE SYSTEM IN ADDITION TO THE TASK INDICATED HEREIN.

ANNUAL TESTING

	Y	N/A	N
C-1.1 Test of Hose per NFPA 1962:			
C-1.2 Test of Hose Nozzle Per NFPA 1962:			

FIVE YEAR INSPECTION

	Y	N/A	N
D-1.1 Internal Inspection of Check Valves: Date: _____			
D-1.2 Internal Inspection of Alarm Check: Date: _____			

FIVE YEAR TESTING

	Y	N/A	N
E-1.1 Pressure Gauges Calibrated: Date: _____			
E-1.2 Pressure Gauges Replaced: Date: _____			
E-2.1 Hydrostatic Test performed: Date: _____			
E-2.2 Water Supplyt Test performed: Date: _____			
E-3.1 Pressure Regulating Type Hose Valves Flow Tested: Date: _____			

(Attach additional pages to record the results of the Flow Test information indicated below which shall be provided for each type of Hose Valve connection including the Roof Manifold, for each Floor and for each Standpipe Riser. The Authority Having Jurisdiction shall be consulted prior to conducting the Flow Test.)

	Y	N/A	N
E-4.1 Volume of Flow: ___ gpm			
E-4.2 Supply Side: ___ psi			
E-4.3 Hose Connection Side: ___ psi			

SAMPLE

(All "NO" answers to be fully explained.)	(AFSA Form 94-108A)
INSPECTOR'S INITIAL _____ OWNER/DESIGNATED REP. INITIAL _____ DATE _____	Page 2 of 2

Figure 11.3 (cont.)

various types of sprinkler systems. Special hazard systems, or very extensive systems, should be covered by a maintenance contract with a local fire protection company.

If a sprinkler system has been in service for 50 years, the sprinkler head must be replaced. A sample of these heads can be submitted to a nationally recognized testing laboratory for field service testing, but a system with this many years of service should have the heads replaced in any event.

After many years in service sprinkler system piping should also be inspected and investigated to determine if the piping has become clogged with corrosion. This evaluation is performed by actually disconnecting sections of the piping and examining the interior with a flashlight. If the piping interior is badly obstructed, it may warrant flushing the system piping using one of two methods — both of which must be performed by a fire protection contractor.

- The hydraulic method consists of flowing water from the yard mains, sprinkler risers, bulk mains, cross mains, and branch lines respectively, in the same direction in which water would flow during a fire.
- The hydropneumatic method uses special equipment and compressed air to blow a charge of about 30 gallons of water from the ends of branch lines back into feed mains and down the riser, washing the foreign material out of an opening at the base of the riser.

Both methods of cleaning out old sprinkler piping are quite expensive and, as mentioned before, must be performed by an experienced automatic sprinkler contractor.

Conclusion

All fire suppression, detection, and alarm systems require a financial investment, but more than that, they are an investment in life safety — of both building occupants and firefighters. Furthermore, they protect an organizational investment in its buildings and their contents. It is imperative that these systems' effectiveness and efficiency be maintained in their original condition as designed and installed by periodic testing, inspection, and maintenance.

Chapter 12

Electrical

Stephen C. Plotner

Maintaining electrical equipment involves working with a variety of materials, and meeting the individual standards of various manufacturers. This sometimes necessitates a separate set of operational, maintenance and repair procedures for each piece of equipment. This chapter focuses on the major components and equipment types often found in facilities' electrical systems. Since there is such a wide variety of equipment available, step-by-step maintenance and repair procedures will not be included here. The reader is referred to *Means Facilities Maintenance and Repair Cost Data* for guidelines, and to the manufacturer of each piece of equipment for more specific information.

The electrician's or electronic technician's responsibilities include a wide range of activities and may cover installation, service, maintenance and repairs of all phases of power distribution, lighting and electronic instrumentation. In larger facilities or in multi-plant situations, there may be many individuals on staff, each performing only part of the task. In smaller facilities, all the duties may be performed by one individual; or the owner or management may contract out electrical maintenance and repair activities.

The focus of this chapter will be to identify, for owners and management, the major components and equipment types in need of maintenance, such as preliminaries to maintenance and repair, lighting systems, motors, batteries, distribution systems and special systems.

Preliminaries to Maintenance and Repair

To implement an effective electrical maintenance and repair program for a facility, it is necessary to gather all known information on the facility's existing electrical systems, set out basic maintenance guidelines, and establish work safety rules.

Information and Records

Before any effective maintenance or repair activities can be performed, the electrical maintenance group or outside electrical contractor needs to know the locations of substations, main feeders, transformers, subfeeders, panelboards, branch circuits, and all electrical equipment, such as motors and lighting. This information should be developed in the form of specifications and record drawings on file which were furnished by the original general building contractor or the electrical subcontractor. If the whereabouts of this information is not presently known, time and resources should be allocated for obtaining such documentation, or much time will be wasted locating items when maintenance or repairs are required. It is important that any subsequent additions/modifications be recorded on the original drawings so that they always reflect current conditions. Included in the drawings should be a single-line diagram of the facility's electrical distribution system, as well as floor plans and a site plan which should depict the following information:

- Location and size of overhead or underground utility company feeders
- Location and size of the building step-down transformers
- Location, type and size of the main switchgear
- Location, type and size of all feeder transformers
- Location, type and size of all raceways, panelboards and conduits
- Sizes, types and numbers of wires for each feeder, branch circuit and circuit
- Location, type and rating of all motors
- Location, type, size and setting of all protective devices
- Location, type and size of all lighting fixtures
- Location and type of all wiring devices

The information in these drawings should be supplemented by an Operations Manual furnished by the original general building contractor or his electrical subcontractor, as well as information such as specifications, recommended maintenance and repair procedures and frequencies, parts lists and any other helpful information for each piece of equipment, device or system furnished by each manufacturer.

Maintenance Guidelines

To prevent electrical breakdowns and thus minimize the amount of repair necessary, follow the general maintenance guidelines listed below:

- Perform routine inspections based on formal procedures at regular intervals.
- Keep the environment adjacent to electrical equipment dry and at controlled humidities according to the manufacturer's instructions.
- Regularly clean electrical apparatus, also according to manufacturer's instructions.
- Maintain connecting electrical parts in tight condition.
- Adhere to all safety precautions relating to the maintenance and repair of the equipment.

If established maintenance schedules are not followed, under-maintenance will result in one or more of the following defects:

- Deterioration of cable insulation from moisture, oil, heat or other causes.
- Excessive or low voltages caused by deficiencies in the system.
- Loss of illumination.
- Corrosion of metals.
- Hot cables caused by unbalanced loading.
- Power failure.

Work Safety

It is extremely important when working on electrical equipment to adhere to the safety rules listed below:

- Do not touch bare wires until a check is made to be certain that the feeder, branch circuit or circuit has been de-energized.

- Shut off all power before working on any piece of equipment.
- Work on energized equipment only when making a test that requires power to be delivered to the equipment.
- Follow the National Electrical Safety Code when working on equipment.

Lighting Systems

Lighting systems use the properties of light reflectance and absorption to obtain the most lighting at the lowest energy cost and, at the same time, deliver a specific level of illumination adequate for those working in the area. This illumination should be maintained at recommended levels to conserve eyesight, improve morale, increase safety, improve housekeeping, decrease fatigue, reduce headaches, and increase production, all of which are directly reflected in lower operating costs. In order to achieve these benefits, it will be necessary to consider lamp types, lighting quality and lamp maintenance.

Lamp Types

Lighting is a significant factor in the operating costs of a facility and, in the last 25 years, much attention has been given to the development of more efficient lamps. Lamps for interior and exterior fixtures can be incandescent, fluorescent or high-intensity discharge (HID). HID lamps include mercury vapor, metal halide and high-pressure sodium (low-pressure sodium lamps are used principally for outdoor applications because of their strong yellow color).

Incandescent bulbs are the least efficient of the above, but their low initial cost keeps them attractive. Also, their "warm" (high red content) color spectrum is pleasing for many applications. One significant advantage of incandescent bulbs is that they light up instantly.

Fluorescent bulbs require the use of ballast coils which make initial fixture costs somewhat higher than the cost of incandescent fixtures. Fluorescent bulbs are, however, more efficient and their color spectrum far more closely imitates natural daylight. A few different fluorescent lamp types are available that offer trade-offs among efficiency, color spectrum and bulb cost.

HID lamps generally are more efficient than fluorescent lamps, but do not generate the same broad light spectrum. HID lamps require special ballasts and require several minutes to warm up before full output is reached.

Mercury vapor lamps put out about 65 lumens per watt and have a blue-green color. Because they have little or no red component in their spectrum, they distort the actual color of objects. They require 3 to 7 minutes after striking to achieve their full brilliance and, since the bulb must cool before it can restrike, it takes about the same amount of time to light up again after a power disruption.

Metal halide lamps are 1.5 to 2 times more efficient than mercury vapor lamps. Their color is a blue-white and they render color slightly better than mercury vapor lamps. They take longer to warm up or restart—about 15 minutes.

High-pressure sodium lamps have an efficiency of about 110 lumens per watt. They have a golden-white color and can distort colors considerably. About 15 minutes are required for a high-pressure sodium lamp to reach full luminance, but it can restart quickly after a momentary interruption—in only 1-1/2 to 2 minutes.

Lighting Quality

Characteristics such as direct glare, reflected glare, brightness ratio, shadows and color are used to describe lighting quality. Both direct glare and reflected glare cause light to shine brightly into the eyes and produce unnecessary strain. Too great a brightness ratio between near and more distant objects can produce fatigue as the eyes shift back and forth to adjust to the varying brightness. By making lighting more diffuse, we can minimize dark shadows which can annoy and tire the eyes. Color of light and color rendition can have a profound effect on light quality.

Lamp burnout clearly reduces the effectiveness of any lighting system, but levels of lighting can begin to deteriorate even before lamp burnout. The amount of illumination initially provided starts to decrease almost as soon as the bulb, or lamp, is put into operation. The rate of decrease is infinitesimal at first, but increases as time passes. This continual decrease of lighting quality is due in large measure to three causes: dirt, lumen depreciation and discoloration. Dirt accumulation on lamps, reflectors, room walls and ceilings will reduce the reflectability of light. Lumen depreciation is the gradual reduction of light output which occurs as lamps age and is caused by heat from within the lamp slowly vaporizing internal components and depositing this material on the inside of the glass bulb. Heat given off by lamps can cause discoloration of reflectors and lenses.

Lamp Maintenance

Careful observation by those who operate and use a facility is needed to discover and report any evidence of defects in lighting systems. Deficiencies should be repaired promptly to prevent progressive deterioration of the system. Maintain the required illumination intensity by keeping lamps, fixtures and reflective areas clean and in good repair, replacing defective lamps, and keeping the voltage steady. The progressive decrease of light caused by the accumulation of dirt necessitates periodic cleaning of lighting equipment, the frequency of which depends on local conditions. The cleaning schedule for a particular lighting system should be determined by a light meter reading after the initial cleaning. When subsequent footcandle readings have dropped 20 to 25 percent, the fixtures should be cleaned again. The exterior surfaces of lighting equipment should be washed with a damp cloth and non-abrasive cleaner, not just wiped off with a dry cloth. Washing reclaims about 5 to 10 percent more light than dry wiping and reduces the possibility of marring or scratching reflective surfaces of fixtures.

Lamp replacement is a critical portion of lighting maintenance. Neglected lamp outages reduce illumination and, if burned-out lamps are not replaced, illumination may drop to unsafe footcandle levels in a short time. In some cases it may be satisfactory and more economical to clean lamp surfaces and fixture interiors only at the time of lamp replacement. Lamp replacement is done by either individual fixture or a group of fixtures.

Individual method: With this approach, burned-out lamps are replaced by an electrician on request. To prevent reduced illumination from lamp outages, follow these maintenance procedures:

1. Instruct employees to report burnouts as soon as they occur.
2. Replace blackened or discolored lamps even though they are still burning.
3. Replace fluorescent lamps as soon as they begin to flicker.
4. Replace any lamp with the same type, wattage and voltage as that of the lamp removed. Lamps of higher or lower wattage than called for on the lighting design plans should not be used.

Group method: With this approach, group replacement of lamps before they burn out is considered the most economical method for replacement in large areas. Whenever possible, group replacement should be performed simultaneously with fixture cleaning. Replacement of lamps by this method is accomplished by installing new lamps in all fixtures in the prescribed area after the old lamps have been in service for 70 to 75 percent of their rated life. The rated life of lamps can be obtained from the manufacturer and is the number of hours elapsed when 50 percent of the lamps in a large test group are burned out and 50 percent are still burning. Proponents of group relamping point out that it may cost as much as ten times more to replace lamps individually than in groups.

Motors

Motors are machines for converting electrical energy into mechanical energy. They are used in appliances, elevators and all kinds of machinery. Motors are available in a multitude of sizes, voltages, types and enclosures. Motor sizes range from fractional HP, single phase to as high as 2000 HP, 3-phase, with voltages ranging from 120 to many thousand Volts AC. The most common motors are fractional HP/110 Volt to 200 HP/600 Volt. The most common housings are drip-proof and totally enclosed.

Motor starters and across-the-line starters are manufactured in a variety of sizes and types. The most common types of across-the-line starters are the following:
- Motor-starting switch with no overload protection
- Single-throw switch with overload protection
- Magnetic switch with thermal overload protection

A *motor-starting switch* is simply a tumbler, rotary, lever or drum switch. It does not provide any protection against overload or inadequate voltage, and it is used for nonreversing small size motors up to 2 HP.

A *single-throw, across-the-line switch* can obtain thermal overload protection by means of thermal cutouts, time-lag fuses, or by a thermal overload release device. In switches equipped with a thermal overload device, the overload trips the holding catch of the switch and allows the switch to open. These starters do not provide protection against inadequate voltage.

Magnetic across-the-line starting switches (motor contactors) are magnetically operated and controlled from either an integral or remote push-button station. They are manufactured with thermal-relay protection against overload and inadequate voltage. Pressing the start button closes the circuit to the operating coil of the magnetic switch. When the stop button is pressed, the operating coil circuit is opened, and the motor is stopped.

Causes of Motor Failure

Good maintenance of motors consists basically of cleanliness and lubrication. Motor failures are caused by contamination, overloading, age, vibration and commutation problems. The most common cause of motor failure is *contamination*. Dirt can coat motor windings and cut down on heat dissipation, or it can block ventilation passages and increase insulation temperatures so that failure ultimately occurs. Dirt also causes wear in moving parts, especially bearings. Moisture in combination with dirt also causes insulation failure by shorting the windings or shorting to ground, and will also cause rust, a deteriorating factor.

An *overloaded* motor will fail earlier in its expected lifetime than one which is not overloaded. Therefore, the easiest way to solve the problem of overloaded motors would be either to eliminate overloads or replace the overloaded motors with larger units. To prevent a sudden overload, periodically check all protective devices such as overload relays, fuses, circuit breakers, field loss relays and voltage relays.

Failures due to *age* can be delayed by thorough periodic cleaning and lubrication, replacing fast-wearing bearings and brushes, and by tightening, turning and overcutting commutators on DC motors.

Failures caused by *vibrations* can take the form of sprung or broken shafts, bearing or seal failures, insulation breakdown, broken electrical connections, damage to the motor's mechanical structure, broken brushes, and a variety of commutator or slip ring problems. The vibration itself can be caused by misalignment, bearing problems or inbalance.

Commutation problems for DC motors (that is, problems with commutators and brushes) are evident in conditions such as rapid brush wear, chipped or broken brushes, burned brushes, copper feathering or drag, commutator burning, and slot or pitch patterns that are overfilming.

An inspection checklist for any motor should include checking for any of the conditions listed above. When practical, in addition to a shut-down or de-energized inspection, a running or energized inspection should be performed where the motor is started, run, and cycled through its load range. Take care in starting motors. Check rotor freedom and lubrication on standby or infrequently run equipment. In humid environments, check records for evidence of regular inspections and maintenance. Otherwise, dry out the windings before starting the motor.

Replacement versus Repair of Electric Motors

Upon failure of an electric motor, a decision must be made whether to replace or to repair it. A motor with unusual electrical or mechanical features will most probably be repaired. This decision is harder to make for standard motors, but repair or replacement should not be based only on the cost difference between a rewind and a new motor. Operating cost penalties as well as the cost of attached features, accessories and modifications should also be considered. Following are items to be considered in the decision to repair or replace a motor:

- Cost
- Characteristics of the motor
- Mounting and connecting conditions
- Mechanical interchangeability
- Available alternatives
- Cost of downtime while waiting for a rewinding
- Nameplate data
- Replacement criteria

Repair prices for smaller sizes of both protected and totally enclosed AC motors can exceed the cost of a new motor. However, in larger HP ratings, the repair cost of a protected motor can be less that 65 percent of a new motor, and for a totally enclosed motor, the repair cost can be less than half of a new drive.

Batteries

As for any other type of electrical equipment, batteries should be placed under a preventive maintenance program. Batteries should be inspected on a periodic basis in order to ensure optimum efficiency. There are two basic types of batteries in common use in commercial and industrial facilities—the dry cell and the wet cell. Both have a finite current-producing life and both are rechargeable to some degree.

Dry Cell Batteries

Dry cell batteries have their electrolyte solution in paste form, packed tightly into a closed sealed container. Being self-contained, they are easily transportable and can be mounted in any position without suffering the loss of electrolyte. The chief disadvantages of dry cell batteries are that they are limited to small sizes and their current-carrying capacity is low. However, they are ideally suited for use in battery-operated hand power tools such as drills and small saws due to their small size and portability. Because dry cell batteries are totally sealed, cleaning their contacts and recharging are the only maintenance activities required.

Wet Cell Batteries

There are three types of wet cell batteries: lead-lead-acid, nickel-iron-alkaline (or Edison) and nickel-cadmium-alkaline (or NiCad) type. Lead-lead-acid types are most commonly used in motor vehicles and industrial trucks. The predictability of wet cell battery life under normal use conditions makes it possible to place them under a preventive maintenance program in order to ensure reliability and long service life. Measures that should be included in a wet cell battery maintenance program are listed below:

- Keep the battery properly charged. Overcharging or undercharging must be avoided. Ideally, the battery should be charged on a *daily basis.*
- Keep water at the proper level in accordance with the manufacturer's recommendations. Check the water level on a *weekly basis.*
- Keep the battery clean and dry and wash down on a *weekly basis.*
- Clean the vent plugs and neutralize on a *monthly basis.*
- Keep battery temperatures at the manufacturer's recommended levels.
- Keep metal objects and tools off batteries to prevent shorts.
- Keep open flames away from the tops of batteries to prevent explosions.
- Add only water to storage batteries. Never add acid, electrolytes, powders or jellies.
- All cells in a battery should take the same amount of water. Any discrepancies could indicate a leaky cell.
- Do not take specific gravity readings from the same cell each time.
- Batteries should be charged as soon as possible after discharge. Do not allow a battery to remain uncharged for more than 24 hours.
- When the difference between cell voltages reaches 0.05 Volts, replace the battery.

Battery Charging Equipment

Battery chargers for dry cell batteries are usually self-contained and not serviceable. Upon failure, they will most likely be replaced due to the fact that the repair cost would greatly exceed the replacement cost. Battery chargers for wet cell batteries can range from very small transformers which incorporate contacts and windings, to larger units which incorporate an added DC generator with bearings and an air-cooling fan. Depending on the size and complexity of the battery charger, the following maintenance checklist should be followed:

- Clean and polish contacts periodically.

- Check the insulation annually for deterioration, by visual inspection and using a megohmmeter (megger).
- Clean the generator windings and fan by blowing them out with air.
- Check the bearing seals to make sure they are intact.
- Lubricate greaseable bearings periodically in accordance with the manufacturer's recommendations.
- Check protective devices such as built-in circuit breakers, timers and charge rate selectors periodically to ensure that they are working properly.

Distribution Systems

Electrical wiring, conduits, ducts, raceways, busways and the various types of junction boxes and enclosures are the streets and intersections of the electrical system. Protection devices such as panelboards and switchgear, fuses and circuit breakers, transformers, grounding system and lightning arrestors behave in a similar fashion to traffic signs and stop lights which control the flow of traffic. Together, they comprise the electrical distribution system of a facility.

Wiring

The main feeders, subfeeders and branch circuits are nothing more than two or more wires, each sheathed in some sort of protective covering such as plastic and/or metal, and run through a conduit or raceway. Particular attention must be given to electrical wiring in a facility. Any loose wires, poor connections, bare conductors, defective convenience outlets or switches, defective attachment cords and any other unsafe conditions must be corrected immediately. All facility wiring should be examined and determined to have been done in a workmanlike manner, with correct circuit arrangements, circuits that are not overloaded, and that all work conforms to the National Electric Code. Wiring systems should be checked frequently to reduce fire hazards, correct defective wiring and eliminate oversized lamps and overloaded circuits. All repairs made to wiring must be made in a permanent manner using approved materials. The following items should be checked when inspecting wires and cables:

- Dirty or other detrimental ambient conditions, poor ventilation and the presence of moisture, grease, oil, chemicals and fumes.
- Improper or unauthorized connections and dangerous temporary connections.
- Damaged wiring devices, defective insulators and cable supports, broken or missing parts or exposed live parts.
- Excessive cable sag and vibration, crowded cable spacing or excessive numbers of conductors in conduits and raceways.
- Evidence of overheating, improper grounds, short circuits, overheated splices and damaged or defective insulation.

Conduit, Boxes, Ducts and Raceways

Conduit in overhead exposed situations is used for power distribution, branch lighting and branch power. The most common types are aluminum, rigid galvanized steel, intermediate metallic conduit (IMC), thin-wall electric metal tubing (EMT) and plastic-coated rigid steel. Conduit in concrete slabs is used for both branch circuit piping and power distribution where the locations of end use will not change, the most common types being Schedule 40 PVC (plastic), rigid galvanized steel and intermediate metallic conduit (IMC). Conduit in trench is used for power distribution, outdoor lighting and communications, the most common types being rigid galvanized steel and rigid PVC, either direct burial or concrete-encased. Flexible metallic conduit, sometimes referred to as "Greenfield" or "Flex," is a single strip of aluminum or galvanized steel that is spiral wound and interlocked to provide a cross-section of high strength and flexibility. It can be covered with a plastic covering and be referred to as "Liquidtight" or "Sealtight." Armored cable is a manufactured product that combines wire inside of flexible conduit and is used where protection is needed but rigid conduit is not practical. Flat conductor cable (FCC), also known as "Undercarpet Power Systems," is used for branch lighting and power circuits, telephone, data and fiber optic systems in many existing buildings in order to rework obsolete wiring systems, and in new buildings where flexibility is paramount, and must be installed under carpet squares.

A box is used in electrical wiring at each junction point, outlet or switch. *Pull boxes* are inserted in a long run of conduit to facilitate the pulling of wire, and they are used where conduit changes direction or wires divide into different directions. *Cabinets* are used where wiring terminates. *Outlet boxes* are used to hold wiring devices such as switches and receptacles, and can also be used as a mount for hanging light fixtures.

Trench duct is a steel trough system set into a concrete floor with a top cover fitted flush with the finished floor. It is used as an accessible underfloor raceway. *Underfloor duct* is used to make power and communication wiring available at numerous locations within a room. It is set in place before the concrete floor is poured.

Busway or *bus duct* is made up of copper bars covered with a metal enclosure. It is primarily used in overhead installations for feeders from the transformer service entrance to various facility locations. Some bus duct is of the more expensive plug-in type, which is designed so that branch circuits can be plugged in anywhere along its length, allowing for future flexibility. A *cable tray* system is a prefabricated metal raceway structure consisting of lengths of ladder, trough or solid-bottom-type tray, usually made of aluminum or galvanized steel, with solid or ventilated covers and associated fittings. *Wireways* are sheet-metal troughs with hinged or removable covers. They are used to house and protect wires and cables, and are used only for exposed situations. *Surface metal raceways* (wiremold) are installed on walls and floors in existing buildings in cases where it would be too costly or difficult to install raceway or conduit within walls.

All the types of conduit, boxes, ducts and raceways referenced in this chapter should be inspected frequently and any found to be damaged due to corrosion or operational abuse should be replaced. All systems of underfloor duct require checks for evidence of oil, water, other liquids, insects or rodents. Bus ducts and cable trays require annual cleaning and removal of oil substances and dirt, and ventilated types should be blown out annually with clean, dry, compressed air.

Fuses, Circuit Breakers, Panelboards and Switchgear
Fuses are used to protect the wiring system and electrical equipment from short circuits or overloads. They have a fusible link, through which the electric current must pass, which will melt from the heat generated by excess current. They are manufactured with either fast-acting or time-delay links, and come in a variety of styles such as plug, cartridge or bolt-on. The fuse size and type must be consistent with the circuit or piece of equipment it is protecting. A blown fuse must be replaced; it cannot be repaired.
The following items must be checked when replacing a fuse or conducting an inspection:

- Insulators: Inspect for breaks, cracks or burns.
- Contact surfaces: Inspect for pitting, burning, alignment and pressure.
- Bolts, nuts, washers, pins and terminal connectors: Inspect for condition and tightness.

As with fuses, *circuit breakers* are used in general distribution and branch circuits to protect downstream wires and equipment from current overload. A circuit breaker is like a fuse in that it is designed to open automatically at a preset ampere rating, but it can usually be reset and put back into service. Most circuit breakers have two modes of tripping: time (thermal) trip, and instant (magnetic) trip. Some breakers also have a third mode: ground fault (shunt coil) trip. Circuit breakers are made in either plug-in or bolt-on designs, and usually as a single-pole which can be ganged into double- or triple-pole devices with a single handle. If a breaker fails to trip, serious damage may result. Smaller circuit breakers in load centers and panelboards are usually maintenance-free, but larger circuit breakers will need to be checked periodically for the following:

- Low bushing oil gauge reading
- Low breaker tank oil level
- Oil or air leaks
- Broken or missing cotter pins, loose hardware or broken porcelain
- Air supply compressor running excessively
- De-energized cabinet heaters
- Excessive corrosion or rust
- Loose gasket covers
- Unusual noise, smoke or temperature

Panelboards are used to group circuit switching and protective devices into one location. They consist of an assembly of bus bars and circuit breakers housed in a metal box enclosure. The box provides space for wiring and includes a trim plate with a cover. Panelboards are available in a variety of configurations, including AC styles up to 600V and DC styles up to 250V. The AC styles can be single- or three-phase.

Switchboards are used in buildings that have larger load requirements than can be handled by a single load center or panelboard. They are modular assemblies of sections or compartments for the main service, auxiliary function, metering and distribution. The service section contains the main (incoming) breaker, which may be rated from 200 to 4000 amps. The auxiliary section is a blank compartment which may be used to facilitate cable pulling. A metering compartment may contain current transformers (CTs), potential transformers (PTs), and relays, as well as meters for amps, volts and watts. The distribution sections may contain any combination of fused disconnect switches, branch circuit breakers rated from 30 to 1200 amps, or motor starters. Bus bars are rated from 200 to 4000 amps and are usually aluminum, but copper bars are an available option.

Transformers

Transformers are used to convert from one voltage to another in an electrical distribution system. They are used in four basic applications: (1) instrument transformers, (2) control transformers, (3) isolating transformers, and (4) power transformers. Their capacities range from fractional Volt-amps (VA) to thousands of kVA. The first two are always single-phase, but the last two can be either single- or three-phase designs. Power transformers may be either dry-type air-cooled or oil- or silicone-based liquid-cooled. Large building service entrance transformers may be pad- or pole-mounted or installed in an underground vault near the service entrance. Transformers require very little, and receive even less, attention than most other electrical equipment. The maintenance required is governed by the size, importance and location within the system, ambient temperatures and the surrounding atmosphere. Dampness, dust and a corrosive atmosphere can all be causes of transformer problems.

Grounding System

In most distribution systems one conductor of the supply, called the "neutral wire," is grounded. In addition, the NEC requires that a grounding conductor be supplied to connect noncurrent-carrying conductive parts to ground. This distinction between the "grounded conductor" and the "grounding conductor" is very important. Grounding protects persons from injury in the event of an insulation failure.

Special Systems   Lightning Protection

Lightning poses two kinds of danger. The first is the lightning strike itself, which can damage structures and electrical distribution systems by passing a very high current for a brief time causing heating, fire and/or equipment failure. The second failure is from induced voltages in electrical cables and wires running near to the lightning's path. These pulses can be very high and can cause damage to sensitive electrical equipment and injury to people.

It is essential that a good, low-resistance ground path be provided. Lightning protection for roof tops can be achieved by a series of copper or aluminum lightning rods or air terminals spaced around the perimeter of the roof and joined together by either copper or aluminum cable. The lightning cable system is connected through a downlead to a ground rod. In addition, any metal bodies on the roof must be bonded to the lightning protection system. A lightning protection system is virtually maintenance-free. However, a visual inspection should be made periodically to make sure that all components are still in place, in sound condition, securely fastened to the building and fully connected—together and to ground. Any discrepancies must be corrected or repaired immediately.

Emergency Lighting System

In an emergency situation, an electrical fault may cause the main lighting system in a facility to lose power. Lighted exit signs and emergency lighting units are placed in stairwells, corridors, lobbies, key entrances and exits, and other critical locations to provide enough light and direction to allow people to vacate the facility in a safe manner. These individual units are connected to a 110 or 277 Volt power source, which keeps their batteries charged. These batteries will be used as the power source in the event of a power failure and will run connected lights and exit signs by low voltage direct current.

Maintenance of battery-powered emergency lighting and exit sign systems consists of inspecting each unit periodically, usually every six months unless otherwise specified by code or regulation. During the test, the test button should be depressed, which will simulate a power failure. The emergency light heads and any low-voltage DC lamps in exit signs which are connected to the battery pack should light up. Any burned-out lamps must be replaced immediately. Wet cell batteries should have the specific gravity of their electrolytes checked as well.

Uninterruptible Power Supply (UPS) Systems

Some facilities are equipped with one or more UPS systems which provides emergency backup battery power during an electrical power failure, as well as smoothing out any power spikes or valleys. A UPS can range in size from fractional kVA for personal computer systems, up to 800 kVA for large computer rooms. The smaller units are usually equipped with dry cell batteries and are plugged into any 110 or 240 Volt circuit, while the larger units are usually equipped with wet cell batteries and are hard-wired to the facility's electrical distribution system and to the equipment they protect.

Maintenance of all UPS units will consist of periodic inspections to make sure that ventilation grilles are kept clean and unobstructed. UPS units which contain wet cell batteries must have their batteries and other critical components checked on a regular basis, usually under a maintenance contract with an outside contractor.

Emergency Power Generators

An emergency power generator can serve up to three purposes. The first is to provide a safe environment for personnel in the event of a power failure through backup power to emergency lights and exit signs, and even full power to complete facility lighting systems. The second is to protect production equipment, sensitive electronic equipment and property from damage through an automatic transfer to emergency backup power. The third is to make continuity of operations possible where stoppage would be detrimental to the profitability of the facility or organization. A system will generally consist of a power failure sensor, a power generator and a transfer switch. Systems are rated in terms of their capacity in kilowatts (KW), from 10 to 1,000 KW, and are available in single-phase or three-phase, and from 120 through 6,000 Volts. Motors in emergency generator sets may be powered by gasoline, propane (or natural) gas or diesel fuel.

Maintenance will consist of a monthly inspection to make sure that the motor's water, oil and fuel and the starter battery's electrolyte levels are in proper quantities and condition. This equipment is idle most of the time, and evaporation, leakage and/or contamination can be detrimental to this equipment, affecting its operation. The manufacturer of an emergency generator set will most likely have a prescribed set of periodic preventive maintenance activities which, when performed, will keep this equipment in top condition and ready in a millisecond's notice for standby power.

Conclusion

This chapter has identified the major systems, components and types of electrical equipment in need of maintenance, whether in a small facility or in a large complex plant. Any facility is packed with thousands of feet, and sometimes miles, of wires, cables and conduit in its electrical distribution system, along with many different types of fuses, circuit breakers, switches and transformers, as well as lighting systems, motors, batteries and special systems. A major investment has been made in this equipment by the owner organization. A substantial commitment of time and resources is also needed to maintain this equipment and to keep the facility operating safely and at peak performance.

Chapter 13

Landscaping

John L. Maas and Al R. Young

Maintenance management for landscaping should include the following goals. First, it should focus on maintaining corporate image as it is expressed in a facility's landscaping. Second, it should protect the owner's investment in a property. Finally, maintenance management and programming should ensure the efficient expenditure of maintenance funds. The methodology presented in this chapter approaches landscape design, construction, and maintenance with these goals in mind.

Phases of Maintenance Programming

Landscape maintenance programming includes four phases:

- Image analysis
- Design
- Implementation
- Management

This discussion provides an overview of these phases and shows their relationship to activities that are typically provided by land planners and landscape architects. The four phases of maintenance programming can be performed by maintenance consultants who are part of a facility's maintenance design team. Figure 13.1 shows activities typically performed by land planners and landscape architects, and the manner in which maintenance programming activities can be integrated into the process of design and construction. The image analysis phase of maintenance programming described in this chapter should be part of the site analysis typically performed in the initial phases of land planning. The design phase should be part of the system documentation step seen in Figure 13.1. Implementation and management can be carried out in the steps illustrated in this figure, and vary according to an owner's approach to planning and budgeting for the maintenance function.

Corporate image is a product of the maintenance function. Landscape maintenance must be focused on this product if it is to be successful. Landscape maintenance serves primarily two objectives:

- Enhancing corporate (or owner) image
- Safeguarding the investment in the landscape

By incorporating maintenance programming into the facility from the outset, the resulting maintenance program can be implemented as part of a strategy that addresses the whole life cycle of the facility. Maintenance

Grounds Management System Design and Implementation

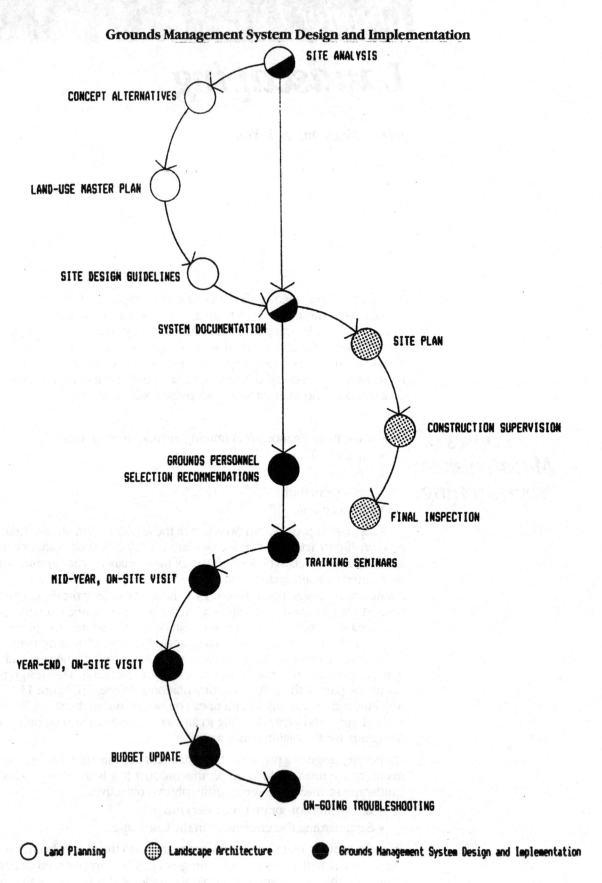

SITE ANALYSIS

CONCEPT ALTERNATIVES

LAND-USE MASTER PLAN

SITE DESIGN GUIDELINES

SYSTEM DOCUMENTATION

SITE PLAN

CONSTRUCTION SUPERVISION

GROUNDS PERSONNEL
SELECTION RECOMMENDATIONS

FINAL INSPECTION

TRAINING SEMINARS

MID-YEAR, ON-SITE VISIT

YEAR-END, ON-SITE VISIT

BUDGET UPDATE

ON-GOING TROUBLESHOOTING

○ Land Planning   ▦ Landscape Architecture   ● Grounds Management System Design and Implementation

Figure 13.1

involves many requirements for which the owner is not necessarily prepared. Maintenance programming, therefore, identifies what must be done to meet the owner's goals within budget requirements.

Image Analysis

Each phase of maintenance programming is the basis for subsequent phases. Image analysis clarifies the image-related objectives on which a facility's landscape maintenance program must focus. The analyses performed during this phase form the foundation of the property's management system. These analyses lead to planning and design activities.

Design

The second phase of maintenance programming synthesizes the results of the image analysis phase into core information for project design. This not only helps put the maintenance program on a firm footing, but can also be used as a reality check as the design progresses.

Maintenance programming necessarily takes a long view of the project. Design decisions are made within the context of the long-term consequences of maintainability and cost. Maintainability is concerned with whether a proposed feature can survive. Cost is concerned with the fiscal implications of maintaining the feature.

Specifications and guidelines developed during the design phase address the feasibility of maintaining the features proposed in the development plan. Labor-hour and cost projections developed during this phase address the actual cost of proposed features. Estimated costs per unit (square feet of turf, number of palm trees, etc.) can be used to establish the fiscal feasibility of maintaining each proposed feature.

Maintenance programming safeguards the interests not only of the owner, but also of design team members, as it puts their work on firmer ground by determining feasibility. Maintenance of the landscape is not only an advertisement for the owner of the property, but the property's designers as well.

Implementation

Implementation of the landscape maintenance program begins as development of the design nears completion. In this way, a smooth transition can be achieved from construction to ongoing management of the facility. Design teams typically focus on completion of construction. Once construction is complete, their work ends because their goal has been achieved. In fact, completion of construction actually initiates the longest and most costly part of the project's life cycle—maintenance of the facility.

Management

The management phase begins as construction concludes. The people who will maintain the landscape and are ultimately responsible for much of the design's success or failure are engaged and trained to ensure that a proper maintenance program is put into place that will uphold the design team's efforts.

The Maintenance Calendar and Budget

The development of the landscape maintenance calendar and budget occurs during the design phase of maintenance programming. These activities parallel master planning and site planning. For new construction, doing all three concurrently helps ensure that:

- Master planning and site planning focus on maintainability as much as on construction.
- Maintenance costs are factors in design decisions.

If master planning and site planning are to focus on maintainability, maintenance must be as important to the design team as climate, topography, and the project's design and installation budget. While site characteristics data is crucial in that construction cannot be completed without it, maintenance is often disregarded because failing to plan for it does not impede construction. Topography is present before and during construction. Maintenance is created as the result of construction.

Maintenance, like any other characteristic of the site, must be incorporated into the design from the outset. As planning progresses, maintenance must emerge with as much clarity as the grading plan, the plant palette, treatment of traffic patterns, and other features of the design. The procedure described in this section — the formulation of the maintenance calendar and budget — is intended to make maintenance as visible, as important, and as completely addressed by the design as any other feature of the facility.

The ability to project maintenance costs for various design options is key to giving maintenance the role it deserves during planning phases. For example, the horticultural feasibility of introducing a particular species to a site is one thing; the expense incurred in maintaining that species is quite a different consideration. When a site's climatic and other conditions do not naturally conform to the horticultural requirements of a species, maintenance must provide a buffer between the species and site conditions. The greater the need for such a buffer, the greater the expense involved in providing it. Maintenance planning must, therefore, consider horticultural and other technical issues, but such issues must be presented in financial terms. Only then is maintenance given appropriate attention.

Information on the financial feasibility of maintaining a design feature almost certainly ensures that maintenance will be as important to the design team as any aspect of the development. If the owner can be made aware of the cost of maintaining various site characteristics, he or she will ensure that maintenance is a key factor in design decision-making.

For existing properties, the time and resource management projections described in this section can be prepared as the basis for establishing or refining a maintenance program. The procedure for generating such projections can be performed at any time in the life cycle of an existing facility.

The procedure described in this section produces a maintenance calendar as well as labor-hour and materials requirements for the facility. The maintenance calendar is a scheduling tool. (See Figure 13.2, a sample summary calendar.) Labor-hour and materials requirements are used to forecast the maintenance budget.

These are the steps in preparing the calendar and budget projections:
1. Group site features according to maintenance tasks.
2. Specify pertinent tasks.
3. Prepare task descriptions and assign local labor and materials costs.
4. Formulate the maintenance calendar.
5. Prepare the maintenance budget/forecast.

These steps are discussed in detail in the following sections.

Grouping Site Features According to Maintenance Tasks

To determine the maintenance tasks required for a facility, the general characteristics of the facility (such as turf, hard-surfaced features, etc.) must be grouped so that tasks related to the maintenance of those characteristics can be defined.

Maintenance Calendar Summary

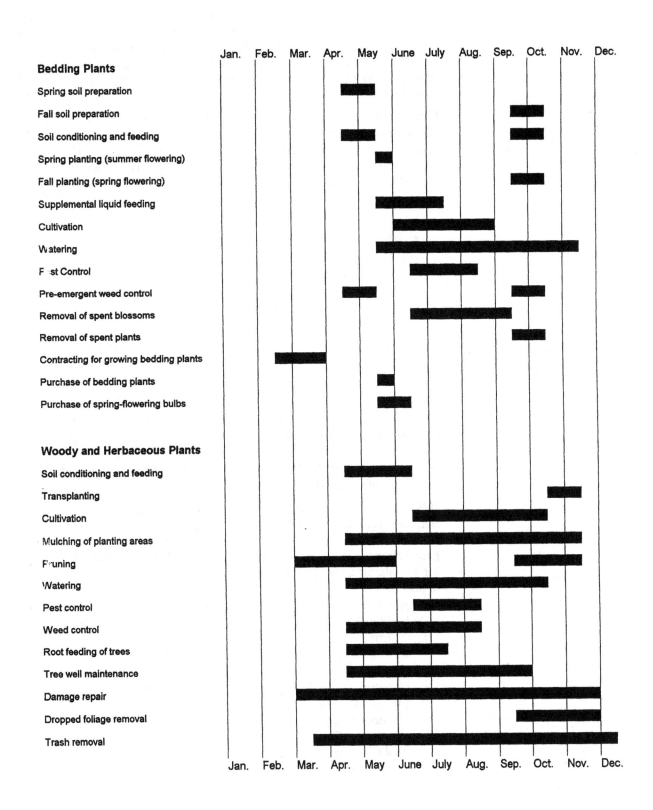

Figure 13.2

One of the primary objectives in grouping site characteristics is to avoid overlap between task groupings. For example, tree pits might seem to "belong" to trees. Thus tree-pit maintenance could be grouped with such tasks as pruning and vertical mulching. Edging turf, however, is usually grouped with turf-related tasks. If the linear feet of tree-pit turf edge is included under *tree maintenance*, care must be taken not to include it also in the edging listed under *turf*. When this kind of ambiguity exists, labor-hour and cost projections are likely to be skewed. Furthermore, the facility's owner could actually wind up paying twice for performance of the same task.

If the site is undeveloped, information about landscape features must be taken from the plans being prepared for the site. If the maintenance program is being prepared for an existing facility, the quantities can be derived from measurement of the site itself. Facilities being renovated or revised necessitate measurement of both existing and planned features.

Site features may be categorized as follows:
- Bedding-plant areas
- Hard-surface areas
- Turf
- Undeveloped areas
- Water features
- Woody and herbaceous plants

Maintaining the above groups of features creates two other tasks that must be part of the facility's time and resource management projections:
- Equipment and irrigation system maintenance
- Supervision

Together, these lists constitute the task groupings on which the management strategy for a facility is based. The first group of site features is derived directly from characteristics of the site (or from the features of the development planned for a site). The second group of characteristics (equipment and supervision) is then derived from the maintenance requirements created by the first list.

Bedding-Plant Areas

Bedding-plant areas include all developed parts of the site unoccupied by turf and hard surfaces. The bedding-plant area includes such features as planter boxes, flower beds, and tree pits. Labor-hour and other projections for maintaining the bedding-plant area are based on the following measurements:
- Number of soil variations or topographic features affecting bedding plants. The purpose of totaling the soil variations is to determine the number of soil tests required to adapt maintenance procedures to site conditions.
- Number of water sources affecting bedding plants. The number of water sources is important only if water used in maintaining the site is non-culinary. For example, brackish water sources require testing to adapt maintenance procedures to site conditions. Where culinary water is available, a standard of water quality can be assumed.
- Square feet of bedding-plant area.

Hard-Surface Areas

Hard-surface areas include roads, sidewalks, outdoor parking areas, plazas—all vehicular and pedestrian traffic areas. Planting areas covered with gravel mulch, mow-curbs, and cemetery headstones are not included in this category.

Maintenance of hard-surface areas usually requires the following minimum measurement of site characteristics:

- Square feet of pedestrian traffic area
- Number of small trash receptacles
- Number of dumpsters
- Square feet of roads and parking lots
- Square feet of steps

This list of quantities, like each of the other lists in this section, must be modified according to site conditions. For example, the list indicates that only the total square feet of pedestrian traffic areas need be recorded; however, a site whose pedestrian walkways are asphalt as well as brick would require one task description for repairing asphalt and another for repairing brick. Thus it would be necessary to measure the square feet of asphalt walkways as well as the square feet of brick walkways. Also, the number of dumpsters, small trash receptacles, etc., is unimportant if maintenance personnel are not responsible for these aspects of a facility.

Turf

Turf includes all lawns at the site. Barren areas intended for turf, or areas where turf has been introduced but worn barren, for example, are included in this category. Areas covered by wild grasses or weeds should be included as undeveloped.

Turf maintenance tasks require the following quantification of site characteristics:

- Number of soil variations or topographic features affecting turf areas. (The purpose of totaling the soil variations is to determine the number of soil tests required to adapt maintenance procedures to site conditions).
- Square feet of turf area.
- Linear feet of turf edges.
- Square feet of turf to be repaired or resodded annually.
- Square feet of turf areas in parcels less than 1,000 square feet.
- Square feet of turf areas in parcels greater than 1,000 square feet.

Undeveloped Areas

If development of the site is to be phased, undeveloped areas should be included in the estimate for this feature. "Undeveloped" refers to the absence of construction or planned landscaping features.

Water

This category includes all ponds, streams, fountains, and other water features.

Woody and Herbaceous Plants

This category includes all trees and shrubs at the site. Note that these quantities are not areal (such as square feet). While turf, hard surfaces, water, and bedding plants all occupy areas, woody and herbaceous plants are located within turf and bedding-plant areas. Thus the ground in which the woody plant is located belongs not to the woody plant, but to the turf or bedding-plant feature in which the woody plant is located.

For example, a tree pit is a bedding-plant area. The square feet of the tree pit area is part of the areal estimate for the bedding-plant feature. The linear feet at the perimeter of a tree pit is part of the linear feet belonging to turf (if the tree pit is in a turf area). If a tree pit is in a parking lot, the pit itself is part of the bedding-plant area. The linear feet at the perimeter of such a pit belong to the linear feet of asphalt at the site.

Maintenance projections for woody and herbaceous plants require the following measurements:

- Linear feet of hedges
- Number of palm trees
- Number of deciduous trees
- Number of evergreen trees
- Square feet of shrubs and vines

Note: Square feet of shrubs and vines represents square feet of the surface area to be trimmed, not the square feet of the area occupied by the shrubs and vines.

Equipment and Irrigation System Maintenance

This group of maintenance tasks includes all equipment required for the maintenance operation. These tasks pertain to the irrigation system, computer equipment, trucks and other vehicles, mowers, edgers, hand tools, etc. Typically, the number of irrigation systems at the site is the only quantity needed. If, however, the nature and complexity of the irrigation system (or systems) requires it, it may be helpful to quantify individual system components to be maintained.

Supervision

Supervision includes management personnel required for the maintenance operation. If the maintenance function will be outsourced, the vendor must factor these costs into his or her bid and operational budget. Thus for outsourced management, these costs need not appear as a line item in the owner's budget. If, however, the maintenance function is staffed by a facility's owner, supervision should appear as a line item in which the labor hours, costs, etc., associated with the management function are projected just like any other maintenance activity.

Specifying Pertinent Tasks

Once the measurement of site characteristics is completed, the maintenance tasks for each group of features should be determined. This section presents a list of tasks typically required for each group of site characteristics.

Each facility has unique requirements. Climatic conditions, soil conditions, plant palette, water conditions, equipment, and other factors determine the list of tasks suited to the requirements of a particular property. Consequently, the tasks listed here should be modified to accommodate actual site conditions, methodologies, and other factors. These principles and information can also be used to add tasks required by a particular facility. The order in which tasks are listed is not significant.

Bedding Plant Area Maintenance Tasks

BP–001:	Soil tests
BP–002:	Water tests
BP–003:	Soil preparation
BP–004:	Soil conditioning and feeding
BP–005:	Cultivation
BP–006:	Irrigation
BP–007:	Pest control

BP–008:	Weed control
BP–009:	Removal of spent plants
BP–010:	Contracting for growing bedding plants
BP–011:	Purchasing of bedding plants
BP–012:	Supplemental feeding
BP–013:	Winter and spring planting (spring flowering)
BP–014:	Trash removal (surface)
BP–015:	Spring planting (summer flowering)
BP–016:	Inspecting for pests and damages

Hard-Surface Area Maintenance Tasks

HS–001:	Sweeping and trash removal (pedestrian)
HS–002:	Trash removal (cans)
HS–003:	Trash removal (dumpsters)
HS–004:	Snow removal (sidewalks)
HS–005:	Snow removal (roads and parking lots)
HS–006:	Snow removal (steps)
HS–007:	Repair
HS–008:	Sweeping and trash removal (parking)

Turf Maintenance Tasks

TF–001:	Soil tests
TF–002:	Mechanical cultivation
TF–003:	Feeding
TF–004:	Edging
TF–005:	Mowing
TF–006:	Irrigation
TF–007:	Pest control
TF–008:	Weed control
TF–009:	Re-sodding
TF–010:	Litter and foliage removal (area < 1,000 S.F.)
TF–011:	Litter and foliage removal (area > 1,000 S.F.)
TF–012:	Pest inspection

Undeveloped Area Maintenance Tasks

UD–001:	Weed control (manual)
UD–002:	Weed control (herbicide)
UD–003:	Trash removal

Water Feature Maintenance Tasks

WT–001:	Cleaning ponds and aerators (no wildlife or aquatic plants)

Woody and Herbaceous Plants Maintenance Tasks

WP–001:	Trimming hedges
WP–002:	Pruning palm trees
WP–003:	Pruning deciduous trees
WP–004:	Pruning evergreen trees
WP–005:	Pruning shrubs and vines
WP–006:	Pest control (trees)
WP–007:	Pest control (shrubs)
WP–008:	Root feeding of trees
WP–009:	Vertical mulching
WP–010:	Damage repair
WP–011:	Transplanting
WP–012:	Winter protection
WP–013:	Inspecting for pests

Equipment Maintenance Tasks

The following tasks pertain to equipment. Note, however, that this list, more than any of the others, is incomplete. While equipment maintenance must be planned with the same care as other aspects of the maintenance operation, the kinds of equipment vary greatly, and particular maintenance tasks are specified by vendors. Consequently, the following list is indicative only of the *kinds* of tasks required:

EQ–001:	Activate sprinkling system
EQ–002:	Deactivate sprinkling system
EQ–003:	Monitor and set controllers
EQ–004:	Monitor system components
EQ–005:	Repair irrigation system

Supervision Tasks

Only two tasks are listed for supervision. As with other aspects of the maintenance operation, the list of supervisory tasks should be tailored to the maintenance operation:

SP–001:	Management
SP–002:	Clerical

Preparing Task Descriptions

Once the maintenance tasks for a facility's landscape have been listed, task descriptions must be developed. Each task description should include the following information:

A. Description of work

B. QA Criteria

C. Materials and equipment list

D. Climatic window for task performance

E. Task performance frequency

F. Total times task is performed

G. Unit of feature

H. Work completion rate per unit

I. Total number of units

J. Labor-hours per total units

K. Total annual task labor-hours

L. Materials and equipment hourly overhead rate

M. Materials and equipment overhead unit cost

N. Labor hourly rate

O. Labor unit cost

P. Laboratory fee

Q. Total unit cost

R. Total annual task cost

These landscape management task descriptions are defined in the sections that follow. These definitions also explain the manner in which derived aspects of a task description are formulated.

A. Description of Work

The description of work in a task description must be written in such a way that it can be used to:

- Give the person who performs the task a precise and thorough description of what must be done.
- Measure the duration of task performance.

The description of work is written primarily for the contractor who performs the task. Well-written descriptions are precise, detailed, and written for the specific needs of a particular facility. Such descriptions:

- Promote compliance with specific standards.
- Reduce the likelihood of either misunderstanding or confrontation.
- Foster a uniform comparison of bid proposals (when maintenance is to be subcontracted).
- Enhance communication and understanding in multilingual work environments.
- Provide a basis for uniform results across diverse sites.
- Form the basis of any landscape management training program.

Consistency in the names of tasks and in the terminology of task documentation promotes clarity of communication within the maintenance organization. Such documentation also helps insulate the maintenance organization from the adverse effects of turnover and inexperience. It must be possible to uniformly replicate the task based on the description. This is essential to the integrity of the labor-hour and cost projections based on the description. Periodically auditing task descriptions by having the task performed also promotes accuracy.

Task descriptions should tell maintenance personnel what to do and how to do it. For example, the following definition of turf edging is inadequate:

- Trim grass along the edge of turf areas such as grass along walls and fences, sidewalks, curbs, tree pits, bedding-plant areas, etc.

The description does not specify whether removal of clippings is included. Tools are significant in determining duration, yet tools are not mentioned. Since maintenance budget projections are based on task descriptions, such descriptions must not overlap. If, for example, cleaning up grass clippings is a task by itself, or is part of a clean-up task, it should not be included in the description of work for turf edging. However, if clean-up is not a separate task, it should be part of turf edging and other tasks where grass clippings must be cleaned up.

B. QA Criteria

The QA criteria should be used by maintenance workers, supervisors, and the facility manager to evaluate the result of each maintenance activity. "Quality" of maintenance must be evaluated from two points of view:

- Compliance with accepted methodologies
- Resulting appearance

QA criteria alert the individuals who are responsible for task completion to the standards by which their efforts will be evaluated. By including such criteria in the task description, everyone involved knows at the outset what the result must be. This tends to preclude misunderstanding and confrontation.

Payment for maintenance services, whether performed in-house or by subcontractors, is usually payment for the way something looks. Consequently, the criteria specified in the task descriptions address not only technique of task performance, but appearance of the result.

C. Materials and Equipment List

The materials and equipment list specifies the tools and other items that are required to complete the task. As the methodology or technology involved in the description of work varies, the list of equipment and tools also varies.

The technology suited to a particular site must be determined by specialists who can evaluate safety, environmental considerations, and other factors. Combining the materials and equipment lists for all maintenance tasks produces a comprehensive list of items needed by the maintenance organization. (For new construction, this list can be used to project purchase or rental costs associated with the landscape.)

D. Climatic Window for Task Performance

The task performance window indicates the beginning and ending dates for the period during which the task can be performed. This part of a task description relates the description to information collected during the analysis of maintenance factors. This climatic information must be compared with the site's soil conditions, water conditions, and species to be maintained to determine when each task can be performed.

For example, turf edging does not begin until temperatures rise to the point at which turf begins to grow. The window of opportunity for performing this task closes when temperatures preclude growth. The beginning and ending dates for task performance vary according to the species to be maintained.

Not all maintenance tasks are so closely related to climatic conditions. Where climate and other site conditions are more or less irrelevant, the fact should be noted in the task descriptions.

E. Task Performance Frequency

Task performance frequency is the periodic frequency with which a task should be performed. For example, a task's climatic window might be April 10 through September 15. During that period, the task might be performed once, twice (or semi-) weekly, every two weeks (or semi-monthly), etc.

F. Total Times Task is Performed

Total times task is performed is simply the total number of times the task must be performed. For example, if the task performance window is April 10 through September 15, and the task performance frequency is weekly, the task is performed 24 times. Were the task performance frequency daily (Monday–Friday), the task would be performed 120 times.

G. Unit of Feature

Unit of feature is the smallest amount of square feet, linear feet, etc., for which the work (specified under *Description of Work*) will be performed. The unit of feature can be any quantity desired. The determinant is how much or how little time seems reasonable to have to specify for work completion rate.

For example, if it is determined that it takes 5 minutes to edge 500 linear feet of turf, 100 linear feet could be specified as the unit of feature. If so, 1 minute would be the work completion rate. If, on the other hand, 1,000 linear feet is specified as the unit of feature, 10 minutes is the work completion rate.

Unit of feature can also be an object. For pruning deciduous trees that are greater than 6 feet tall, the unit of feature might be a single tree.

Note: Unit of feature is not the total square footage or linear footage, or number of units of a particular feature.

H. Work Completion Rate per Unit

Work completion rate per unit is the number of minutes, per unit of feature, necessary to complete the description of work. Various sources are available for this kind of information. The best source is, of course, a measurement of the performance of the task as specified in the Description of Work.

I. Total Number of Units

The total number of units is the quantity of a feature that must be maintained at a site or facility. For example, the unit might be 500 linear feet (for turf edging). If a site includes 100,000 linear feet of turf edge, the total number of units of turf edge for that site is 200. If the unit of turf edge were 1,000 linear feet, the total number of units of turf edge at the site would be 100.

J. Labor-Hours per Total Units

Labor-hours per total units is the product of work completion rate per unit times total number of units, divided by 60:

$$J = (H \times I)/60$$

Letters in the formula correspond to headings in this chapter's definition of task description elements. J, for example, is the letter designating "labor hours per total units." H corresponds to "work completion rate per unit." I represents the "total number of units."

K. Total Annual Task Labor-Hours

The total annual task labor-hours is the product of the total number of times the task is performed, multiplied by labor-hours per total units:

$$K = F \times J$$

Again, the letters in the formula correspond to the task description headings.

L. Materials and Equipment Hourly Overhead Rate

The overhead in materials and equipment is stated as an hourly rate because labor (the other cost associated with each maintenance task) is best expressed as an hourly rate. By translating materials and equipment into an hourly rate, task cost can be derived simply as a product of the task completion rate times the cost of labor and the cost of materials.

M. Materials and Equipment Overhead Unit Cost

Materials and equipment overhead unit cost is the product of work completion rate per unit times the materials and equipment hourly overhead rate. This cost is the expenditure in materials and equipment overhead necessary to perform the task for one unit of feature. Using the letters assigned to the parts of the task description, this is the formula for calculating materials and equipment overhead unit cost:

$$M = H \times L$$

For example, if the work completion rate per unit is 15 minutes, the materials and equipment overhead unit cost is the amount spent on materials and equipment to perform the task for 15 minutes.

The materials and equipment hourly overhead rate states how much money is required to operate the equipment (required by the task) for a single hour. The materials and equipment overhead task cost states how much it will cost to operate the equipment for the number of minutes required to complete the description of work for a single unit of feature.

N. Labor Hourly Rate

The labor hourly rate is the wage per hour for the person performing the task.

O. Labor Unit Cost

The labor unit cost is similar to materials and equipment overhead unit cost, except that the labor unit cost is the amount spent on labor to perform the task for a single unit of feature. If, for example, the work completion rate per unit is 15 minutes, the labor unit cost is the amount spent on labor to perform the task for 15 minutes.

The cost of labor should include lost time from holidays and should account for the fact that labor estimates are inaccurate if it is assumed that a worker works 60 minutes for each labor-hour. Lost time is calculated at the rate of 4.26 minutes per hour. To include lost time in the labor cost, add another 7.1% to the cost. In other words, for every dollar of labor estimated, add 7.1 cents.

This 7.1% is added to the labor task cost instead of the work completion rate, because the materials cost is based on the minutes for task completion. Adding the 7.1% to the minutes for task completion would needlessly increase the materials and equipment overhead task cost.

P. Laboratory Fee
Usually, laboratory fees pertain only to tasks such as soil or water analyses that require laboratory testing. The amount entered on the work sheet is the total fee per unit.

Q. Total Unit Cost
The total unit cost is the sum of the unit costs for labor, materials and equipment, and fees:

$$Q = M + O + P$$

R. Total Annual Task Cost
Total annual task cost is the product of total annual task labor-hours times total unit cost. For example, if the annual total of labor-hours for a particular task is 100 and the total unit cost is $34.50, then the total annual task cost for the task is $3,450. The formula for deriving total unit cost is:

$$R = K \times Q$$

Formulating the Maintenance Calendar
The next step is to formulate the maintenance calendar, which is simply the summary of items D, E, and F for each task description.

Preparing the Maintenance Budget/Forecast
The maintenance budget or forecast summarizes items Q (total unit cost) and R (total annual task cost) from each task description. Figure 13.3 shows an excerpt from an annual landscape maintenance budget. All the information concerning the three tasks in the excerpt is derived from the kind of task description discussed above.

In Figure 13.4, a graphic summary of an annual landscape maintenance budget presents labor-hours and costs that show an owner where his or her maintenance dollar is being spent.

Finally, in Figure 13.5, the landscape maintenance calendar is related to the annual maintenance budget to provide a monthly indication of site characteristics targeted by a facility's maintenance program.

Designing for Cost-Effective Maintenance

Landscape construction creates a significant long-term financial obligation for a facility's owner. The owner's expenses for routine maintenance of the landscape are determined by the degree to which design and construction focus on maintainability. Again, because the development team tends to focus primarily on the successful completion of construction, the cost of maintenance is often overlooked.

Excerpt from a Sample Annual Landscape Maintenance Budget

Turf

	Unit Definition	Total Units	Labor per Unit (min.)	Labor per Total Units (hr.)	Hourly Wage	Hourly Overhead	As Req	Jan	Feb	March	April	May	June	July	Aug	Sept	Oct	Nov	Dec	Annual Total	Percent of Grand Total
25 Trimming turf edges (39,600 L.F.)																					
Freq.	33 L.F.	1200	1	–	–	–	0	4	4	4	5	4	5	4	5	4	4	5	4	52	
Labor-hours	–	–	–	20	–	–	0	80	80	80	100	80	100	80	100	80	80	100	80	1040	2.25%
Cost	–	–	–	–	10.58	4.50	.00	1206.40	1206.40	1206.40	1508.00	1206.40	1508.00	1206.40	1508.00	1206.40	1206.40	1508.60	1206.40	15683.20	2.24%
26 Mechanical cultivation of active burial sites (1,611,720 S.F.)																					
Freq.	300 S.F.	5372	1	–	–	–	0	0	0	0	1	0	0	0	0	0	0	0	0	1	
Labor-hours	–	–	–	90	–	–	0	0	0	0	90	0	0	0	0	0	0	0	0	90	.20%
Cost	–	–	–	–	10.58	4.50	.00	.00	.00	.00	1357.20	.00	.00	.00	.00	.00	.00	.00	.00	1357.20	.19%
27 Trimming turf around markers (72,000 L.F.)																					
Freq.	75 L.F.	960	1	–	–	–	0	4	4	4	5	4	5	4	5	4	4	5	4	52	
Labor-hours	–	–	–	16	–	–	0	64	64	64	80	64	80	64	80	64	64	80	64	832	1.80%
Cost	–	–	–	–	10.58	6.50	.00	1093.12	1093.12	1093.12	1366.40	1093.12	1366.40	1093.12	1366.40	1093.12	1093.12	1366.40	1093.12	14210.56	2.03%
28 Feeding (74 acres, 3,223,440 S.F.)																					
Freq.	500 S.F.	6447	1	–	–	–	0	0	0	0	1	0	0	0	0	0	0	0	0	1	
Labor-hours	–	–	–	107	–	–	0	0	0	0	107	0	0	0	0	0	0	0	0	107	.23%
Cost	–	–	–	–	10.58	30.00	.00	.00	.00	.00	4342.06	.00	.00	.00	.00	.00	.00	.00	.00	4342.06	.62%
29 Mowing (with 36-in. mower, 5 mph, 3,223,440 S.F.)																					
Freq.	1452 S.F.	2220	1	–	–	–	0		4	4	5	4	5	4	5	4	4	5	4	48	
Labor-hours	–	–	–	37	–	–	0	0	148	148	185	148	185	148	185	148	148	185	148	1776	3.85%
Cost	–	–	–	–	10.58	4.50	.00	.00	2231.84	2231.84	2789.80	2231.84	2789.80	2231.84	2789.80	2231.84	2231.84	2789.80	2231.84	26782.08	3.87%
30 Watering																					
Freq.	hds, vlvs	3200	10	–	–	–	1	0	0	0	0	0	0	0	0	0	0	0	0	1	
Labor-hours	–	–	–	533	–	–	533	0	0	0	0	0	0	0	0	0	0	0	0	533	1.16%
Cost	–	–	–	–	10.58	12.50	12301.64	.00	.00	.00	.00	.00	.00	.00	.00	.00	.00	.00	.00	12301.64	1.76%
31 Pest (gopher) control																					
Freq.	site	1	960	–	–	–	0	0	1	0	1	1	0	0	1	0	0	1	0	4	
Labor-hours	–	–	–	16	–	–	0	0	16	0	0	16	0	0	16	0	0	16	0	64	.14%
Cost	–	–	–	–	10.58	2.50	.00	.00	209.28	.00	.00	209.28	.00	.00	209.28	.00	.00	209.28	.00	837.12	.12%

Figure 13.3

Maintenance as a Long-Term Financial Obligation

As a facility ages, the cost of routine maintenance can far surpass the initial investment in landscape design and construction. See Figure 13.6. This table projects the cost of routine landscape maintenance for a 19.2-acre corporate headquarters complex. The landscaping of the campus is relatively simple (e.g., there are no water features). The projection is in fixed dollars: neither inflation nor escalating maintenance and materials costs are included in the calculation.

Initial landscape design and construction costs for the campus were just under $2,000,000. According to the projections for the facility illustrated in Figure 13.5, the owner will spend approximately $23,955 per acre each year in routine landscape maintenance. This amounts to an annual grounds maintenance budget of approximately $459,936.

Allocation of Maintenance Costs/Labor-Hours

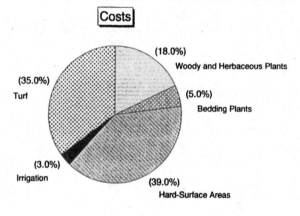

Costs

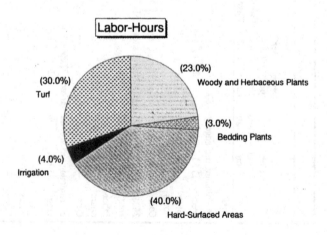

Labor-Hours

Figure 13.4

212

These projections forecast that:

- By the end of the 5th year in the facility's life cycle, more will have been spent on landscape maintenance than was spent on design and construction.
- By the end of the 9th year, more than twice as much will have been spent on maintenance than was spent on design and construction; that is, for every $1.00 spent originally on design and construction, the owner will have spent $2.15 on maintenance.
- The initial expenditure of $1,920,000 has created a financial obligation (in terms of future maintenance requirements) that, by the end of 10 years, will cost the owner nearly $4.6 million.
- For the owner to spend that $4.6 million efficiently, the initial $1.9 million spent on design and construction must focus on maintainability. Control of maintenance costs begins with project design because maintenance costs cannot be reduced below the problems and inefficiencies that are designed and built into a project.

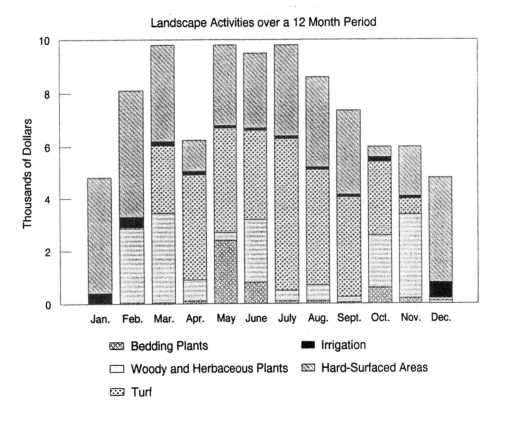

Landscape Activities over a 12 Month Period

Figure 13.5

- The landscaping has created a financial obligation amounting to $9.1 million over 20 years. An investment of such magnitude must be managed not only to control costs, but to help maximize the investment.

Figure 13.7 graphically presents the projections in Figure 13.6. The graph compares the significance of funds spent on design and construction to the significance of funds spent on maintenance. For example, over the 20 years shown, routine maintenance represents 82 percent of the owner's total landscape costs.

Reducing Routine Maintenance Expenses

The importance of landscape maintenance planning can also be illustrated by its projected impact on the maintenance of other types of facilities. For example, a hospital was spending $130,610 annually to maintain

Maintenance Cost Projections for a 19-Acre Complex				
Maintenance Year	Running Total Spent on Landscape Maintenance	Cumulative Running Total Spent on Landscape (including initial design/const.)	Percent of Cumulative Running Total Spent on Landscape Maintenance	Percent of Cumulative Running Total Spent Originally on Landscape Design/Const.
Year 0	$ 0	$ 1,920,000	0%	100%
Year 1	459,936	2,379,936	19%	80%
Year 2	919,872	2,839,872	32%	67%
Year 3	1,379,808	3,299,808	41%	58%
Year 4	1,839,744	3,759,744	48%	51%
Year 5	2,299,680	4,219,680	54%	45%
Year 6	2,759,616	4,679,616	58%	41%
Year 7	3,219,552	5,139,552	62%	37%
Year 8	3,679,488	5,599,488	65%	34%
Year 9	4,139,424	6,059,424	68%	31%
Year 10	4,599,360	6,519,360	70%	29%
Year 11	5,059,296	6,979,296	72%	27%
Year 12	5,519,232	7,439,232	74%	25%
Year 13	5,979,168	7,899,168	75%	24%
Year 14	6,439,104	8,359,104	77%	22%
Year 15	6,899,040	8,819,040	78%	21%
Year 16	7,358,976	9,278,976	79%	20%
Year 17	7,818,912	9,738,912	80%	19%
Year 18	8,278,848	10,198,848	81%	18%
Year 19	8,738,784	10,658,784	81%	18%
Year 20	9,198,720	11,118,720	82%	17%

Figure 13.6

37.5 acres of landscaping. Maintenance programming for the facility reduced the annual budget from $130,610 to $97,957—a savings of 25%. The significance of this savings is illustrated in Figure 13.8.

A maintenance program for another facility—a cemetery—recommended an annual grounds management budget of $700,255. This was only a 6.2% reduction from the $746,839 already being spent on grounds maintenance, but the savings over a number of years would be considerable. Figure 13.9 illustrates (again in fixed dollars) the significance of this savings over the period shown. Ideally, such savings are built into the development of the landscaping program from the outset.

Controlling Landscape Maintenance Costs Using Long-Range Planning

In addition to the maintenance programming tools described in the foregoing sections, the following list suggests ways in which landscape maintenance costs can be controlled during construction:

- Avoid loopholes in design specifications.
- Be specific in handling existing conditions. Prepare a list of existing features to be preserved and establish penalties (fines) for damage or destruction.

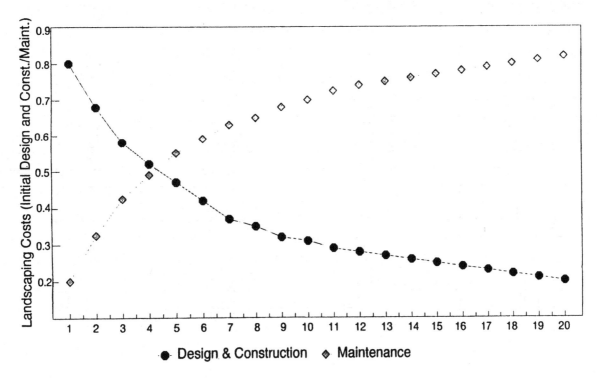

Maintenance Cost Projections

Figure 13.7

- Avoid unauthorized changes in plant palette by the contractor.
- Avoid changes by the contractor through channels other than the supervising designer or his or her agent.
- Do not accept lower quality merchandise. Specifications should, for example, stipulate that labels from seed packages be supplied as part of the documentation of compliance.
- Ensure that the agreement with vendors includes a guarantee period so that maintenance personnel are not burdened with maintaining features that are not completed according to specifications.
- Ensure that specifications require testing and certification of all systems before acceptance of the work.

	Savings Achieved Through Maintenance Programming for a Hospital		
Maintenance Year	Cumulative Running Total for Client's Budget (fixed dollar)	Cumulative Running Total for LMC's Budget (fixed dollar)	Cumulative Running Total of Savings (fixed dollar)
Year 1	$ 130,610	$ 97,957	$ 32,652*
Year 2	261,220	195,915	65,305
Year 3	391,830	293,873	97,957
Year 4	522,441	391,830	130,610
Year 5	653,051	489,788	163,262
Year 6	783,661	587,746	195,915
Year 7	914,272	685,704	228,568
Year 8	1,044,882	783,661	261,220
Year 9	1,175,492	881,619	293,873
Year 10	1,306,103	979,577	326,525
Year 11	1,436,713	1,077,535	359,178
Year 12	1,567,323	1,175,492	391,830
Year 13	1,697,934	1,273,450	424,483
Year 14	1,828,544	1,371,408	457,136
Year 15	1,959,154	1,469,366	489,788
Year 16	2,089,765	1,567,323	522,441
Year 17	2,220,375	1,665,281	555,093
Year 18	2,350,985	1,763,239	587,746
Year 19	2,481,596	1,861,197	620,399
Year 20	2,612,206	1,959,154	653,051

*25% annual savings

Figure 13.8

- Ensure that as-built documents are updated daily by the contractor. For example, if an irrigation line is repositioned three feet to the south of where it is laid out on the plan, the revision must be noted in the as-built documents while the trench is open. Otherwise, no one will be able to remember the exact location of the pipe.

The as-built documents are extremely important to the landscape maintenance function because they most accurately describe the facility. If an irrigation pipe is moved and the move is not documented, the owner must pay for two excavations instead of one (not to mention the likelihood of damage repair).

Part of the as-built documentation should be photographic. Videotape, for example, is an excellent way to document the nature and location of landscape features. Videotaping the location of buried features can employ a tape measure included in the shot to show precise location relative to other features.

Cumulative Savings from Maintenance Programming for a Cemetery			
Maintenance Year	Cumulative Running Total for Client's Budget (fixed dollar)	Cumulative Running Total for LMC's Budget (fixed dollar)	Cumulative Running Total of Savings (fixed dollar)
Year 1	$ 746,839	$ 700,255	$ 46,584*
Year 2	1,493,678	1,400,510	93,168
Year 3	2,240,517	2,100,765	139,752
Year 4	2,987,356	2,801,020	186,336
Year 5	3,734,195	3,501,275	232,920
Year 6	4,481,034	4,201,530	279,504
Year 7	5,227,873	4,901,785	326,088
Year 8	5,974,712	5,602,040	372,672
Year 9	6,721,551	6,302,295	419,256
Year 10	7,468,390	7,002,550	465,840
Year 11	8,215,229	7,702,805	512,424
Year 12	8,962,068	8,403,060	559,008
Year 13	9,708,907	9,103,315	605,592
Year 14	10,455,746	9,803,570	652,176
Year 15	11,202,585	10,503,825	698,760
Year 16	11,949,424	11,204,080	745,344
Year 17	12,696,263	11,904,335	791,928
Year 18	13,443,102	12,604,590	838,512
Year 19	14,189,941	13,304,845	885,096
Year 20	14,936,780	14,005,100	931,680

*6.2% annual savings

Figure 13.9

Reducing the Number of High-Maintenance Items

The following items represent some of the most typical labor-intensive landscape maintenance tasks. These items are important to cover in a landscape maintenance schedule and budget. Accounting for related labor in this way may also point out the high cost of maintaining certain features, and may prompt consideration of appropriate alternative plant materials, irrigation systems, or new landscape design and construction.

- Individual curvilinear lawn areas smaller than 500 square feet.
- Large annual flower beds that require yearly purchase of plants, planting, weeding, and removal of plants.
- Perennials that require staking because they are taller than 18 to 24 inches and have soft stems that bend to the ground under rain or overhead irrigation.
- Ground covers that are readily infested by weeds and other plants.
- Open soil areas that require frequent weeding.
- Rock formations unable to maintain designed plant material, which develop into unsightly weed patches.
- Thorny shrubs requiring frequent selective pruning, or uniform hedges requiring repeated trimming.
- Climbing (thorny) roses requiring yearly untying from a support, pruning and refastening, along with frequent pest control and removal of spent flowers.
- Trees that exceed an approximate height of 35 feet and require special climbing equipment and specially trained personnel, along with higher insurance rates.
- High (superficial) rooting trees and shrubs that present an obstacle to normal maintenance of lawn or ground covers.
- Manmade landscape features that require above-normal mechanical and other care.

There is usually more than one way to solve a high-maintenance landscape problem. Here are two examples.

- A small individual curvilinear shady lawn area requiring weekly mowing and edging with a small lawnmower was planted with an aggressively growing ground cover planting. Maintenance labor for this area was subsequently reduced by approximately 70% per year.
- A previously neglected tall deciduous hedge requiring at least two trimmings per year was cut to 4 feet, and the width reduced to approximately 2.5 feet. Parts were removed where not needed and used to replace other damaged sections. The remodeled hedge received frequent liquid feeding with a low nitrogen/phosphorous fertilizer until new growth was well established. Labor-intense maintenance was reduced by 50%.

The first step is analyzing maintenance cost overruns. Next, investigate and decide on horticulturally sound maintenance-oriented solutions that are aesthetically acceptable, while preserving the overall established design.

Conclusion

Effective planning and budgeting for landscape maintenance involves attention to the four phases of landscape maintenance programming, as explained at the beginning of this chapter. First, the facility's image must be analyzed, and an appropriate, affordable design created and implemented. If the design incorporates direct consideration of the facility's landscape maintenance budget (along with its needed physical features and aesthetic requirements), then ongoing maintenance schedules and budgets should be simply a matter of competent, responsible management, and a host of costly problems can be avoided.

Part 3

Estimating and Budgeting Maintenance Costs

Part 3

Introduction

With an understanding of a facility's overall financial picture, and familiarity with the requirements of the major building components assessed in the facility audit, we are ready to begin assigning costs to planned maintenance and repair. Chapter 14 is an introduction to *Means Facilities Maintenance & Repair Cost Data*, one source of information on maintenance tasks, crews, frequencies, and costs. This chapter, together with Chapters 15–17, includes guidance on how to apply the information contained in this cost data resource for different purposes, from developing an annual maintenance program, to creating a cost estimate for a particular repair item. Several examples show how the information is extracted and used in practical applications.

Chapter 18 addresses the long-term cost implications of materials or systems selection in both new construction and renovations, where initial installation costs are compared to future expenditures in maintenance and repair. This chapter also presents tips on making a case for a capital improvement to address a continuing or growing maintenance and/or repair problem.

Deferred maintenance is inevitably a major issue in facilities planning. The author of Chapter 19 offers guidance on how to identify and prioritize items in this category. Chapter 20 addresses "Reserve Funds," a topic that may be of particular interest to owners and managers of apartments and condominiums, but which is applicable to many other types of facilities as well. Whether a reserve fund is necessary, and how it is set up and monitored, are among the issues addressed.

With a grasp of basic cost concerns and estimating procedures, it should be possible to proceed with a maintenance and repair plan and estimate. At this point, it is helpful to review the issues of particular significance to specific types of facilities, such as those covered in Part 4.

Chapter 14

Using Means Facilities Maintenance & Repair Cost Data

Phillip R. Waier, P.E.

Estimating facilities maintenance and repair costs has always been a challenge. Facility managers are asked to perform many different estimates, often with little background information and even less time to do it. Following are some examples of items that have to be estimated.

- Repair or replacement of a building component.
- The value of deferred maintenance for an entire facility.
- The annual maintenance budget.
- Life cycle costs, to justify a request for funds.
- Funds required to address the deficiencies identified during a recent building audit.

In response to this need Means has developed *Facilities Maintenance & Repair Cost Data.* This publication was designed as a complete reference and cost data source for facility managers, owners, engineers and contractors, as well as anyone who manages real estate.

A complete facilities management program starts with a building audit. (See the Introduction to Part 2 for more on building audits.) The audit describes the structures, their characteristics and major equipment, and apparent deficiencies. An audit is not a prerequisite to using any information contained in the *Facilities Maintenance & Repair Cost Data* book — it just provides a base line from which to begin the budgeting process. The list of equipment identified by the audit forms the basis for a preventive maintenance (PM) program. The building deficiencies identified fuel the maintenance budget. The audit also provides facility size and usage criteria on which the general maintenance (GM), or cleaning program, will be based.

The building information gathered in the audit can also be used as the basis for benchmarking studies among facilities. The size and usage criteria facilitate comparisons of utility usage, general and preventive maintenance, and repair costs. Together, the five sections of *Facilities Maintenance & Repair Cost Data* provide a framework and definitive data for developing a complete program for all facets of facilities maintenance.

For the occasional user, the cost data provides:

- A reference for time and material requirements for maintenance and repair tasks to fill in where an organization lacks its own historical data.

- Preventive maintenance (PM) checklists with labor-hour standards that can be used to benchmark a facility's program.
- A quick reference for general maintenance (cleaning) productivity rates, to analyze proposals from outside contractors.
- An explanation of life cycle cost analysis, as a tool to assist in making the right budget decisions.
- Data to validate line items in budgets.

For the frequent user, the *Facilities Maintenance & Repair Cost Data* supplies all of the above, plus:

- Guidance for getting the audit under way.
- A foundation for zero-based budgeting of maintenance and repair, preventive maintenance, and general maintenance.
- Cost data that can be used to estimate preventive maintenance, deferred maintenance (repairs), and general maintenance (cleaning) programs.
- Detailed task descriptions for establishing a complete PM program.

Facilities Maintenance & Repair Cost Data is published in loose-leaf form, organized into the following clearly defined sections.

Section 1: Maintenance & Repair (M&R)

The *Maintenance & Repair* section is a listing of common maintenance tasks performed at facilities. The tasks include removal and replacement, repair, and refinishing. This section is organized according to the 12-division UniFormat system. The UniFormat was originally developed by the General Services Administration as a systems approach to organizing building construction and has since become widely used in the industry. The UniFormat divisions are as follows:

Division 1 — Foundations
Division 2 — Substructures
Division 3 — Superstructures
Division 4 — Exterior Closure
Division 5 — Roofing
Division 6 — Interior Construction
Division 7 — Conveying Systems
Division 8 — Mechanical
Division 9 — Electrical
Division 10 — General Conditions & Profit
Division 11 — Special Construction
Division 12 — Site Work

The purpose of Section 1 is to provide cost data and an approximate frequency of occurrence for each maintenance and repair task. The data is also used to estimate labor-hours and material costs for in-house staff, or total cost including overhead and profit for an outside contractor. The frequency listing facilitates preparation of a zero-based budget.

Under the "System Description" column, each task including its components is described. In some cases additional information (such as "2% of total roof area") appears in parentheses. The next column, "Frequency," refers to how often one can expect, and therefore should estimate, this task will have to be performed. Labor-hours have been enhanced where appropriate by 30% to include setup and delay time. In some cases, the list of steps comprising the task includes most of the potential repairs that could be made on a system. For those tasks, individual items, as appropriate, can be selected from the all-encompassing list.

The data in this section can be applied to many facilities estimating problems such as:

- Providing an estimate for a current repair or replacement project.
- Preparing a deferred maintenance backlog budget.
- Estimating the replacement cost for items identified in the reserve study.
- Estimating costs for several options in a value engineering study.
- Estimating the costs to comply with new codes and regulations.
- Providing benchmarks for maintenance tasks.

See Chapter 15 for additional information on using *Means Facilities Maintenance & Repair Cost Data*.

Section 2: Preventive Maintenance (PM)

The Preventive Maintenance section provides the framework for a complete PM program. The establishment of a program is frequently hindered by the lack of a comprehensive list of equipment, actual PM steps, and budget documentation. Once the comprehensive list of equipment is obtained as part of the building audit or from existing records, this section fulfills the remaining needs.

The facility manager, plant engineer, or owner can use the schedules provided in *Facilities Maintenance & Repair Cost Data* to establish labor-hours and a budget. The maintenance contractor can use these PM checklists as the basis for a comprehensive maintenance proposal or as an estimating aid when bidding PM contracts. Selected pages, with preventive maintenance steps, can be photocopied and assembled into a customized PM program. Copies of individual sheets can also be distributed to maintenance personnel to identify the procedures required.

The hours listed are predicated on work being performed by experienced technicians familiar with the PM tasks and equipped with the proper tools and materials. The PM section lists the tasks and their frequency, whether weekly, monthly, quarterly, semi-annually, or annually. The frequency of those procedures is based on non-critical usage (e.g., in "normal use" situations, versus facilities such as surgical suites or computer rooms that demand absolute adherence to a limited range of environmental conditions.) The labor-hours to perform each item on the checklist are listed.

Beneath the table is cost data for the PM schedule. The data is shown both annually and annualized, to provide the facility manager with an estimating range. If all tasks on the schedule are performed once a year, the *annually* line should be used in the PM estimate. The *annualized* data is used when all items on the schedule are performed at the frequency shown. See Chapter 16, "Preventive & Predictive Maintenance Estimating," for additional information.

Section 3: General Maintenance (GM)

This section provides labor-hour estimates and costs to perform day-to-day cleaning. The data is used to estimate cleaning times, compare and assess estimates submitted by cleaning companies, or to budget in-house staff. The information is divided into *Interior* and *Exterior* maintenance. A common maintenance laborer *(Clam)* is used to perform these tasks. See Chapter 17 "General Maintenance Estimating," for additional information.

Section 4: Facilities Audits

The Facilities Audit is generally the basis for preparation of the maintenance & repair, preventive maintenance, and general maintenance estimates. The audit provides the following:

- A list of deficiencies for a current maintenance & repair budget or future deferred maintenance budget.
- A list of equipment and other items used as the basis for a preventive maintenance budget.
- A list of the facilities including size, usage, and space distribution, required to prepare a general maintenance estimate.

The introduction to facilities audits in Section 2 of *Facilities Maintenance & Repair Cost Data* explains the rationale for the audit, and the steps required to complete it successfully. Section 4 of the book also provides forms for listing all of the organization's facilities, and a separate detail form for the specifics of each facility. Checklists are provided for major building components. Each checklist should be accompanied by a blank audit form, used to record the audit findings, prioritize the deficiencies, and estimate the cost to remedy these deficiencies.

Section 5: Reference

This section contains data that complement or support information contained in the other sections. Included are an overview of life cycle cost analysis, a listing of equipment rental rates, a table of travel costs, crews, city cost indexes for adjusting costs to specific locations, and standard abbreviations.

Life Cycle Costing: This section provides definitions and basic equations for performing life cycle analysis. This analysis forms a part of a complete value engineering study. A sample problem demonstrates the methodology.

Equipment Rental Rates: Lists rental rates and hourly operating costs for equipment frequently used in facilities maintenance. These rates may be used to estimate supplemental equipment needed by in-house staff or outside contractors.

Travel Cost Tables: This chart provides labor-hour costs for round-trip travel between a base of operation and the project site. The costs shown in *Facilities Maintenance & Repair Cost Data* do not contain travel time. The preparation of a maintenance and repair budget or PM budget for a campus facility or locations spread across a geographic area requires the inclusion of travel time.

Crews: This section lists all crews referenced in this book. A crew is composed of more than one trade classification and/or the addition of power equipment to any trade classification. Power equipment is included in the cost of the crew. Costs are shown with bare labor rates and with both the in-house mark-ups and the installing contractor's overhead and profit added. For each, the total crew cost per eight-hour day and the composite cost per labor-hour are listed.

City Cost Indexes: Obviously, costs vary depending on the regional economy. You can adjust the "national average" costs in this book to 305 major cities throughout the U.S. and Canada by using the data in this section. The city cost index is a percentage ratio of cost at a specific location compared to the national average cost. In other words, these index figures represent relative construction factors (or multipliers) for material costs and installation costs, as well as the weighted average for Total In Place costs for each UniFormat division. Estimates calculated using this publication should be adjusted to a specific location. In 1996, the index ranges from .745 to 1.36; therefore, the correction for your location could be significant.

Abbreviations: A listing of the abbreviations used throughout this book, along with the terms they represent, is included.

Design Assumptions

This book is designed to be as easy to use as possible. To that end, we have established models based on typical facility scenarios and requirements.

1. Unlike any other Means publication, *Facilities Maintenance & Repair Cost Data* was designed for estimating tasks in a wide range of existing, diverse environments. Because of this diversity, the level of accuracy of the data is +/–20%.

2. Material prices have been established based on a national average.

3. Labor costs are based on a 30-city national average of union wage rates.

4. Except where major equipment or component replacement is described, the projects in this book are small; therefore, material prices and labor-hours have been enhanced to reflect the increased costs of small-scale work.

5. The PM frequencies are based on non-critical applications. Increased frequencies are required for critical environments.

Conclusion   *Means Facilities Maintenance & Repair Cost Data* is a tool that can be readily applied to the tasks of planning, budgeting, and managing facility maintenance. Together with other tools, such as the audit survey, this resource can help facility and maintenance managers assign rational dollar values to their organization's facility requirements.

Chapter 15

Maintenance & Repair Estimating

Phillip R. Waier, P.E. and Stephen C. Plotner

Maintenance and Repair (M&R) Estimating is the most common type of estimate performed at a facility. Tasks in this category include removal and replacement of defective parts and equipment, and repair and refinishing of deteriorated materials. Estimating this work begins with the development of the annual operating and maintenance budget. This work is based on the frequency and scope of the maintenance work to be performed during the year. Historical records and trends are the most valuable tools available when preparing the budget. From historical data the overall cost, or budget, for each area of work is developed. An alternate to this is the zero-based budget. Instead of predicating next year's budget on the current budget plus an inflation factor, zero-based budgeting requires starting from zero and building the budget based on the historical or published data on the frequency of failure of the building elements, times the estimated cost to repair or replace these elements.

Repair work is identified during the course of the year in the form of trouble reports or maintenance requests. Simple, inexpensive repairs may not require a formal estimate. More involved or costly repairs may necessitate a detailed estimate for planning and approval. A cost estimate may also be done in order to determine whether an item is repaired or replaced. Cost estimates play a similar role when evaluating the feasibility or desirability of major modifications or improvements to a facility. This chapter addresses the information required, cost data available and established procedures to compile a maintenance and repair estimate.

M&R work can be accomplished by an in-house staff or outsourced to outside contractors. The decision whether or not to outsource usually depends on the size of the facility and the importance of the maintenance to the organization's primary mission. For example, the mission of a college or university is to educate; therefore, the choice of who performs the maintenance function is not of central importance. The mission of Ford Motor company is to produce cars; therefore, any maintenance related to car production will probably be accomplished in-house. The maintenance of the building in which the car production takes place may or may not be contracted out. Another factor, as stated above, is the size of the facility. When larger organizations have their own maintenance staff, it is usually comprised of at least one skilled tradesperson from each of the major trades: carpenter, electrician, plumber and HVAC technician. As a result of the facility's larger size, there is a reasonable expectation that the staff will be kept busy throughout the year. In a smaller facility it is more difficult to continually maintain a sufficient work level; therefore, outside services are often contracted. Most larger organizations employ both in-house staff and outside contractors. Usually, projects above a certain size or level of complexity are outsourced. This leaves the in-house staff to handle day-to-day repairs. This is an oversimplification of the outsourcing criteria, which is presented in depth in Chapter 3, "Maintenance Outsourcing."

Incurred Costs

M&R estimates begin with an understanding of all the costs incurred. The costs are categorized as *direct* and *indirect*. The direct expenses are related to performing work at the specific location, including materials, labor, and equipment. Indirect work elements are other activities that facilitate the direct work. While not performed directly on the facility, indirect costs are part of an effective program. The usual indirect expenses are related to the contractors' overhead or in-house administrative costs, such as administrative salaries, supervision, office supplies, telephones, office space, transportation, and other expenses. In a maintenance organization there is an expanded list of indirect expenses. These work elements include:

- Work Identification
- Cost Estimating
- Purchasing, Supplies and Inventory Control
- Cost Accounting and Control
- Scheduling
- Work Tracking and Monitoring
- Facility and Equipment Histories
- Engineering

A complete maintenance and repair estimate must include all of these costs, which are generally expressed as a percentage of direct maintenance costs. In large organizations this percentage can range from 0 to more than 50%, depending on the company and its accounting methods. Some companies, for instance, do not charge supervision, administration, or maintenance vehicles to the maintenance budget. When using *Means Facilities Maintenance and Repair Cost Data*, you will note two *Total Cost* columns. One is the *Total In-House* and the second is the *Total with Overhead and Profit* for outside contractors. The markups are different when applied to bare material, labor, and equipment to calculate each column.

Items 8 and 9 in the "How to Use the Maintenance and Repair Assemblies Cost Tables" (Figure 15.1) define the markups applied to the bare costs. The labor markup for in-house staff includes Workers' Compensation and Average Fixed Overhead (FICA, Federal and State Unemployment, Builder's Risk and Public Liability) only. If your company charges administration, supervision, vehicles, and other expenses to the maintenance organization, then an additional overhead markup must be added to account for these costs. This markup would be expressed as a percentage of direct maintenance costs.

Sources of Maintenance and Repair Estimating Information

As stated earlier, one of the best sources for cost information is the facility's own current data. The invoice you just received from a contractor for a particular maintenance task or the completed, priced, in-house work order for that task are the best indicator of what a subsequent similar task will cost. Keep in mind that there is always room for improvement (lower costs) with the application of better management practices.

How to Use the Maintenance & Repair Assemblies Cost Tables

The following is a detailed explanation of a sample Maintenance & Repair Assemblies Cost Table. Next to each bold number that follows is the item being described with the appropriate component of the sample entry following in parenthesis.

Total system costs as well as the individual component costs are shown. In most cases, the intent is for the user to apply the total system costs. However, changes and adjustments to the components or partial use of selected components is also appropriate. In particular, selected equipment system tables in the mechanical section include complete listings of operations that are meant to be chosen from, rather than used in total.

System/Line Numbers (5.1-105-0700)

Each Maintenance & Repair Assembly has been assigned a unique identification number based on the UniFormat classification system.

UniFormat Division

5.1 105 0700

Means Subdivision
Means Major Classification
Means Individual Line Number

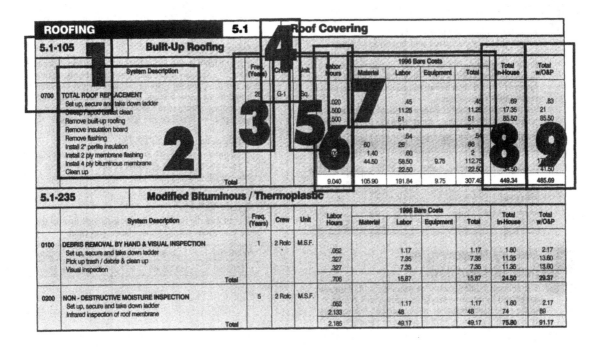

ROOFING — 5.1 — Roof Covering

5.1-105 — Built-Up Roofing

	System Description	Freq. (Years)	Crew	Unit	Labor Hours	1996 Bare Costs Material	Labor	Equipment	Total	Total In-House	Total w/O&P
0700	TOTAL ROOF REPLACEMENT	25	G-1	Sq.							
	Set up, secure and take down ladder				.020		.45		.45	.69	.83
	Sweep / spud surface clean				.500		11.25		11.25	17.35	21
	Remove built-up roofing				.500		51		51	85.50	85.50
	Remove insulation board										
	Remove flashing						.54		.54		
	Install 2" perlite insulation					.60	26		86		
	Install 2 ply membrane flashing					1.40	.60		2		
	Install 4 ply bituminous membrane					44.50	58.50	9.75	112.75	34.50	41.50
	Clean up						22.50		22.50		
	Total				9.040	105.90	191.84	9.75	307.49	449.34	485.69

5.1-235 — Modified Bituminous / Thermoplastic

	System Description	Freq. (Years)	Crew	Unit	Labor Hours	1996 Bare Costs Material	Labor	Equipment	Total	Total In-House	Total w/O&P
0100	DEBRIS REMOVAL BY HAND & VISUAL INSPECTION	1	2 Rofc	M.S.F.							
	Set up, secure and take down ladder				.052		1.17		1.17	1.80	2.17
	Pick up trash / debris & clean up				.327		7.35		7.35	11.35	13.60
	Visual inspection				.327		7.35		7.35	11.35	13.60
	Total				.706		15.87		15.87	24.50	29.37
0200	NON - DESTRUCTIVE MOISTURE INSPECTION	5	2 Rofc	M.S.F.							
	Set up, secure and take down ladder				.052		1.17		1.17	1.80	2.17
	Infrared inspection of roof membrane				2.133		48		48	74	89
	Total				2.185		49.17		49.17	75.80	91.17

I – 4

Figure 15.1

2 System Description

A one-line description of each system is given, followed by a description of each component that makes up the system. In selected items, a percentage factor for planning annual costs is also given.

3 Frequency (28)

The projected frequency of occurrence in years for each system is listed. For example, 28 indicates that this operation should be budgeted for once every 28 years. If the frequency is less than 1, then the operation should be planned for more than once a year. For example, a value of .5 indicates a planned occurrence of 2 times in a given year.

4 Crew (G-1)

The "Crew" column designates the typical trade or crew used to install the item, although other trades may be used for minor portions of the work. If an installation can be accomplished by one trade and requires no power equipment, that trade and the number of workers are listed (for example, "2 Rofc"). If an installation requires a composite crew, a crew code designation is listed (for example, "G-1"). You'll find full details on all composite crews in the Crew Listings.

• For a complete list of all trades utilized in this book and their abbreviations, see the last page of the book.

Crews

Crew No.	Bare Costs		In House Costs		Incl. Subs O&P		Cost Per Labor-Hour		
Crew G-1	Hr.	Daily	Hr.	Daily	Hr.	Daily	Bare Costs	In House	Incl. O&P
1 Roofer Foreman	$24.55	$ 196.40	$37.85	$ 302.80	$45.45	$ 363.60	$20.95	$32.29	$38.79
4 Roofers, Composition	22.55	721.60	34.75	1112.00	41.75	1336.00			
2 Roofer Helpers	15.95	255.20	24.60	393.60	29.55	472.80			
1 Application Equipment		160.90		160.90		177.00			
1 Tar Kettle/Pot		34.60		34.60		38.05	3.49	3.49	3.84
56 M.H., Daily Totals		$1368.70		$2003.90		$2387.45	$24.44	$35.78	$42.63

5 Unit of Measure for Each System (Sq.)

The abbreviated designation indicates the unit of measure, as defined by industry standards, upon which the price of the component is based. In this example, "Cost per Sq." is the unit of measure for this system or "assembly."

6 Labor-Hours (9.040)

The "Labor-Hours" figure represents the number of labor-hours required to install one unit of work. To find out the number of labor-hours required for a particular task, multiply the quantity of the item times the number of labor-hours shown. For example:

Quantity	x	Productivity Rate	=	Duration
30 Sq.	x	9.040 Labor-Hours/ Sq.	=	271 Labor-Hours

7 Bare Costs:

Material (105.90)

This figure for the unit material cost for the line item is the "bare" material cost with no overhead and profit allowances included. *Costs shown reflect national average material prices for the current year and generally include delivery to the facility. Small purchases may require an additional delivery charge. No sales taxes are included.*

Labor (191.84)

The unit labor cost is derived by multiplying bare labor-hour costs for Crew G-1 and other trades by labor-hour units. In this case, the bare labor-hour cost for the G-1 crew is found in the Crew Section. (If a trade is listed or used for minor portions of the work, the hourly labor cost—the wage rate—is found on the last page of the book).

Equip. (Equipment) (9.75)

Equipment costs for each crew are listed in the description of each crew. The unit equipment cost is derived by multiplying the bare equipment hourly cost by the labor-hour units.

Total (307.49)

The total of the bare costs is the arithmetic total of the three previous columns: mat., labor, and equip.

Material	+	Labor	+	Equip.	=	Total
$105.90	+	$191.84	+	$9.75	=	$307.49

I–5

Figure 15.1 (cont.)

234

8 Total In-House Costs (449.34)

"Total In-House Costs" include suggested mark-ups to bare costs for the direct overhead requirements of in-house maintenance staff. The figure in this column is the sum of three components: the bare material cost plus 10% (for purchasing and handling small quantities); the bare labor cost plus workers' compensation and average fixed overhead (per the labor rate table on the last page of the book or, if a crew is listed, from the crew listings); and the bare equipment cost.

9 Total Costs Including O & P (485.69)

"Total Costs Including O&P" include suggested mark-ups to bare costs for an outside subcontractor's overhead and profit requirements. The figure in this column is the sum of three components: the bare material cost plus 25% for profit; the bare labor cost plus total overhead and profit (per the labor rate table on the last page of the book or, if a crew is listed, from the crew listings); and the bare equipment cost plus 10% for profit.

1-6

Figure 15.1 (cont.)

Another good source of cost information is historical data from your facility, with a correction for inflation. If the removal and replacement of a 300-ton cooling tower in 1990 cost $21,000, then using historical economic data or Means Historical Cost Indexes, this cost can be converted to 1996 dollars. See more about the Historical Cost Indexes later in this chapter.

A third source of cost information is Means cost data. Depending on the amount of information available, the degree of accuracy, and amount of time available to complete the estimate, there are several applicable publications.

- The most accurate and time-consuming type of estimate is the *unit price estimate*. This estimate is used when complete construction documentation is available. It is usually used for new construction or major remodeling. The accuracy of a unit price estimate is approximately $+/-10\%$. *Means Facilities Construction Cost Data* is the best source of costs for this type of estimate.

- Since the lack of time and documentation is sometimes a problem for the facility/maintenance manager, the next best estimate is an assemblies or systems estimate. This estimate is faster and easier to complete because it is based on more general construction parameters. The accuracy of an assemblies estimate is $+/-15\%$ for new construction or major remodeling, and $+/-20\%$ for true maintenance work. *Means Assemblies Cost Data* is the best source of costs for construction or remodeling, while *Means Facilities Maintenance & Repair Cost Data* is best for maintenance estimating.

- For maintenance planning and budgeting *Means Facilities Maintenance & Repair Cost Data* is the most appropriate.

Other sources of information are trade organizations that provide information on labor-hours and vendor catalogs with material/equipment costs. This type of information will generally provide enough data to calculate a unit price estimate. Research of this type can be very informative, but it is also time-consuming.

Productivity

Productivity is the number of units of a specified task that a tradesperson or crew can install in an 8-hour day. The productivity of maintenance and repair work is different than new construction. When new construction work is performed, there is unrestricted access for materials and labor; and noise, work coordination, and service disruptions are seldom a problem. Also, the amount of mobilization, demobilization, and set-up time are minimal compared to the total labor-hours. Maintenance work, conversely, has a high ratio of mobilization, set-up and demobilization time to actual repair time. *Means Facilities Maintenance & Repair Cost Data* has accounted for the coordination and set-up time by adding 30% to the new construction labor-hours. This percentage was derived from studies conducted by the Army Corps of Engineers Construction Engineering Research Laboratory and confirmed by other research groups.

Travel Time

Travel time is not included in Means cost data because of the high variability. Travel in a facility's environment has three possible scenarios.

1. Travel within a high-rise building; that is, between the maintenance area and a specific location. In this case, the minimum travel is via elevator, and no vehicle is required.

2. Travel within a multi-building campus; that is, travel between the central maintenance area and a particular site. A vehicle may be necessary, and travel time is short.
3. Travel within a multi-building district, such as schools in a school district. In this case, travel between the central maintenance area and each building requires a vehicle and possibly considerable travel time.

The Table in Figure 15.2 was calculated to account for travel costs. This table is based on national average wage rates for the common maintenance trades. Note: This table does not include vehicle costs. The cost of service vehicles, if applicable, must be included in the overhead percentage.

Here are some shortcuts to account for travel time in your budget or estimate. If all the trips are short, such as around a campus, add .50 hours round trip to each service call. Travel within a school district may be longer and more varied. For estimating purposes, use the travel time to the average location and apply it to all service calls in the district.

Site Visits

The routine maintenance or repair call is seldom viewed by anyone other than the assigned tradesperson. There are so many calls that it is impossible for a supervisor to visit each site and perform an estimate or assess the extent of the work. There are usually parameters set within an organization that establish an approval process for routine maintenance requiring more than one day to complete. A site visit may be part of that process. Emergencies and safety-related problems are usually exempt from the approval process. Generally, there are three circumstances under which a site visit will be conducted. They are:

1. When requested by mechanic
2. When a job exceeds a certain value/complexity
3. When recurring problems warrant complete evaluation of the equipment

Supervisory site visits are not estimated separately, but are included as part of the overhead markup.

Using Means Cost Data

Means cost data is used by M&R contractors, general contractors, facility/maintenance managers, and owners. The data is used to:

- provide budget estimates for projects
- prepare complete M&R budgets for facilities
- value the corrective action required as the result of an audit
- value the deferred maintenance backlog
- evaluate proposals submitted by outside contractors

The data is also used as a benchmark for in-house staff. The data is arranged according to the 12-division UniFormat (systems format) introduced in Chapter 14, "Using *Means Facilities Maintenance and Repair Cost Data.*"

The presentation of the data is straightforward for ease of use. There are, however, several ways in which it can be applied. Performing an estimate may be as simple as just turning to the appropriate page and retrieving a price for the work (which includes the removal and replacement of a particular item such as a door and frame with hardware). In other instances, it may be necessary to modify an existing assembly or combine assemblies. Following are some examples of the modification process.

Reference Section

The following table is used to estimate the cost of in-house staff or contractor's personnel to travel to and from the job site. The amount incurred must be added to the in-house total labor cost or contractor's billing application to calculate the total cost with travel.

In-House Staff—Travel Times versus Costs

Travel Time versus Costs	Crew Abbr. Rate with Mark-ups	Skwk	Clab	Clam	Asbe	Bric	Carp	Elec	Pord	Plum	Rofc	Shee	Stpi
		$35.45	$27.40	$18.45	$38.75	$35.25	$34.90	$36.55	$30.55	$38.15	$34.55	$37.80	$38.45
Round Trip Travel Time (hours)	Crew Size												
.50	1	17.73	13.70	9.23	19.38	17.63	17.45	18.28	15.28	19.08	17.28	18.90	19.23
	2	35.45	27.40	18.45	38.75	35.25	34.90	36.55	30.55	38.15	34.55	37.80	38.45
	3	53.18	41.10	27.68	58.13	52.88	52.35	54.83	45.83	57.23	51.83	56.70	57.68
.75	1	26.59	20.55	13.84	29.06	26.44	26.18	27.41	22.91	28.61	25.91	28.35	28.84
	2	53.18	41.10	27.68	58.13	52.88	52.35	54.83	45.83	57.23	51.83	56.70	57.68
	3	79.76	61.65	41.51	87.19	79.31	78.53	82.24	68.74	85.84	77.74	85.05	86.51
1.00	1	35.45	27.40	18.45	38.75	35.25	34.90	36.55	30.55	38.15	34.55	37.80	38.45
	2	70.90	54.80	36.90	77.50	70.50	69.80	73.10	61.10	76.30	69.10	75.60	76.90
	3	106.35	82.20	55.35	116.25	105.75	104.70	109.65	91.65	114.45	103.65	113.40	115.35
1.50	1	53.18	41.10	27.68	58.13	52.88	52.35	54.83	45.83	57.23	51.83	56.70	57.68
	2	106.35	82.20	55.35	116.25	105.75	104.70	109.65	91.65	114.45	103.65	113.40	115.35
	3	159.53	123.30	83.03	174.38	158.63	157.05	164.48	137.48	171.68	155.48	170.10	173.03
2.00	1	70.90	54.80	36.90	77.50	70.50	69.80	73.10	61.10	76.30	69.10	75.60	76.90
	2	141.80	109.60	73.80	155.00	141.00	139.60	146.20	122.20	152.60	138.20	151.20	153.80
	3	212.70	164.40	110.70	232.50	211.50	209.40	219.30	183.30	228.90	207.30	226.80	230.70

Installing Contractors—Travel Times versus Costs

Travel Time versus Costs	Crew Abbr. Rate with O & P	Skwk	Clab	Clam	Asbe	Bric	Carp	Elec	Pord	Plum	Rofc	Shee	Stpi
		$43.60	$33.60	$23.05	$47.80	$43.30	$42.75	$45.65	$37.80	$47.50	$41.75	$46.80	$47.85
Round Trip Travel Time (hours)	Crew Size												
.50	1	21.80	16.80	11.53	23.90	21.65	21.38	22.83	18.90	23.75	20.88	23.40	23.93
	2	43.60	33.60	23.05	47.80	43.30	42.75	45.65	37.80	47.50	41.75	46.80	47.85
	3	65.40	50.40	34.58	71.70	64.95	64.13	68.48	56.70	71.25	62.63	70.20	71.78
.75	1	32.70	25.20	17.29	35.85	32.48	32.06	34.24	28.35	35.63	31.31	35.10	35.89
	2	65.40	50.40	34.58	71.70	64.95	64.13	68.48	56.70	71.25	62.63	70.20	71.78
	3	98.10	75.60	51.86	107.55	97.43	96.19	102.71	85.05	106.88	93.94	105.30	107.66
1.00	1	43.60	33.60	23.05	47.80	43.30	42.75	45.65	37.80	47.50	41.75	46.80	47.85
	2	87.20	67.20	46.10	95.60	86.60	85.50	91.30	75.60	95.00	83.50	93.60	95.70
	3	130.80	100.80	69.15	143.40	129.90	128.25	136.95	113.40	142.50	125.25	140.40	143.55
1.50	1	65.40	50.40	34.58	71.70	64.95	64.13	68.48	56.70	71.25	62.63	70.20	71.78
	2	130.80	100.80	69.15	143.40	129.90	128.25	136.95	113.40	142.50	125.25	140.40	143.55
	3	196.20	151.20	103.73	215.10	194.85	192.38	205.43	170.10	213.75	187.88	210.60	215.33
2.00	1	87.20	67.20	46.10	95.60	86.60	85.50	91.30	75.60	95.00	83.50	93.60	95.70
	2	174.40	134.40	92.20	191.20	173.20	171.00	182.60	151.20	190.00	167.00	187.20	191.40
	3	261.60	201.60	138.30	286.80	259.80	256.50	273.90	226.80	285.00	250.50	280.80	287.10

Figure 15.2

The maintenance manager wants to estimate the cost to repair a 2,000 MBH oil boiler. The technician sent to investigate the problem indicated the blower motor is defective. The Table of Contents of the Maintenance & Repair section of *Facilities Maintenance & Repair Cost Data* directs him to assembly 8.3-406. Assembly 8.3-406-2010 (shown in Figure 15.3) addresses the 2,000 MBH boiler. The *Total In-House* repair cost is $942.45. This amount includes removal and replacement of many items, including the fireye, oil pump, etc. Since only the blower motor is defective, the removal and replacement lines can be broken out to estimate that cost. The *Total In-House* cost to remove and replace is $35.50 + $213 = $248.50. This amount includes the standard markups on material, labor and equipment but does not include sales tax, travel time or city cost adjustment factors.

In another case a maintenance manager has to estimate the removal and replacement of fire-rated gypsum drywall that is two layers of 5/8″ thick gypsum. The metal studs are undamaged. The Table of Contents directs him to Assembly 6.5-230-0040 (see Figure 15.4), which prices the removal and replacement of one layer. Since the removal and replacement of two layers is required, it is necessary to multiply the drywall installation cost times two. We assume this wall has been a maintenance problem, and it has to be painted frequently. Now vinyl wall covering is being considered, and a cost is needed. Refer to Assembly 6.5-440-0020 (Figure 15.5). Note that the unit of measure varies among the assemblies; therefore, it is always necessary to check the units. In the above example the costs were converted to S.F.

These are just two examples of how the data can be manipulated.

Putting It All Together

The following is a summary of items that must be considered for inclusion in a maintenance and repair estimate. Separate checklists are provided for in-house work and for an outside contractor. In each case, assume you have completed the first part of the estimate using bare material, labor, and equipment costs.

In-House
1. Add appropriate in-house markups to the bare costs: 10% to material, 37% to labor and 0% to equipment.
2. Add marked-up travel costs to the labor subtotal.
3. Add additional markups to new labor subtotal for in-house overhead, if applicable.
4. Add sales tax to bare material costs.
5. Apply city cost adjustment factors.

The first of the sample estimates later in this chapter incorporates these five items.

Contractor's Work
1. Add appropriate contractors' markups to the bare costs: 25% to material, 68% to labor, and 10% to equipment.
2. Add marked-up travel costs to the labor subtotal.
3. Add sales tax to bare material costs.
4. Apply city cost adjustment factors.

System Description	Freq. (Years)	Crew	Unit	Labor Hours	1996 Bare Costs				Total In-House	Total w/O&P
					Material	Labor	Equipment	Total		
1060 REPLACE BOILER, OIL, 250 MBH	30	Q-7	Ea.							
Remove boiler				24.465		705		705	900	1,125
Replace boiler, oil, 250 MBH				48.930	2,200	1,425		3,625	4,225	4,975
Total				73.395	2,200	2,130		4,330	5,125	6,100
2010 REPAIR BOILER, OIL, 2000 MBH	7	1 Stpi	Ea.							
Remove / replace burner blower				.600		18.20		18.20	23	29
Remove burner blower bearing				1		30.50		30.50	38.50	48
Replace burner blower bearing				2	25	60.50		85.50	104	127
Remove burner blower motor				.976		28.50		28.50	35.50	44.50
Replace burner blower motor				1.951	129	57		186	213	250
Remove burner fireye				.195		5.90		5.90	7.50	9.30
Replace burner fireye				.300	85.50	9.10		94.60	106	121
Remove burner ignition transformer				.286		8.65		8.65	11	13.65
Replace burner ignition transformer				.571	75	17.30		92.30	104	121
Remove burner ignition electrode				.200		6.05		6.05	7.70	9.55
Replace burner ignition electrode				.400	10.20	12.10		22.30	26.50	32
Remove burner oil pump				.400		12.10		12.10	15.35	19.10
Replace burner oil pump				.800	150	24		174	195	225
Remove burner nozzle				.167		5.05		5.05	6.40	8
Replace burner nozzle				.333	12	10.10		22.10	26	31
Repair controls				.600		18.20		18.20	23	29
Total				10.779	486.70	323.25		809.95	942.45	1,117.10
2060 REPLACE BOILER, OIL, 2000 MBH	30	Q-7	Ea.							
Remove boiler				57.762		1,675		1,675	2,125	2,625
Replace boiler, oil, 2000 MBH				148	13,000	4,300		17,300	19,700	23,000
Total				205.762	13,000	5,975		18,975	21,825	25,625

Figure 15.3

6.5-210 Plaster

	System Description	Freq. (Years)	Crew	Unit	Labor Hours	1996 Bare Costs				Total In-House	Total w/O&P
						Material	Labor	Equipment	Total		
0010	REPAIR PLASTER WALL - (2% OF WALLS)	13	1 Carp	S.Y.							
	Remove damage				.468		9.25	3.51	12.76	16.35	19.60
	Replace two coat plaster finish, incl. lath				.752	6.85	17.30	.62	24.77	31	37.50
	Total				1.220	6.85	26.55	4.13	37.53	47.35	57.10
0030	REFINISH PLASTER WALL	4	1 Pord	S.Y.							
	Prepare surface				.033		.81		.81	1.08	1.33
	Paint / seal surface				.072	.36	1.62		1.98	2.56	3.12
	Place and remove mask and drops				.033		.81		.81	1.08	1.33
	Total				.138	.36	3.24		3.60	4.72	5.78
0040	REPLACE PLASTER WALL	75	2 Carp	S.Y.							
	Set up, secure and take down ladder				.130		3.28		3.28	4.55	5.55
	Remove material				.468		9.25	3.51	12.76	16.35	19.60
	Replace two coat plaster wall, incl. lath				.752	6.85	17.30	.62	24.77	31	37.50
	Total				1.350	6.85	29.83	4.13	40.81	51.90	62.65

6.5-230 Drywall

	System Description	Freq. (Years)	Crew	Unit	Labor Hours	1996 Bare Costs				Total In-House	Total w/O&P
						Material	Labor	Equipment	Total		
0010	REPAIR 5/8" DRYWALL - (2% OF WALLS)	20	1 Carp	S.Y.							
	Remove damage				.087		1.72	.65	2.37	3.03	3.63
	Replace 5/8" drywall, taped and finished				.194	1.89	4.86		6.75	8.80	10.60
	Total				.281	1.89	6.58	.65	9.12	11.83	14.23
0030	REFINISH DRYWALL	4	1 Pord	S.Y.							
	Prepare surface				.033		.81		.81	1.08	1.33
	Paint / seal surface				.072	.36	1.62		1.98	2.56	3.12
	Place and remove mask and drops				.033		.81		.81	1.08	1.33
	Total				.138	.36	3.24		3.60	4.72	5.78
0040	REPLACE 5/8" DRYWALL	75	2 Carp	S.Y.							
	Set up, secure and take down ladder				.130		3.28		3.28	4.55	5.55
	Remove drywall				.087		1.72	.65	2.37	3.03	3.63
	Replace 5/8" drywall, taped and finished				.194	1.89	4.86		6.75	8.80	10.60
	Total				.411	1.89	9.86	.65	12.40	16.38	19.78

Figure 15.4

6.5-440 Vinyl Wall Covering

System Description	Freq. (Years)	Crew	Unit	Labor Hours	1996 Bare Costs				Total In-House	Total w/O&P
					Material	Labor	Equipment	Total		
0020 REPLACE MEDIUM WEIGHT VINYL WALL COVERING	15	1 Carp	C.S.F.							
Set up and secure scaffold				.880		22.50		22.50	31	38
Remove vinyl wall covering				1.156		27		27	36	44.50
Install vinyl wall covering				2.391	77	56		133	160	188
Remove scaffold				.880		22.50		22.50	31	38
Total				5.307	77	128		205	258	308.50

6.5-450 Wallpaper

System Description	Freq. (Years)	Crew	Unit	Labor Hours	1996 Bare Costs				Total In-House	Total w/O&P
					Material	Labor	Equipment	Total		
0010 REPAIR WALLPAPER - (2% OF WALLS)	8	1 Carp	S.Y.							
Remove damaged wallpaper				.104		2.43		2.43	3.25	4
Repair / replace wallpaper				.175	3.87	4.05		7.92	9.65	11.50
Total				.279	3.87	6.48		10.35	12.90	15.50
0020 REPLACE WALLPAPER	20	1 Carp	S.Y.							
Set up and secure scaffold				.130		3.28		3.28	4.55	5.55
Remove damaged wallpaper				.104		2.43		2.43	3.25	4
Replace wallpaper				.175	3.87	4.05		7.92	9.65	11.50
Remove scaffold				.130		3.28		3.28	4.55	5.55
Total				.539	3.87	13.04		16.91	22	26.60

6.5-510 Plywood Paneling

System Description	Freq. (Years)	Crew	Unit	Labor Hours	1996 Bare Costs				Total In-House	Total w/O&P
					Material	Labor	Equipment	Total		
0010 REPAIR PLYWOOD PANELING - (2% OF WALLS)	10	1 Carp	C.S.F.							
Remove damaged paneling				1.156		23		23	32	39
Replace paneling				4.952	100	125	6	231	289	345
Total				6.108	100	148	6	254	321	384

Figure 15.5

The second of the two examples shown later in this chapter incorporates these four items into an actual estimate. The labor mark-ups are shown in Figure 15.6. The labor mark-up for the SKWK (Skilled Workers Average) has been used in these examples.

The preceding is the checklist for completing an estimate of in-house or contracted work. Each of these items should be addressed in the estimate. At the beginning of this chapter it was stated that the degree of accuracy of an assemblies M&R estimate is $+/-20\%$. This should be taken into account before finalizing the estimate.

The estimates presented in the Sample Estimate section later in this chapter demonstrate the application of these points.

City Cost Factors

The cost data in all Means books are based on national average material, labor, and equipment prices. The data derived from the books can be adjusted to a particular location. This is done using the city cost indexes. The index is a percentage ratio of the desired city cost to the national average cost of the same item at a stated time period. In other words, the index figures represent relative construction factors (or multipliers) for material and installation costs. A sample of the index for San Antonio and other Texas cities is shown in Figure 15.7.

Index figures for both material and installation are based on the 30-city average of 100. The index for each division is computed from representative material and labor quantities for that division. The weighted average for the city is a weighted total of the components listed above it, but does not include productivity differences between trades or cities.

The most common use of the index is to modify national average prices listed in the book to a specific location. Remember, the index is not a fixed number but a ratio. It is a percentage ratio of a building component's cost at any stated time, relative to the national average. Put in the form of an equation:

$$\frac{\text{Index for Specific City}}{100} \times \text{National Average Cost} = \text{Specific City Cost}$$

Historical Cost Factors

Historical cost indexes are used to adjust costs from one year to another. On many occasions an estimate is completed for a project, then the project is abandoned for lack of funds or interest. Then four years later the project is resurrected, and a current estimate is requested. The expedient way to provide this new estimate is by applying the historical cost index to the old estimate. The index is the ratio of cost at any time to cost at a specific time. Means' index is based on January 1, 1993 = 100. The historical cost index is shown in Figure 15.8.

A project in 1992 was estimated to cost $1,000,000. The project was not built at that time. What will the project cost in 1996?

$$\frac{\text{Present Historical Index}}{\text{Former Historical Index}} \times \text{Former Cost} = \text{Present Cost}$$

The answer to this question is:

$$\frac{108.3}{99.4} \times \$1,000,000 = \$1,089,537$$

This index is a valuable tool when an existing estimate needs to be updated.

In-house & Contractor's Overhead & Profit

Below are the **average** percentage mark-ups applied to base labor rates to calculate a typical in-house direct labor cost or contractor's billing rate.

In-House Labor Rate = Hourly Base Rate + Workers' Compensation + Avg. Fixed Overhead.

Contractor's Labor = Hourly Base Rate + Workers' Compensation + Avg. Fixed Overhead + Overhead + Profit.

Column A: Labor rates are based on union wages averaged for 30 major U.S. cities. Base rates including fringe benefits are listed hourly. These figures are the sum of the wage rate and employer-paid fringe benefits such as vacation pay, employer-paid health and welfare costs, pension costs, plus appropriate training and industry advancement funds costs.

Column B: Workers' Compensation rates are the national average of state rates established for each trade.

Column C: Column C lists average fixed overhead figures for all trades. Included are Federal and State Unemployment costs set at 7.3%; Social Security Taxes (FICA) set at 7.65%; Builder's Risk Insurance costs set at 0.34%; and Public Liability costs set at 1.55%. All the percentages except those for Social Security Taxes vary from state to state as well as from company to company.

Columns D and E: Percentage mark-ups in Columns D and E are applied to the mark-ups of columns B and C to calculate the contractor's billing rate. They are based on the presumption that the contractor has annual billing of $500,000 and up. The overhead percentages may increase with smaller annual billing. The overhead percentages may vary greatly among contractors depending on a number of factors, such as the contractor's annual volume, logistical support costs, and staff requirements. The figures for overhead and profit will also vary depending on the type of work, the job location and prevailing economic conditions.

Column F: Column F lists the total in-house markup of Columns B plus C.

Column G: Column G is Column A (hourly base labor rate) multiplied by the percentage in Column F (in-house overhead percentage).

Column H: Column H is the total of Column A (hourly base labor rate) plus Column G (in-house overhead).

Column I: Column I lists the total contractor's mark-up of Columns B, C, D, and E.

Column J: Column J is Column A (hourly base labor rate) multiplied by the percentage in Column I (contractor's O&P).

Column K: Column K is total of Column A (hourly base labor rate) plus Column J (total contractor's O&P).

Abbr.	Trade	A — Base Rate Incl. Fringes Hourly	B — Workers' Comp. Ins.	C — Average Fixed Overhead	D — Overhead	E — Profit	F — Rate with In-House Mark-Up %	G — Amount	H — Hourly	I — Rate with Contractor's O&P %	J — Amount	K — Hourly
Skwk	Skilled Workers Average (35 trades)	$25.95	20.2%	16.8%	16%	15%	37.0	$ 9.60	$35.55	68.0	$17.65	$43.60
	Helpers Average (5 trades)	19.25	21.4				38.2	7.35	26.60	69.2	13.30	32.55
	Foreman Average, Inside ($.50 over trade)	26.45	20.2				37.0	9.80	36.25	68.0	18.00	44.45
	Foreman Average, Outside ($2.00 over trade)	27.95	20.2				37.0	10.35	38.30	68.0	19.00	46.95
Clab	Common Building Laborers	19.80	21.9				38.7	7.65	27.45	69.7	13.80	33.60
Clam	Common Maintenance Laborer	14.85	7.4				24.2	3.60	18.45	55.2	8.20	23.05
Asbe	Asbestos Workers	28.55	19.7				36.5	10.40	38.95	67.5	19.25	47.80
Boil	Boilermakers	30.05	17.7				34.5	10.35	40.40	65.5	19.70	49.75
Bric	Bricklayers	25.90	19.4				36.2	9.40	35.30	67.2	17.40	43.30
Brhe	Bricklayer Helpers	20.00	19.4				36.2	7.25	27.25	67.2	13.45	33.45
Carp	Carpenters	25.20	21.9				38.7	9.75	34.95	69.7	17.55	42.75
Cefi	Cement Finishers	24.35	12.8				29.6	7.20	31.55	60.6	14.75	39.10
Elec	Electricians	29.30	8.0				24.8	7.25	36.55	55.8	16.35	45.65
Elev	Elevator Constructors	30.05	9.6				26.4	7.95	38.00	57.4	17.25	47.30
Eqhv	Equipment Operators, Crane or Shovel	26.75	12.9				29.7	7.95	34.70	60.7	16.25	43.00
Eqmd	Equipment Operators, Medium Equipment	25.70	12.9				29.7	7.65	33.35	60.7	15.60	41.30
Eqlt	Equipment Operators, Light Equipment	24.70	12.9				29.7	7.35	32.05	60.7	15.00	39.70
Eqol	Equipment Operators, Oilers	21.90	12.9				29.7	6.50	28.40	60.7	13.30	35.20
Eqmm	Equipment Operators, Master Mechanics	27.55	12.9				29.7	8.20	35.75	60.7	16.70	44.25
Glaz	Glaziers	24.90	16.0				32.8	8.15	33.05	63.8	15.90	40.80
Lath	Lathers	24.95	13.5				30.3	7.55	32.50	61.3	15.30	40.25
Marb	Marble Setters	25.65	19.4				36.2	9.30	34.95	67.2	17.25	42.90
Mill	Millwrights	26.55	13.2				30.0	7.95	34.50	61.0	16.20	42.75
Mstz	Mosaic & Terrazzo Workers	25.25	11.0				27.8	7.00	32.25	58.8	14.85	40.10
Pord	Painters, Ordinary	22.95	16.8				33.6	7.70	30.65	64.6	14.85	37.80
Psst	Painters, Structural Steel	23.95	62.5				79.3	19.00	42.95	110.3	26.40	50.35
Pape	Paper Hangers	23.30	16.8				33.6	7.85	31.15	64.6	15.05	38.35
Pile	Pile Drivers	25.35	33.6				50.4	12.80	38.15	81.4	20.65	46.00
Plas	Plasterers	24.20	17.4				34.2	8.30	32.50	65.2	15.80	40.00
Plah	Plasterer Helpers	20.15	17.4				34.2	6.90	27.05	65.2	13.15	33.30
Plum	Plumbers	30.05	10.2				27.0	8.10	38.15	58.0	17.45	47.50
Rodm	Rodmen (Reinforcing)	27.75	36.3				53.1	14.75	42.50	84.1	23.35	51.10
Rofc	Roofers, Composition	22.55	37.4				54.2	12.20	34.75	85.2	19.20	41.75
Rots	Roofers, Tile & Slate	22.60	37.4				54.2	12.25	34.85	85.2	19.25	41.85
Rohe	Roofers, Helpers (Composition)	16.96	37.4				54.2	8.65	24.60	85.2	13.60	29.55
Shee	Sheet Metal Workers	28.95	13.8				30.6	8.85	37.80	61.6	17.85	46.80
Spri	Sprinkler Installers	31.30	10.4				27.2	8.50	39.80	58.2	18.20	49.50
Stpi	Steamfitters or Pipefitters	30.30	10.2				27.0	8.20	38.50	58.0	17.55	47.85
Ston	Stone Masons	25.90	19.4				36.2	9.40	35.30	67.2	17.40	43.30
Sswk	Structural Steel Workers	27.85	46.4				63.2	17.60	45.45	94.2	26.25	54.10
Tilf	Tile Layers	25.05	11.0				27.8	6.95	32.00	58.8	14.75	39.80
Tilh	Tile Layers Helpers	20.30	11.0				27.8	5.65	25.95	58.8	11.95	32.25
Trlt	Truck Drivers, Light	20.35	17.0				33.8	6.90	27.25	64.8	13.20	33.55
Trhv	Truck Drivers, Heavy	20.70	17.0				33.8	7.00	27.70	64.8	13.40	34.10
Sswl	Welders, Structural Steel	27.85	46.4				63.2	17.60	45.45	94.2	26.25	54.10
Wrck	*Wrecking	19.80	44.8				61.6	12.20	32.00	92.6	18.35	38.15

*Not included in Averages.

Figure 15.6

TENNESSEE

DIV. NO.	BUILDING SYSTEMS	CHATTANOOGA			JACKSON			JOHNSON CITY			KNOXVILLE			MEMPHIS		
		MAT.	INST.	TOTAL	MAT.	INST.	TOTAL	MAT.	INST.	TOTAL	MAT.	INST.	TOTAL	MAT.	INST.	TOTAL
1-2	FOUND/SUBSTRUCTURES	99.1	71.1	81.2	98.5	59.9	73.8	87.1	70.8	76.6	93.3	69.2	77.9	96.3	75.5	83.0
3	SUPERSTRUCTURES	96.8	76.5	87.7	95.1	67.1	82.5	90.9	77.9	85.1	96.0	77.1	87.6	96.4	81.4	89.7
4	EXTERIOR CLOSURE	86.3	61.2	74.3	94.1	35.4	66.0	103.0	61.8	83.3	75.9	61.8	69.2	83.3	67.5	75.8
5	ROOFING	101.2	60.5	83.7	100.6	45.2	76.8	95.7	61.3	80.9	93.8	60.5	79.5	100.5	65.2	85.3
6	INTERIOR CONSTRUCTION	98.6	61.9	83.6	97.5	43.5	75.4	95.5	65.6	83.3	90.9	65.6	80.6	98.3	65.1	84.7
7	CONVEYING	100.0	74.4	92.8	100.0	70.5	91.7	100.0	71.4	91.9	100.0	71.4	91.9	100.0	76.2	93.3
8	MECHANICAL	100.0	58.8	81.8	100.1	51.7	78.7	99.9	62.4	83.3	99.9	62.4	83.3	100.0	68.8	86.2
9	ELECTRICAL	105.2	76.0	85.7	100.6	43.5	62.5	92.4	46.5	61.8	102.1	68.0	79.3	100.6	80.4	87.1
11	EQUIPMENT	100.0	63.1	97.6	100.0	46.0	96.5	100.0	65.1	97.7	100.0	66.2	97.8	100.0	67.0	97.9
12	SITE WORK	110.0	96.0	99.5	110.1	93.8	97.9	121.7	84.7	94.0	98.2	84.7	88.1	108.1	96.7	99.6
1-12	WEIGHTED AVERAGE	98.0	69.6	84.3	98.1	53.9	76.7	96.9	65.4	81.7	94.1	68.6	81.8	97.0	74.9	86.3

TENNESSEE / TEXAS

DIV. NO.	BUILDING SYSTEMS	NASHVILLE			ABILENE			AMARILLO			AUSTIN			BEAUMONT		
		MAT.	INST.	TOTAL	MAT.	INST.	TOTAL	MAT.	INST.	TOTAL	MAT.	INST.	TOTAL	MAT.	INST.	TOTAL
1-2	FOUND/SUBSTRUCTURES	88.4	73.1	78.6	96.7	64.6	76.2	98.6	68.9	79.6	86.9	72.8	77.9	95.0	77.0	83.5
3	SUPERSTRUCTURES	95.7	80.4	88.8	97.3	66.9	83.7	98.0	68.2	84.7	96.7	74.1	86.6	97.7	76.7	88.3
4	EXTERIOR CLOSURE	78.9	66.9	73.2	88.7	61.0	75.4	90.3	58.1	74.9	85.5	66.9	76.6	92.3	77.3	85.1
5	ROOFING	93.9	64.9	81.4	91.5	64.6	80.0	94.3	59.9	79.5	87.9	67.2	79.0	92.7	73.6	84.5
6	INTERIOR CONSTRUCTION	95.3	66.3	83.5	98.8	62.5	84.0	98.8	63.2	84.2	95.4	69.1	84.6	103.0	79.4	93.4
7	CONVEYING	100.0	75.8	93.2	100.0	72.3	92.2	100.0	71.3	91.9	100.0	70.9	91.8	100.0	76.0	93.2
8	MECHANICAL	99.9	67.5	85.6	100.0	50.9	78.3	100.0	61.5	83.0	100.1	66.9	85.4	100.2	70.8	87.2
9	ELECTRICAL	103.5	67.9	79.7	97.0	55.3	69.2	97.0	63.5	74.6	95.6	72.8	80.4	97.3	83.0	87.7
11	EQUIPMENT	100.0	66.7	97.8	100.0	60.6	97.4	100.0	67.9	97.9	100.0	73.4	98.3	100.0	81.8	98.8
12	SITE WORK	97.0	92.0	93.2	102.8	82.6	87.6	102.0	83.7	88.3	94.7	83.5	86.3	103.9	84.1	89.1
1-12	WEIGHTED AVERAGE	95.1	71.9	83.9	97.3	61.6	80.1	97.8	65.3	82.1	95.3	71.3	83.7	98.7	77.6	88.5

TEXAS

DIV. NO.	BUILDING SYSTEMS	CORPUS CHRISTI			DALLAS			EL PASO			FORT WORTH			HOUSTON		
		MAT.	INST.	TOTAL	MAT.	INST.	TOTAL	MAT.	INST.	TOTAL	MAT.	INST.	TOTAL	MAT.	INST.	TOTAL
1-2	FOUND/SUBSTRUCTURES	102.4	64.1	77.9	100.4	72.0	82.2	98.2	61.1	74.5	95.4	70.6	79.6	100.9	77.8	86.1
3	SUPERSTRUCTURES	98.7	75.8	88.4	95.8	67.3	91.3	97.5	63.3	82.2	97.1	71.4	85.6	100.2	87.4	94.5
4	EXTERIOR CLOSURE	84.9	59.0	72.5	85.9	67.3	77.0	89.0	64.0	77.1	90.1	64.4	77.8	89.7	77.2	83.8
5	ROOFING	89.9	61.5	77.7	89.7	71.6	81.9	91.5	59.2	77.6	92.3	67.8	81.7	86.1	76.5	82.0
6	INTERIOR CONSTRUCTION	102.5	58.0	84.3	102.3	68.9	88.7	98.8	50.0	78.9	101.3	69.1	88.2	107.2	76.7	94.7
7	CONVEYING	100.0	73.2	92.4	100.0	75.3	93.0	100.0	73.7	92.6	100.0	74.5	92.8	100.0	77.7	93.7
8	MECHANICAL	100.1	51.3	78.5	99.8	68.2	85.8	100.0	54.5	79.9	100.0	64.6	84.3	100.1	75.7	89.3
9	ELECTRICAL	92.5	69.3	77.0	98.6	73.9	82.1	97.0	63.7	74.8	97.0	69.6	78.8	96.3	80.0	85.4
11	EQUIPMENT	100.0	58.6	97.3	100.0	70.4	98.1	100.0	45.5	96.5	100.0	69.9	98.0	100.0	81.8	98.8
12	SITE WORK	128.3	78.1	90.7	125.5	78.5	90.3	103.2	83.2	88.2	102.2	82.8	87.7	130.3	80.9	93.3
1-12	WEIGHTED AVERAGE	98.3	64.1	81.8	98.6	72.9	86.2	97.5	61.2	79.9	97.9	69.4	84.1	100.3	79.3	90.2

TEXAS

DIV. NO.	BUILDING SYSTEMS	LAREDO			LUBBOCK			ODESSA			SAN ANTONIO			WACO		
		MAT.	INST.	TOTAL	MAT.	INST.	TOTAL	MAT.	INST.	TOTAL	MAT.	INST.	TOTAL	MAT.	INST.	TOTAL
1-2	FOUND/SUBSTRUCTURES	82.0	66.4	72.0	102.0	67.8	80.2	97.1	65.8	77.1	81.2	71.5	75.0	91.2	68.5	76.7
3	SUPERSTRUCTURES	95.3	66.9	82.6	100.8	78.3	90.7	97.5	67.0	83.9	95.6	70.8	84.5	96.2	70.3	84.6
4	EXTERIOR CLOSURE	89.9	59.3	75.3	89.9	56.1	73.8	88.7	54.5	72.4	89.8	63.1	77.1	90.9	64.0	78.1
5	ROOFING	86.6	62.3	76.1	86.0	65.3	77.1	91.5	58.3	77.3	86.6	67.3	78.3	92.3	63.9	80.1
6	INTERIOR CONSTRUCTION	95.0	57.6	79.7	105.6	60.0	87.0	98.8	57.9	82.1	95.0	63.2	82.0	101.3	56.1	82.8
7	CONVEYING	100.0	70.7	91.7	100.0	72.8	92.3	100.0	71.0	91.8	100.0	72.6	92.2	100.0	74.5	92.8
8	MECHANICAL	100.0	60.4	82.5	99.9	55.7	80.4	100.0	54.7	80.0	100.0	73.3	88.2	100.0	63.1	83.7
9	ELECTRICAL	95.6	71.1	79.3	95.5	62.6	73.6	97.0	62.7	74.1	95.6	71.1	79.3	97.0	63.6	74.8
11	EQUIPMENT	100.0	58.5	97.3	100.0	64.0	97.7	100.0	60.9	97.5	100.0	65.5	97.8	100.0	57.0	97.2
12	SITE WORK	94.8	83.3	86.2	129.3	80.2	92.6	102.8	83.1	88.1	94.6	87.1	89.0	101.4	82.8	87.5
1-12	WEIGHTED AVERAGE	95.2	65.2	80.7	100.0	64.8	83.0	97.3	62.1	80.3	95.2	70.5	83.3	97.6	65.7	82.2

TEXAS / UTAH

DIV. NO.	BUILDING SYSTEMS	WICHITA FALLS			LOGAN			OGDEN			PROVO			SALT LAKE CITY		
		MAT.	INST.	TOTAL	MAT.	INST.	TOTAL	MAT.	INST.	TOTAL	MAT.	INST.	TOTAL	MAT.	INST.	TOTAL
1-2	FOUND/SUBSTRUCTURES	92.3	67.6	76.5	94.4	77.9	83.9	93.4	77.9	83.5	95.4	77.5	83.9	96.5	77.9	84.6
3	SUPERSTRUCTURES	96.4	69.0	84.1	101.7	74.2	89.4	101.9	74.2	89.5	100.3	74.3	88.7	102.0	74.3	89.6
4	EXTERIOR CLOSURE	91.0	63.4	77.8	120.7	67.9	95.5	107.3	67.9	88.5	122.1	67.9	96.2	113.2	67.9	91.6
5	ROOFING	92.3	64.6	80.3	104.8	73.4	91.3	103.6	73.4	90.6	106.9	73.4	92.5	105.5	73.4	91.7
6	INTERIOR CONSTRUCTION	101.3	65.3	86.6	91.9	63.8	80.4	91.4	63.8	80.1	93.7	64.0	81.6	91.4	64.0	80.2
7	CONVEYING	100.0	70.1	91.6	100.0	81.5	94.8	100.0	81.5	94.8	100.0	81.5	94.8	100.0	81.5	94.8
8	MECHANICAL	100.0	59.5	82.1	99.8	72.5	87.8	99.8	72.5	87.8	99.8	72.5	87.8	99.8	72.5	87.8
9	ELECTRICAL	101.8	64.0	76.6	97.2	71.9	80.3	97.2	71.9	80.3	97.6	76.2	83.3	98.0	76.2	83.4
11	EQUIPMENT	100.0	63.7	97.6	100.0	64.7	97.7	100.0	64.7	97.7	100.0	64.9	97.7	100.0	64.9	97.7
12	SITE WORK	101.7	82.6	87.4	101.9	97.3	98.4	88.9	97.3	95.2	98.3	95.3	96.1	88.3	97.2	95.0
1-12	WEIGHTED AVERAGE	98.0	66.0	82.5	100.9	73.4	87.7	98.8	73.4	86.6	101.3	74.0	88.1	99.8	74.2	87.4

VERMONT / VIRGINIA

DIV. NO.	BUILDING SYSTEMS	BURLINGTON			RUTLAND			ALEXANDRIA			ARLINGTON			NEWPORT NEWS		
		MAT.	INST.	TOTAL	MAT.	INST.	TOTAL	MAT.	INST.	TOTAL	MAT.	INST.	TOTAL	MAT.	INST.	TOTAL
1-2	FOUND/SUBSTRUCTURES	103.8	78.8	87.8	103.4	78.8	87.6	97.5	81.5	87.3	98.6	80.4	86.9	98.9	69.3	80.0
3	SUPERSTRUCTURES	104.7	71.1	89.7	105.5	71.1	90.1	95.8	90.1	93.3	94.7	89.2	92.3	96.1	79.7	88.7
4	EXTERIOR CLOSURE	109.3	55.8	83.8	105.2	55.8	81.6	94.9	76.6	86.2	106.9	75.6	92.0	95.1	58.3	77.5
5	ROOFING	99.3	67.7	85.7	99.3	67.7	85.7	93.5	83.1	89.1	94.1	82.2	89.0	93.6	55.8	77.3
6	INTERIOR CONSTRUCTION	101.3	63.0	85.7	101.3	63.0	85.6	95.0	80.3	89.0	93.8	79.6	88.0	94.6	59.6	80.3
7	CONVEYING	100.0	75.3	93.0	100.0	75.3	93.0	100.0	91.4	97.6	100.0	90.9	97.4	100.0	82.8	95.1
8	MECHANICAL	100.0	70.5	86.9	100.0	70.5	86.9	100.1	83.5	92.8	100.1	80.9	91.6	100.1	63.0	83.7
9	ELECTRICAL	106.7	55.9	72.8	106.7	55.9	72.8	98.6	93.5	95.2	95.1	91.4	92.7	98.6	65.3	76.4
11	EQUIPMENT	100.0	64.1	97.7	100.0	64.1	97.7	100.0	78.7	98.6	100.0	77.5	98.5	100.0	61.5	97.5
12	SITE WORK	96.4	101.4	100.2	96.3	101.4	100.1	123.6	83.1	93.2	132.4	82.7	95.2	110.7	83.8	90.5
1-12	WEIGHTED AVERAGE	102.9	68.5	86.2	102.5	68.5	86.0	97.8	84.9	91.5	98.8	83.5	91.4	97.5	67.5	83.0

Figure 15.7

City Adjustments

City to City: To convert known or estimated costs from one city to another, cost indexes can be used as follows:

$$\text{Unknown City Cost} = \text{Known Cost} \times \frac{\text{Unknown City Index}}{\text{Known City Index}}$$

For example: If the building cost in Boston, Mass., is $1,500,000, how much would a duplicated building cost in Los Angeles, Calif.?

$$\text{L.A. Cost} = \$1,500,000 \times \frac{\text{(Los Angeles) } 113.3}{\text{(Boston) } 118.8} = \$1,430,500$$

The table below lists both the Means City Cost Index based on January 1, 1993 = 100 as well as the computed value of an index based on January 1, 1996 costs. Since the January 1, 1996 figure is estimated, space is left to write in the actual index figures as they become available thru either the quarterly "Means Construction Cost Indexes" or as printed in the "Engineering News-Record." To compute the actual index based on January 1, 1996 = 100, divide the Quarterly City Cost Index for a particular year by the actual January 1, 1996 Quarterly City Cost Index. Space has been left to advance the index figures as the year progresses.

Historical Cost Indexes

Year	Historical Cost Index Jan. 1, 1993 = 100 Est.	Actual	Current Index Based on Jan. 1, 1996 = 100 Est.	Actual	Year	Historical Cost Index Jan. 1, 1993 = 100 Actual	Current Index Based on Jan. 1, 1996 = 100 Est.	Actual	Year	Historical Cost Index Jan. 1, 1993 = 100 Actual	Current Index Based on Jan. 1, 1996 = 100 Est.	Actual
Oct 1996					July 1981	70.0	64.6		July 1963	20.7	19.1	
July 1996					1980	62.9	58.1		1962	20.2	18.7	
April 1996					1979	57.8	53.4		1961	19.8	18.3	
Jan 1996	108.3		100.0	100.0	1978	53.5	49.4		1960	19.7	18.2	
July 1995		107.6	99.4		1977	49.5	45.7		1959	19.3	17.8	
1994		104.4	96.4		1976	46.9	43.3		1958	18.8	17.4	
1993		101.7	93.9		1975	44.8	41.4		1957	18.4	17.0	
1992		99.4	91.8		1974	41.4	38.2		1956	17.6	16.3	
1991		96.8	89.4		1973	37.7	34.8		1955	16.6	15.3	
1990		94.3	87.1		1972	34.8	32.1		1954	16.0	14.8	
1989		92.1	85.1		1971	32.1	29.6		1953	15.8	14.6	
1988		89.9	83.0		1970	28.7	26.5		1952	15.4	14.2	
1987		87.7	81.0		1969	26.9	24.8		1951	15.0	13.9	
1986		84.2	77.8		1968	24.9	23.0		1950	13.7	12.7	
1985		82.6	76.3		1967	23.5	21.7		1949	13.3	12.3	
1984		82.0	75.7		1966	22.7	21.0		1948	13.3	12.3	
1983		80.2	74.0		1965	21.7	20.0		1947	12.1	11.2	
▼ 1982		76.1	70.3		▼ 1964	21.2	19.6		▼ 1946	10.1	9.3	

To find the **current cost** from a project built previously in either the same city or a different city, the following formula is used:

$$\text{Present Cost (City X)} = \frac{\text{Present Index (City X)}}{\text{Former Index (City Y)}} \times \text{Former Cost (City Y)}$$

For example: Find the construction cost of a building to be built in San Francisco, CA, as of January 1, 1996 when the identical building cost $500,000 in Boston on July 1, 1968.

To Project Future Construction Costs: Using the results of the last five years average percentage increase as a basis, an average increase of 2.5% could be used.

The historical index figures above are compiled from the Means Construction Index Service.

$$\text{Jan. 1, 1996 (San Francisco)} = \frac{\text{(San Francisco) } 108.3 \times 126.3}{\text{(Boston) } 24.9 \times 118.8} = \$500,000 = \$2,312,000$$

Figure 15.8

Sample Estimates

Estimate No. 1

The maintenance manager at a San Antonio, Texas facility has to estimate the removal and replacement of 1,000 S.F. of fire-rated gypsum drywall that is comprised of two layers of 5/8" thick gypsum. The work will be done by in-house tradesmen. Two carpenters will be assigned to the project. The new drywall will be covered with vinyl wall covering. The metal studs and other drywall face are undamaged. Assembly 6.5-230-0040, Figure 15.4, is used to price the drywall removal and replacement. This assembly prices the removal and replacement of one layer. Since the removal and replacement of two layers is required, multiply the cost times two. Refer to assembly 6.5-440-0020, Figure 15.5, for the price to install vinyl wall covering. (Note that the unit of measure varies among the assemblies; therefore, it is always necessary to check the units.) The facility has an additional overhead expense for administration and vehicles of 30%. The travel time to the job site from the maintenance shop is one hour round-trip. The sales tax is 6.25%. See Figure 15.9, the Cost Analysis Sheet, for this estimate.

Estimate No.2

Figure 15.10 depicts the ground floor of a 100 ft. x 100 ft. 6-story office building located in a suburban office park. A pipe joint in the fire sprinkler distribution system failed in the concealed space above the first floor ceiling, causing extensive water damage before the system's main valve could be closed. The repair work must be scheduled and the facility manager needs to estimate its cost, as performed by an outside contractor. An inspection or audit of the office areas revealed the following damage:

Carpet: Water flooded the ground floor—primarily in Suite B, but some water also soaked into Suite A. A decision has been made to replace the carpet in all the individual offices (nominal size is 12' x 12') and the central common area of Suite B, and just the three individual offices and corridor area in Suite A.

Walls: All the walls in the carpet-soaked area also soaked up water and must have the 5/8" drywall removed and replaced on both sides of interior partitions and the inside of exterior walls. The dividing wall between Suites A and B is a rated wall with 2 layers of 5/8" drywall. The metal studs and track will remain for all partitions.

Painting: All the new drywall will need to be painted, along with the fourth wall that does not require new drywall in some individual offices. The new paint can terminate at an inside or outside corner in the central common office areas in both Suites A and B.

Rubber Base: The rubber cove base on water-soaked walls will obviously be thrown out with the wet drywall and must be replaced. Because the old color cannot be matched exactly, the base in individual offices must be removed and replaced, including on the fourth wall that requires no new drywall. The old and new base can meet at an inside or outside corner in the central common office areas of both Suites A and B.

Doors: All six solid wood core doors and their frames must be removed from the water-soaked walls in order to remove and replace the drywall efficiently. Four of the doors and frames can be reused, but two doors have warped and delaminated and must be replaced. The old doors will not need to be refinished, but the new doors and frames must be painted.

Ceilings: The damage to the suspended acoustical ceilings was limited primarily to one individual office in Suite B, with some collateral damage to two adjoining areas. Although the other remaining ceilings are in good condition, they must be removed and replaced in all the individual offices

COST ANALYSIS

PROJECT: Repair Estimate - Replace Gypsum Drywall   CLASSIFICATION: Interior Construction - Wall Finishes   SHEET NO: 1 of 1

LOCATION: San Antonio, Texas   ARCHITECT: ESTIMATE NO: 1501

TAKE OFF BY: PRW   QUANTITIES BY: PRW   PRICES BY: SCP   EXTENSIONS BY: SCP   DATE: February 1996   CHECKED BY: PRW

DESCRIPTION	SOURCE/DIMENSIONS				QUANTITY	UNIT	MATERIAL COST	MATERIAL TOTAL	LABOR COST	LABOR TOTAL	EQUIPMENT COST	EQUIPMENT TOTAL	SUBCONTRACT COST	SUBCONTRACT TOTAL	TOTAL COST	TOTAL TOTAL
Drywall																
Remove & replace 1st layer	FM&R	6.5	230	0040	111	S.Y.	1.89	210	9.86	1,094	0.65	72				
Remove & replace 2nd layer	FM&R	6.5	230	0040	111	S.Y.	1.89	210	9.86	1,094	0.65	72				
Vinyl wall covering																
Set up and secure scaffold	FM&R	6.5	440	0020	10	C.S.F.			22.50	225						
Install vinyl wall covering	FM&R	6.5	440	0020	10	C.S.F.	77.00	770	56.00	560						
Remove scaffold	FM&R	6.5	440	0020	10	C.S.F.			22.50	225						
Travel time (1 hour round trip)																
Drywall (2 men x 3 trips)					3	Trips			69.80	209						
Vinyl wall covering (1 man x 5 trips)					5	Trips			34.90	175						
TOTAL BARE COSTS								1,190		3,583		144				
Markups for in-house work:																
Materials (6.25% for Texas sales tax)								74								
Labor (38.7% for In-House Carpenters)										1,387						
Additional administrative overhead (30%)										1,075						
Equipment (0%)												0				
SUBTOTALS								1,264		6,044		144				
City cost adjustments (for San Antonio, TX):																
Materials (95.0%)								x 0.95								
Labor (63.2%)										x 0.632						
TOTALS								1,201		3,820		144				5,165

Figure 15.9

248

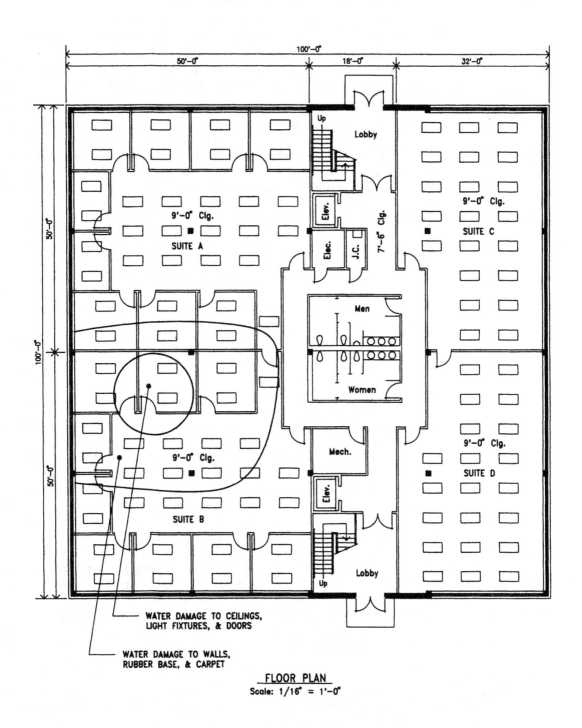

WATER DAMAGE TO CEILINGS,
LIGHT FIXTURES, & DOORS

WATER DAMAGE TO WALLS,
RUBBER BASE, & CARPET

FLOOR PLAN
Scale: 1/16" = 1'-0"

Figure 15.10

and in some of the common areas that sustained water damage to the walls. This must be done in order to remove and replace the drywall efficiently. Some old ceiling materials will need to be salvaged to be used in blending in at the new drywall with the remaining ceilings in the central common areas in both Suites A and B.

Light Fixtures: Only the two light fixtures in the individual office in Suite B sustained enough water damage to require removal and replacement with new fixtures. All the others in the individual affected offices and some common areas will need to be removed, along with the suspended ceilings, saved, and then reinstalled in their previous locations after the new suspended ceiling grid is in place.

Demolition: It has been determined that a 30 C.Y. open-top construction dumpster can be located outside one of the windows of Suite B, and that the window glass can be removed daily in order to keep the hauling of the debris to a distance no greater than approximately 50 feet. It is estimated that the dumpster will be emptied about three times in the course of the project.

Cleanup: After all the repairs are completed, the entire carpeting in both suites must be vacuumed. In addition, all the interior wall surfaces will need to be dusted, the 5' high continuous band of windows will need to be washed inside, and all the window sills will need to be dusted.

All these repair activities, as well as their units of measure and quantities, have been listed in Figure 15.11, a Cost Analysis form. The source of cost data, whether from *Means Facilities Maintenance & Repair Cost Data* (FM&R) or *Means Facilities Construction Cost Data* (FCCD), was then entered onto the form, along with the line number for each repair activity. These sources provided the bare material, labor, and equipment unit costs for each activity, which were also entered. (Alternatively, the contractor's or facility manager's own historical data could be used to calculate these bare unit costs.) Finally, the total material, labor, and equipment costs were calculated for each activity by multiplying the quantity of units times the unit cost. The remainder of the estimate is extended similarly. The summation of each column is the bare material, labor, and equipment cost for the project. Now that the totals of the bare costs have been calculated, it is necessary to add the labor markup plus indirect costs. These are usually expressed as a percentage of the bare costs. The facility, if using in-house labor for this project, would add 0% to the material and equipment, and 37% (Workers' Comp + FICA + Medicare + Federal & State Unemployment + Insurance) to the labor, plus a markup for overhead as appropriate. (Note: Because various tradesmen will be used for this project, the mark-ups for Skilled Workers Average will be used as shown in Column F on the "Installing Contractor's Overhead & Profit" chart in Figure 15.6). A contractor would add 25% to bare material cost, 68% to labor, and 10% to equipment as shown on the estimate.

Conclusion

Maintenance and repair work is a major, and sometimes difficult to estimate, component in the overall maintenance budget. An organization's own historical data can serve as a basis for future projections, supported by objective cost data such as published by R.S. Means, Co. Predictive and preventive maintenance play a key role in controlling repair costs, as shown in the following chapter.

COST ANALYSIS

PROJECT: Repair water damage
CLASSIFICATION: XYZ Office Building, 100'x100', 6-Story, Ground Floor
LOCATION: Anytown, USA
ARCHITECT: As shown
TAKE OFF BY: SCP QUANTITIES BY: SCP PRICES BY: SCP EXTENSIONS BY: SCP

DESCRIPTION	SOURCE/DIMENSIONS				QUANTITY	UNIT	MATERIAL COST	MATERIAL TOTAL	LABOR COST	LABOR TOTAL	EQUIPMENT COST	EQUIPMENT TOTAL	SUBCONTRACT COST	SUBCONTRACT TOTAL	TOTAL COST	TOTAL TOTAL
Demolition																
30 CY dumpster	FCCD	020	620	0800	3	Wks					330.00	990				
50-foot haul, out window	FCCD	020	620	2040	90	CY			13.20	1,188						
Rubber cove base - replace	FMR	6.2	160	0010	468	LF	0.56	262	1.18	552						
Solid core wood door & frame																
Replace with new	FMR	6.4	420	1030	2	EA	174.00	348	55.80	112	10.72	21				
Reuse - remove & replace	FMR	6.4	420	1030	4	EA			55.80	223						
Refinish door & frame	FMR	6.4	420	1020	2	EA	2.84	6	19.90	40						
Drywall																
Replace (1 side)	FMR	6.5	230	0040	454	SY	1.89	858	9.86	4,476	0.65	295				
Add 2nd layer (1 side)	FMR	6.5	230	0040	107	SY	1.89	202	9.86	1,055	0.65	70				
Refinish drywall																
New	FMR	6.5	230	0030	454	SY	0.36	163	3.24	1,471						
Existing	FMR	6.5	230	0030	54	SY	0.36	19	3.24	175						
Acoustical ceiling (2'x4')																
Remove grid & tiles	FCCD	020	702	1250	1,500	SF			0.26	390						
New grid & tiles	FCCD	095	106	0810	1,500	SF	1.21	1,815	0.53	795						
Replace carpet	FMR	6.6	910	0020	334	SY	7.55	2,522	6.46	2,158						
Light fixtures (2'x4' recessed fluorescent)																
Replace with new	FMR	9.6	130	0030	2	EA	58.00	116	58.74	117						
Reuse - remove & replace	FMR	9.6	130	0030	16	EA			58.74	940						
Cleaning																
Vacuum carpet, 16" upright vacuum	FMR	018	510	0100	5	MSF			4.75	24	1.00	5				
Dust walls, backpack vacuum	FMR	018	540	0060	10	MSF			2.64	27						
Wash windows, over 12 SF each	FMR	018	560	0030	1	MSF			9.30	9						
Dust windowsill	FMR	018	560	0060	200	LF			0.01	2						
TOTAL BARE COSTS								6,312		13,755		1,381				
Markups:																
Materials (25%)								1,578								
Labor (68%, average of 35 trades)										9,353						
Equipment (10%)												138				
Sales tax (5% on materials)								316								
SUBTOTALS								8,205		23,108		1,519				
City Cost Adjustment																
TOTALS																32,832

Figure 15.11

Chapter 16

Preventive & Predictive Maintenance Estimating

Phillip R. Waier, P.E., William R. Steele, and Stephen C. Plotner

Two generally accepted programs for prolonging the life of equipment and avoiding emergency breakdowns are Preventive Maintenance (PM) and Predictive Maintenance. Predictive Maintenance is also referred to as *condition monitoring*. Each program has its advocates, and the programs are not mutually exclusive. Many PM programs have failed because they are too costly to maintain, or it has taken too long to show the economic benefits. Savings can be realized in some areas by converting from preventive maintenance to predictive. Predictive maintenance is a higher-tech approach focused on identifying specific characteristics that signal imminent failure through condition monitoring. Using the tools of predictive maintenance such as vibration analysis, the maintenance department is often able to reduce the frequency of inspection and thus realize savings. This chapter includes a description of each program and an example for estimating the cost of preventive maintenance.

Preventive Maintenance

The objective of Preventive Maintenance is to prevent premature failure of facility building envelopes, interiors, components, systems and equipment and, at the same time, to increase productivity of company operations and to lower the maintenance costs of doing business. PM includes periodic facility and equipment inspections to discover conditions which may lead to breakdown or failure. Certain PM activities also include minor repairs or replacement of minor building or equipment components to ensure the upkeep of the plant or facility by correcting defects while they are still in a minor stage. The installation of a PM program is an investment that will require total commitment from company owners, executives, managers and facility maintenance staff, as well as the budgetary resources necessary to implement the program. Some of the major returns on the investment of instituting a PM program include:

- Less production down time
- Less overtime pay for personnel for emergency maintenance situations
- Fewer large scale and repetitive repairs
- Lower repair cost for simple repairs discovered before major breakdowns
- Better quality control and fewer product rejections
- Less standby equipment needed

- Lower unit cost of manufacture
- Better spare parts control
- Decreased breakdown maintenance
- Fewer safety hazards
- Increased life expectancy of buildings and equipment
- Fewer breakdowns due to a systematic lubrication schedule
- Reduction of power requirements and utility expenses

Unfortunately, some building owners and managers are not cognizant of these and other benefits of an effectively planned and implemented PM program. Instead they hope that breakdowns and failures will not occur. Sometimes, owners and managers think that warranties will take care of various building components and equipment, not realizing that warranties require PM during the term of the warranty. Sometimes it is difficult to sell a PM program to the owners, executives or managers of an organization because, although breakdowns and failures are always more expensive to repair or correct than the performance of a PM program, they are reluctant to allocate sufficient funds for preventive maintenance unless there are obvious visible signs of imminent and costly failure.

Establishing and implementing a PM program is beneficial to any size organization, company or facility, whether it is a large complex manufacturing plant with a sizable in-house facility maintenance staff that is well trained and equipped, a simple facility occupied and/or managed by the owner who will contract out most or all PM activities, or even a retail facility that is part of a corporate chain that has a corporate facility and/or maintenance department which will perform PM activities and/or oversee activities that are contracted out.

Much has been written about preventive maintenance programs. While it is not the purpose of this chapter to explain these programs in depth, some of the more important factors will be highlighted. Regardless of the size or complexity of the facility, there are general steps to be employed when planning and estimating an effective preventive maintenance program. These steps will include identifying PM needs, developing a PM program, scheduling PM functions, simplifying paperwork and information storage, and identifying costs. For illustrative purposes only, the chapter will examine the preventive maintenance requirements of a medium-sized facility with no in-house maintenance staff.

Identifying Preventive Maintenance Needs

If the maintenance manager were to thoroughly examine each element of a facility, a PM activity could be derived which, if implemented, would enhance the performance of that element and, therefore, improve the overall operation of that facility. However, not all of these enhancements would provide benefits that justify the cost of the associated PM activity. The field of possible PM tasks should be narrowed to that minimum level which maximizes the ratio of benefits to cost. There are several methods by which appropriate PM activities can be identified.

Impact Analysis

One method for determining which PM tasks are essential involves analyzing the impact of not performing a particular task. Simply because a system or piece of equipment is critical does not imply that all PM tasks performed on it will prevent failure. Each element, piece of equipment, and system should be examined and classified by its impact on the operation of the facility, and only those which are critical to the continued safe and efficient operation of the facility should receive high priority consideration for preventive maintenance.

Failure Analysis

PM activities may be identified by examining the maintenance history of the facility. In particular, each failure of a facility system or piece of equipment offers an opportunity to determine whether preventive maintenance could have averted or delayed the failure.

Manufacturers' Recommendations

A primary source for identifying PM activities is a list of procedures recommended by equipment manufacturers. When entire systems or major system components are mass produced, the manufacturer usually recommends service procedures and frequencies. Any warranties provided with such equipment usually depend on the completion of recommended preventive maintenance. While such warranties are in force, these tasks are, in fact, required. After the warranties expire, the maintenance manager may consider the desirability of continuing these activities at the recommended frequency.

Analogous Equipment

When components of mechanical and electrical equipment are examined, many similarities are revealed. PM procedures which have been developed for one piece of equipment may be readily adaptable to a similar piece of equipment. Such analogies between equipment components allow for easy development of PM procedures when a manufacturer fails to provide any instructions.

Developing a Preventive Maintenance Program

Once a comprehensive list has been made of what items to inspect, a checklist should be developed that highlights all the points to be inspected and/or repaired on any one piece or type of equipment or property. (Figure 16.1 is an example.) The checklist for each piece of equipment or system should contain procedures for inspection, calibration, lubrication and replacement of wear parts. The list should be complete and yet allow for the inclusion of general information such as the day of inspection, name of inspector, and any other specific comments. There are no readily available standard or universal answers as to how often to inspect equipment or systems. This information must be developed by the maintenance department, taking into consideration such factors as the age of the facility, types and condition of equipment, types and cycle frequencies of operations, environmental factors such as moisture, temperature, wind, contaminants and ultraviolet rays, as well as poor design and/or construction. To develop a checklist and timetable for each piece of equipment, consult with:

- Historical records of the facility
- Manufacturer of the equipment
- *Means Facilities Maintenance & Repair Cost Data*
- Employees of the facility
- Manufacturer's associations
- Trade associations
- Insurance agent

If contracting out the PM activities for facility systems or equipment, the developed checklist and timetable will become the work specifications for the contractor and the basis of a contract, whether it be long-term fixed price, one-time fixed price, cost reimbursement, fixed cost and retainer,

PREVENTIVE MAINTENANCE ORDER

Description
_____

Equipment Data
Name	Model No.	Manufacturer

P.M. Priority	Frequency	Due Date

LOCATION
Building	Room	Other

TOOLS REQUIRED

_____

MATERIALS REQUIRED
Quantity	Description	Stock Location

_____

SAFETY PROCEDURES

_____

MAINTENANCE PROCEDURES

_____

COMPLETION DATA
Date Completed	Completed By	Man-hours Expended	Foreman Initials

Craftsman Remarks
_____

Figure 16.1

or unit price. The responsibility for achieving proper performance of a system or piece of equipment still depends on the maintenance manager's ability to predict and describe the required maintenance steps, and on his/her ability to ensure quality workmanship from the contractor.

Keep in mind that once a checklist and timetable have been established, they should be reviewed on a regular basis for possible updating or improvement.

Scheduling Preventive Maintenance Functions

Again, the most effective and efficient schedule is one that fits into the operations of the facility, and causes a minimum of down time. There are three categories of PM activities that should be scheduled:

- Routine upkeep.
- Periodic inspections and replacement of parts.
- Contingent work to be performed at indefinite intervals where equipment is down for other reasons.

In scheduling PM activities for the first two categories above, the following objectives should be kept in mind:

- Handle them on the day shift if possible to minimize overtime.
- Distribute them as evenly over the year as possible. Make use of slack seasons whenever possible.
- Keep production or operating down time to a minimum.

The total scheduling for a PM program involves stipulating a definite recurring day and time for the maintenance and repair activities for each facility system or piece of equipment. The exact type of scheduling system depends on the resources of the facility and can be as simple as an erasable writing board or a magnetic status board, or as complicated as a computerized system. An impressive amount of computer software is available for administering and scheduling PM activities. However, once the schedule is established, it should be somewhat flexible so as to allow for emergency maintenance.

Simplifying Paperwork and Information Storage

Probably the biggest argument against establishing a PM program is the expected quantity of paperwork involved. Too many forms or ill-conceived forms make procedures unnecessarily complex and can overwhelm the facility staff. A PM program's paperwork usually consists of inspection checklists, work orders, equipment logs and schedule tools, such as those mentioned in the previous section. Inspection checklists are needed for each facility system or piece of equipment being inspected. The key document in the PM program is the work order. Although it is uniquely numbered, costed and tracked through the PM program, like the other forms, it must be flexible enough to be used for different systems and equipment. All of these forms must be kept simple and must be easy to use by all personnel.

Information from PM program activities such as inspections, work orders, service calls, emergency repairs and routine repetitive maintenance must be recorded and stored in an accessible format. The system may be as simple as a file folder for each piece of equipment and system, or it could be more sophisticated, using a computer database with spreadsheets and reports. For smaller facilities, manual systems can work very effectively. In larger facilities, computerized systems can be effective, but can also be time-consuming to maintain because in some cases the system may prompt the user to provide more data than is truly required to administer the PM program efficiently.

Identifying Costs

Estimating and tracking costs serves as the foundation for most of the managerial tools used in directing a PM program. The accuracy of budgeting for a PM program depends on how well the facility maintenance manager can predict the costs. Some of the procedures for cost estimating include:

- Determining the scope of work.
- Determining the required types and quantities of materials needed to perform the task.
- Determining the amount of labor and crew equipment needed to perform the task.
- Determining unit prices for the materials, labor and equipment.
- Determining indirect or overhead expenses.
- Summarizing all costs.

Utilizing Standards and Standard Formats

When making an estimate, all figures should be recorded on standard forms. Separate sheets should be used for the material takeoff quantity, pricing of labor and equipment amounts, and summarizing. The most effective format is one that is developed by the facility or organization for its specific needs. R.S. Means recommends standard forms, such as those used in the sample estimate later in this chapter.

The best source of cost information for any PM program is your own current data. If the facility or organization does not have elemental work time units or standards developed for estimating, this information can be requested of any contractor performing PM activities at your facility. The accuracy of a standard is a function of the degree of analysis of a job process. Some of the major types are listed below in increasing order of accuracy in terms of analysis:

1. **Published averages or ratios:** This method utilizes published cost data. The reader is referred to the latest edition of *Means Facilities Maintenance & Repair Cost Data* and *Means Facilities Construction Cost Data* for such data and related topics. This data is based on the total cost for doing work based on a designated unit, such as square foot or performed activity.

2. **Specific company job standard (historical records):** For this particular method, the time, labor, materials and equipment to perform the specific PM activity are noted from past experience and recorded as company historical records which can be consulted when future, similar jobs must be done. It should be noted that these figures should be updated from time to time for reasons of inflation, changes in wage rates and material prices, and changes in productivity rates. Also, if one or more aspects of the new PM activity are somewhat different from the one on record, appropriate corrections should be made.

3. **Activity or task standards:** Every job can be broken down into a series of activities which can be further divided into elements. It is possible through such means as time-and-motion studies and methods-time measure to obtain times and productivity rates for each one of these elements.

Sample Estimate

The best way to explain PM estimating is through an example. The regional facility manager of a retail chain must develop a preventive maintenance budget for each of his 20 identical 80,000 S.F. warehouse-type retail facilities, each consisting of a 200' x 400' one-story building with a flat roof and an eave height of 22 feet. Since he has no in-house maintenance staff, he will contract out all PM activities. The following building elements, systems and equipment have already been identified as needing preventive maintenance:

Exterior Enclosure

- (2) 8' x 10' electric roll-up overhead doors at the rear loading dock.
- (8) 12' x 8' electric sliding bi-parting doors at the front and side vestibules.

Interior Construction

- (4) 8' x 8' roll-up fire doors at door openings in a fire separation wall between the retail sales floor and the receiving and backstock areas.

Mechanical

- (1) 2-zone wet pipe fire suppression sprinkler system with a 6" supply.
- (1) 6-inch backflow preventer on the sprinkler system.
- (6) private yard hydrants for fire department use.
- (1) multi-zone annunciated fire alarm system.
- (1) 100-gallon gas-fired hot water heater.
- (10) 15-ton centrifugal-type roof-top air-cooled packaged units for cooling and heating.
- (3) urinals in public and employee rest rooms.
- (8) vacuum breaker-type toilets in public and employee rest rooms.
- (8) lavatories in public and employee rest rooms.

Electrical

- (20) dry-cell emergency light packages with dual light heads.
- (1) intrusion alarm system with a control panel and several keypads, equipped with an auto-dialer to call the alarm company's central office and the local police department.

Special Construction

- (2) drinking fountains.

Means Facilities Maintenance & Repair Cost Data contains a Preventive Maintenance (PM) section which provides the framework for a complete PM program. As shown on a sample page in Figure 16.2, this section lists the tasks and their frequency (whether weekly, monthly, quarterly, semi-annually or annually), associated with each system that has been identified as being in need of periodic preventive maintenance. The recommended frequencies of those tasks is based on noncritical usage (e.g., "normal use" situations, versus systems in facilities such as surgical suites or computer rooms that demand absolute adherence to a limited range of environmental conditions). The labor-hours to perform each item on the checklist are listed in the next column. Beneath the table is cost data, which is updated quarterly, for the PM schedule. The cost data is shown both annually and annualized, to provide the facility manager or owner with an estimating range. If all the tasks listed on the schedule are performed only once per year, the *annually* line should be used in the PM estimate. The *annualized* cost data should be used when all the tasks listed on the schedule are performed at the recommended frequency shown.

Getting back to the sample estimate, the regional facility manager is committed to performing all listed tasks at their recommended frequency, hence he will be using the *annualized* cost data. All the building elements, systems, and equipment that are listed above have been entered onto a Cost

MECHANICAL — PM8.4-810 Package Unit, Air Cooled

PM Components	Labor-hrs.	W	M	Q	S	A
System PM8.4-810-1950						
Package unit, air cooled, 3 tons through 24 tons						
1 Check with operating or area personnel for deficiencies.	.035			✓	✓	✓
2 Check tension, condition, and alignment of belts; adjust as necessary.	.029			✓	✓	✓
3 Lubricate shaft and motor bearings.	.047			✓	✓	✓
4 Replace air filters.	.055			✓	✓	✓
5 Clean electrical wiring and connections; tighten loose connections.	.120					✓
6 Clean coils, evaporator drain pan, blowers, fans, motors and drain piping as required.	.385					✓
7 Perform operational check of unit; make adjustments on controls and other components as required.	.077			✓	✓	✓
8 During operation of unit, check refrigerant pressure; add refrigerant as necessary.	.135			✓	✓	✓
9 Check compressor oil level; add oil as required.	.033					✓
10 Clean area around unit.	.066			✓	✓	✓
11 Fill out maintenance checklist and report deficiencies.	.022			✓	✓	✓
Total labor hours/period				.466	.466	1.004
Total labor hours/year				.932	.466	1.004

			Cost Each				
			1996 Bare Costs			Total	Total
Description	Labor-hrs.	Material	Labor	Equip.	Total	In-House	w/O&P
1900 Package unit, air cooled, 3 thru 24 ton, annually	1.004	31.50	30.50		62	73.50	87.50
1950 Annualized	2.402	50	72.50		122.50	147	177

PM Components	Labor-hrs.	W	M	Q	S	A
System PM8.4-810-2950						
Package unit, air cooled, 25 tons through 50 tons						
1 Check with operating or area personnel for deficiencies.	.035			✓	✓	✓
2 Check tension, condition, and alignment of belts; adjust as necessary.	.029			✓	✓	✓
3 Lubricate shaft and motor bearings.	.047			✓	✓	✓
4 Replace air filters.	.078			✓	✓	✓
5 Clean electrical wiring and connections; tighten loose connections.	.120					✓
6 Clean coils, evaporator drain pan, blowers, fans, motors and drain piping as required.	.385					✓
7 Perform operational check of unit; make adjustments on controls and other components as required.	.130			✓	✓	✓
8 During operation of unit, check refrigerant pressure; add refrigerant as necessary.	.272			✓	✓	✓
9 Check compressor oil level; add oil as required.	.033					✓
10 Clean area around unit.	.066			✓	✓	✓
11 Fill out maintenance checklist and report deficiencies.	.022			✓	✓	✓
Total labor hours/period				.679	.679	1.217
Total labor hours/year				1.358	.679	1.217

			Cost Each				
			1996 Bare Costs			Total	Total
Description	Labor-hrs.	Material	Labor	Equip.	Total	In-House	w/O&P
2900 Package unit, air cooled, 25 thru 50 ton, annually	1.217	44	37		81	95.50	113
2950 Annualized	3.254	70	98.50		168.50	202	243

Figure 16.2

Analysis form in Figure 16.3, along with their quantities. The system number (e.g., PM8.4-810-1950 for the 15-ton air-cooled packaged units) has been entered onto the form, along with the *annualized* bare cost data for material, labor and equipment. (Alternatively, the facility manager's own historical data could be used to calculate these bare costs and labor-hours). Finally, the total material, labor and equipment costs were calculated for each activity by multiplying the quantity of units by the unit costs. The remainder of the estimate is extended similarly. The summation of each column is the bare material, labor and equipment cost for the PM program. Now that the totals of the bare costs have been calculated, it is necessary to add the labor markup plus indirect costs. These are usually expressed as a percentage of the bare costs. The facility manager, if using in-house labor for this PM program, would add 0% to the material and equipment and 37.0% (Workers' Comp. + FICA + Medicare + Federal & State Unemployment + Insurance) to the labor, plus a markup for overhead as appropriate. (Note: for budgetary purposes, the *Skilled Workers Average* (35 Trades) was used as shown on the *Installing Contractor's Overhead & Profit* chart in Figure 16.4. This chart appears on the last page of *Means Facilities Maintenance & Repair Cost Data*. In this sample estimate, because the facility manager will contract out all PM activities, 25% was added to the bare material cost and 68.0% to labor as shown on the Cost Analysis form in Figure 16.3. If anything had been listed in the *Equipment* column, the facility manager would have added 10% to the bare cost of equipment in this example. The columns to the far right of the Cost Analysis form are used to record the annualized labor-hours per unit for each system, and to extend out the total labor-hours for each system. The sum of the total labor-hours is the amount of labor-hours that will be expended throughout the year for all preventive maintenance activities.

Summary

Planned, or preventive, maintenance is a most effective way to minimize emergency maintenance. A sound PM program is essential to preclude more expensive failure of equipment, installed systems, or other facility components. The elements of a facility which can benefit from periodic PM should be identified, and the cost of such maintenance should be weighed against the impact costs of not performing that maintenance. For the PM activities which are determined to be worth performing, a PM program should be formulated which utilizes a checklist of inspection, calibration, lubrication and worn parts replacement procedures for each PM activity. These activities should then be scheduled in a periodic fashion according to determined frequencies. As the PM program is implemented, a record of costs, labor-hours and productivity should be kept to form a database which will help facilitate the formation of PM budgets each year.

A PM program should be reviewed periodically to determine its effectiveness, and that effectiveness should be measured by the performance of the facility and of the equipment maintained. It is also measured by the relative cost of executing the entire PM program. Sometimes, in an effort to ensure that no equipment or system failures occur, a degree of over-maintenance can creep into any program. If discretionary PM activities are continually deferred with no apparent effect on facility and equipment performance, they should be dropped. If inspections rarely note discrepancies, with or without PM, consideration should be given to deleting these PM activities or revising their frequency. In addition to examining the PM program for wasted efforts or over-maintenance, the failure record of equipment should be examined periodically to identify possible "holes"

COST ANALYSIS

PROJECT: Annualized Preventive Maintenance
CLASSIFICATION: Retail Store, 80,000 SF (200'x400'), 1-Story
LOCATION: Anytown, USA
ARCHITECT:
TAKE OFF BY: SCP   QUANTITIES BY: SCP   PRICES BY: SCP   EXTENSIONS BY: SCP

DESCRIPTION	SOURCE/DIMENSIONS				QUANTITY	UNIT	MATERIAL COST	MATERIAL TOTAL	LABOR COST	LABOR TOTAL	EQUIPMENT COST	EQUIPMENT TOTAL	SUBCONTRACT COST	SUBCONTRACT TOTAL	LABOR-HOURS /UNIT	LABOR-HOURS TOTAL
Exterior Closure																
Electric overhead door	FMR	PM 4.2	120	1950	2	EA	41.00	82	103.00	206					4.070	8.140
Electric sliding entrance door	FMR	PM 4.2	130	1950	8	EA	26.00	208	61.50	492					2.448	19.584
Interior Construction																
Overhead fire door	FMR	PM 6.4	210	3950	4	EA	12.00	48	79.50	318					2.856	11.424
Mechanical																
Wet pipe extinguisher system	FMR	PM 8.2	170	1950	1	EA	110.00	110	355.00	355					11.341	11.341
6" Backflow preventer	FMR	PM 8.5	110	2950	1	EA			14.85	15					0.494	0.494
Yard hydrant	FMR	PM 8.2	280	1950	6	EA	5.00	30	19.95	120					0.637	3.822
Fire alarm system	FMR	PM 8.2	270	1950	1	EA	100.00	100	345.00	345					11.050	11.050
Gas-fired water heater	FMR	PM 8.3	910	1950	1	EA	40.00	40	52.00	52					1.719	1.719
15-ton rooftop air-cooled packaged unit	FMR	PM 8.4	810	1950	10	EA	50.00	500	72.50	725					2.402	24.020
Urinal	FMR	PM 8.5	050	1950	3	EA			6.85	21					0.228	0.684
Toilet, vacuum breaker type	FMR	PM 8.5	050	2950	8	EA			6.85	55					0.228	1.824
Lavatory	FMR	PM 8.5	050	4950	8	EA			9.05	72					0.348	2.784
Electrical																
Dry-cell emergency light	FMR	PM 9.3	130	2950	20	EA	33.00	660	10.80	216					0.370	7.400
Security/intrusion alarm system	FMR	PM 9.3	150	1950	1	EA	33.00	33	112.00	112					3.832	3.832
Special Construction																
Drinking fountain	FMR	PM 11.2	130	1950	2	EA	10.00	20	16.10	32					0.620	1.240
TOTAL BARE COSTS								1,831		3,136		0		0		109.358
Markups:																
Materials (25%)								458								
Labor (68.0%)										2,132						
Equipment (10%)																
SUBTOTALS								2,289		5,268		0		0		
City Cost Adjustment																
TOTALS																

Figure 16.3

In-house & Contractor's Overhead & Profit

Below are the average percentage mark-ups applied to base labor rates to calculate a typical in-house direct labor cost or contractor's billing rate.

In-House Labor Rate = Hourly Base Rate + Workers' Compensation + Avg. Fixed Overhead.

Contractor's Labor = Hourly Base Rate + Workers' Compensation + Avg. Fixed Overhead + Overhead + Profit.

Column A: Labor rates are based on union wages averaged for 30 major U.S. cities. Base rates including fringe benefits are listed hourly. These figures are the sum of the wage rate and employer-paid fringe benefits such as vacation pay, employer-paid health and welfare costs, pension costs, plus appropriate training and industry advancement funds costs.

Column B: Workers' Compensation rates are the national average of state rates established for each trade.

Column C: Column C lists average fixed overhead figures for all trades. Included are Federal and State Unemployment costs set at 7.3%; Social Security Taxes (FICA) set at 7.65%; Builder's Risk Insurance costs set at 0.34%; and Public Liability costs set at 1.55%. All the percentages except those for Social Security Taxes vary from state to state as well as from company to company.

Columns D and E: Percentage mark-ups in Columns D and E are applied to the mark-ups of columns B and C to calculate the contractor's billing rate. They are based on the presumption that the contractor has annual billing of $500,000 and up. The overhead percentages may increase with smaller annual billing. The overhead percentages may vary greatly among contractors depending on a number of factors, such as the contractor's annual volume, logistical support costs, and staff requirements. The figures for overhead and profit will also vary depending on the type of work, the job location and prevailing economic conditions.

Column F: Column F lists the total in-house markup of Columns B plus C.

Column G: Column G is Column A (hourly base labor rate) multiplied by the percentage in Column F (in-house overhead percentage).

Column H: Column H is the total of Column A (hourly base labor rate) plus Column G (in-house overhead).

Column I: Column I lists the total contractor's mark-up of Columns B, C, D, and E.

Column J: Column J is Column A (hourly base labor rate) multiplied by the percentage in Column I (contractor's O&P).

Column K: Column K is total of Column A (hourly base labor rate) plus Column J (total contractor's O&P).

Abbr.	Trade	A Base Rate Incl. Fringes Hourly	B Workers' Comp. Ins.	C Average Fixed Overhead	D Overhead	E Profit	F Rate with In-House Mark-Up %	G Amount	H Hourly	I Rate with Contractor's O&P %	J Amount	K Hourly
Skwk	Skilled Workers Average (35 trades)	.$25.95	20.2%	16.8%	16%	15%	37.0	$ 9.60	$35.55	68.0	$17.65	$43.60
	Helpers Average (5 trades)	19.25	21.4				38.2	7.35	26.60	69.2	13.30	32.55
	Foreman Average, Inside ($.50 over trade)	26.45	20.2				37.0	9.80	36.25	68.0	18.00	44.45
	Foreman Average, Outside ($2.00 over trade)	27.95	20.2				37.0	10.35	38.30	68.0	19.00	46.95
Clab	Common Building Laborers	19.80	21.9				38.7	7.65	27.45	69.7	13.80	33.60
Clam	Common Maintenance Laborer	14.85	7.4				24.2	3.60	18.45	55.2	8.20	23.05
Asbe	Asbestos Workers	28.55	19.7				36.5	10.40	38.95	67.5	19.25	47.80
Boil	Boilermakers	30.05	17.7				34.5	10.35	40.40	65.5	19.70	49.75
Bric	Bricklayers	25.90	19.4				36.2	9.40	35.30	67.2	17.40	43.30
Brhe	Bricklayer Helpers	20.00	19.4				36.2	7.25	27.25	67.2	13.45	33.45
Carp	Carpenters	25.20	21.9				38.7	9.75	34.95	69.7	17.55	42.75
Cefi	Cement Finishers	24.35	12.8				29.6	7.20	31.55	60.6	14.75	39.10
Elec	Electricians	29.30	8.0				24.8	7.25	36.55	55.8	16.35	45.65
Elev	Elevator Constructors	30.05	9.6				26.4	7.95	38.00	57.4	17.25	47.30
Eqhv	Equipment Operators, Crane or Shovel	26.75	12.9				29.7	7.95	34.70	60.7	16.25	43.00
Eqmd	Equipment Operators, Medium Equipment	25.70	12.9				29.7	7.65	33.35	60.7	15.60	41.30
Eqlt	Equipment Operators, Light Equipment	24.70	12.9				29.7	7.35	32.05	60.7	15.00	39.70
Eqol	Equipment Operators, Oilers	21.90	12.9				29.7	6.50	28.40	60.7	13.30	35.20
Eqmm	Equipment Operators, Master Mechanics	27.55	12.9				29.7	8.20	35.75	60.7	16.70	44.25
Glaz	Glaziers	24.90	16.0				32.8	8.15	33.05	63.8	15.90	40.80
Lath	Lathers	24.95	13.5				30.3	7.55	32.50	61.3	15.30	40.25
Marb	Marble Setters	25.65	19.4				36.2	9.30	34.95	67.2	17.25	42.90
Mill	Millwrights	26.55	13.2				30.0	7.95	34.50	61.0	16.20	42.75
Mstz	Mosaic & Terrazzo Workers	25.25	11.0				27.8	7.00	32.25	58.8	14.85	40.10
Pord	Painters, Ordinary	22.95	16.8				33.6	7.70	30.65	64.6	14.85	37.80
Psst	Painters, Structural Steel	23.95	62.5				79.3	19.00	42.95	110.3	26.40	50.35
Pape	Paper Hangers	23.30	16.8				33.6	7.85	31.15	64.6	15.05	38.35
Pile	Pile Drivers	25.35	33.6				50.4	12.80	38.15	81.4	20.65	46.00
Plas	Plasterers	24.20	17.4				34.2	8.30	32.50	65.2	15.80	40.00
Plah	Plasterer Helpers	20.15	17.4				34.2	6.90	27.05	65.2	13.15	33.30
Plum	Plumbers	30.05	10.2				27.0	8.10	38.15	58.0	17.45	47.50
Rodm	Rodmen (Reinforcing)	27.75	36.3				53.1	14.75	42.50	84.1	23.35	51.10
Rofc	Roofers, Composition	22.55	37.4				54.2	.12.20	34.75	85.2	19.20	41.75
Rots	Roofers, Tile & Slate	22.60	37.4				54.2	12.25	34.85	85.2	19.25	41.85
Rohe	Roofers, Helpers (Composition)	15.95	37.4				54.2	8.65	24.60	85.2	13.60	29.55
Shee	Sheet Metal Workers	28.95	13.8				30.6	8.85	37.80	61.6	17.85	46.80
Spri	Sprinkler Installers	31.30	10.4				27.2	8.50	39.80	58.2	18.20	49.50
Stpi	Steamfitters or Pipefitters	30.30	10.2				27.0	8.20	38.50	58.0	17.55	47.85
Ston	Stone Masons	25.90	19.4				36.2	9.40	35.30	67.2	17.40	43.30
Sswk	Structural Steel Workers	27.85	46.4				63.2	17.60	45.45	94.2	26.25	54.10
Tilf	Tile Layers	25.05	11.0				27.8	6.95	32.00	58.8	14.75	39.80
Tilh	Tile Layers Helpers	20.30	11.0				27.8	5.65	25.95	58.8	11.95	32.25
Trlt	Truck Drivers, Light	20.35	17.0				33.8	6.90	27.25	64.8	13.20	33.55
Trhv	Truck Drivers, Heavy	20.70	17.0				33.8	7.00	27.70	64.8	13.40	34.10
Sswl	Welders, Structural Steel	27.85	46.4				63.2	17.60	45.45	94.2	26.25	54.10
Wrck	*Wrecking	19.80	44.8				61.6	12.20	32.00	92.6	18.35	38.15

*Not included in Averages.

Figure 16.4

in the program, and consideration should be given to adding PM activities to the program or increasing their frequency. or deficiencies

An effective PM program ensures the continuous operation of a facility, system or piece of equipment. It protects the owner's investment and prevents unexpected failures of building systems that could disrupt facility activities and operations. An effective PM program is not easy to develop or to implement, but once installed it can be a major asset to any organization.

Predictive Maintenance or Condition Monitoring

Condition Monitoring, also known as *Predictive Maintenance*, is the philosophy that maintenance should be performed based on the condition of the system or equipment, rather than on a time (or interval) basis. Condition Monitoring involves continuous or periodic monitoring and diagnosis of equipment and components in order to forecast failures. In some cases, it is the most effective maintenance that can be performed.

Condition Monitoring is analogous to a physician monitoring and diagnosing a patient's condition. While the physician cannot predict when you will die, he or she can monitor your condition and provide you with a plan to maintain and improve your health which will, hopefully, extend your life. *Predictive* implies *positive* control of machinery condition. Condition Monitoring will improve the overall health of machines, including extending life and diagnosing most impending failures. Other names for the approach are used in various industries or agencies; for example, NASA uses the term *Predictive Testing and Inspection* (PT&I).

Condition Monitoring has been emerging as an important maintenance tool as more people have become aware of its benefits, and as the costs associated with monitoring the condition of systems and equipment have decreased. Prior to the 1950s, there was little discussion regarding the maintenance of systems and equipment. Most maintenance was intuitive or breakdown (reactive), and there was little examination of the relationship between failures and maintenance. In the 1960s, the airline industry set out to improve the effectiveness of maintenance in order to reduce costs and increase reliability without sacrificing safety. *Reliability-Centered Maintenance*, a book by Stanley Nowlan and Howard Heap, was the first detailed discussion on the subject and is the basis for modern Reliability-Centered Maintenance (RCM) programs. A key element of RCM is the understanding that time-based maintenance is sometimes not the most effective maintenance method. Time-based maintenance may introduce problems into otherwise healthy machines or, in extreme cases, result in premature overhaul or replacement. Based on this understanding, it became apparent that, when possible, a time-based inspection of systems and equipment would result in more effective utilization of maintenance resources. In the eighties, advances in technology and proven results in the aerospace, military, utilities, and process industries have raised the awareness of time-based inspections, which have become known as Condition Monitoring or Predictive Maintenance.

Condition Monitoring is a subset of periodic maintenance and is forecast-oriented. Inspection techniques that are nonintrusive/nonrestrictive are always preferred, in order to avoid introducing problems. In addition, Condition Monitoring generally involves data collection devices, data analysis, and computer databases to store and trend information.

Proven Condition Monitoring Techniques

Vibration Analysis

When people think of Condition Monitoring, they often think of vibration analysis. The technology and techniques have been developing for over 30 years, and over 78% of all manufacturing or processing plants use vibration analysis. Vibration analysis of rotating machines such as motors, pumps, fans, and gears is widely accepted as a viable technique to identify changing conditions. Reduced costs of test equipment and data management (primarily computers), availability of training, and development of computer-based expert systems are all contributing to this acceptance.

The technique measures machinery movement (vibration), typically through the use of an accelerometer, and examines the vibration spectrum to identify and trend frequencies of interest. Some frequencies are associated with a machine's design, regardless of its condition. For example, a healthy fan or rotary compressor may have a frequency that is equal to the machine speed times the number of fan blades. The vibration analysts may monitor this frequency to note changes in the amplitude that indicate a degrading condition. Other frequencies, for example, those associated with rolling element bearings, may be a sign of bearing damage and will alert the analysts to the start of bearing failure. It is common for electric motor problems, such as broken rotor bars or stator eccentricity, to be seen in vibration associated with electrical line frequency.

The vibration data may be collected with a portable device for periodic monitoring, or a continuous monitoring system may be installed for costly or critical systems. Analysis of the vibration data requires a detailed understanding of machinery operations and of vibration analysis techniques.

Infrared Thermography

Infrared Thermography (IRT) is the application of infrared detection instruments to identify pictures of temperature differences (thermogram). The test instruments used are noncontact, line-of-sight, thermal measurement and imaging systems. Because IRT is a noncontact technique, it is especially attractive for identifying hot/cold spots in energized electrical equipment, large surface areas such as boilers, roofs, and building walls, and other areas where "stand off" temperature measurement is necessary.

IRT inspections are identified as either *qualitative* or *quantitative*. The quantitative inspection is concerned with the accurate measurement of the temperature of the item of interest. The qualitative inspection concerns relative differences, hot and cold spots, and deviations from normal or expected temperature ranges. Qualitative inspections are significantly less time-consuming than quantitative because the thermographer is not concerned with highly accurate temperature measurement. What the thermographer does obtain is highly accurate temperature differences (T) between like components. For example, a typical motor control center will supply three-phase power, through a circuit breaker and controller, to a motor. Current flow through the three-phase circuit should be uniform, which means that the components within the circuit should have similar temperatures, one to the other. Any uneven heating, perhaps due to dirty or loose connections, would quickly be identified with IRT imaging system.

IRT can be utilized to identify degrading conditions in electrical systems such as transformers, motor control centers, switchgear, switchyards, or power lines. In mechanical systems, IRT can identify blocked flow conditions in heat exchangers, condensers, transformers, cooling radiators, and pipes. It can also be used to verify fluid level in large containers such as fuel storage tanks, and degraded refractory in boilers and furnaces.

Lubricating Oil Analysis

Lubricating oil analysis is performed on in-service machines to monitor and trend emerging conditions, confirm problems identified through other means such as vibration, and to troubleshoot known problems. Tests have been developed to address indicators of the machine mechanical wear condition, the lubricant condition, and to determine if the lubricant has become contaminated. Contaminated oil can contain water, dirt, or oil that was meant for another application. If not corrected, these contaminants can quickly lead to machine damage and failure. Lube condition trending, such as depletion of additives, can identify when the oil should be changed. Material, such as metal or seal particles, can identify machine damage before catastrophic failure, allowing for less costly repair.

Other techniques include Ultrasonic Noise Testing (identifies arcing in electrical equipment and leaks in gas systems), Ultrasonic Thickness Measurement (used on pipe wall), Flow Measurement, Valve Operation Analysis, Corrosion Monitoring, Process Parameter Analysis, and Insulation Resistance Trending of motors and circuits.

Analysis Techniques

Pattern Recognition

Often machines exhibit recognizable operation patterns. Deviations from the pattern or norm are indications of changes that may identify the onset of failure. For example, the infrared thermography inspections discussed earlier are seeking to identify unexpected thermal patterns.

Test Against Limits or Ranges

This should be done for parameters or conditions that do not follow continuous trends or repeatable patterns. This is useful in instrument calibration.

Relative Comparison of Data

Look for change as related to earlier data or from another baseline (such as similar equipment). This requires stable plant/equipment conditions.

Statistical Process Analysis (also called "Parameter Control Monitoring")

This type of analysis generally uses process or maintenance data that already exists or is collected. It applies statistical techniques to process or maintenance data in order to identify deviation from the norm.

Correlation Analysis

The most powerful technique is the one that uses data from multiple sources, related technologies, or different analysts. The cross-reference chart in Figure 16.5 illustrates how several of the techniques discussed earlier can be related to confirm the condition diagnosis.

Many companies are benefiting from Condition Monitoring. A few of the success stories:

- Allied-Signal's Chesterfield Plant reduced unscheduled repairs from 33% to 4% through use of vibration analysis.
  AIPE Facilities, March/April 1991
- Companies save $6 to $10 for each dollar spent on Condition Monitoring. One DuPont plant saved $22 for each dollar spent. Another DuPont plant increased up-time from 50% to 86% through use of Condition Monitoring. In a study by Mobil Oil, the company documented a 62% reduction in rotating equipment failures through Condition Monitoring.
  Maintenance Technology, February 1994

- Houston Lighting & Power, using infrared thermography inspection in a predictive maintenance program to reduce maintenance costs, ". . . realized a potential savings totaling $13 million by averting forced outages and increasing plant efficiency."
Power Engineering, June 1994

- Finding and repairing 324 air leaks saves company $52,000/year.
Maintenance Technology, January 1994

- Extending oil drain intervals enables company to save on oil, machine downtime, and labor hours.
Industrial Maintenance & Plant Operation, January 1996

Predictive Maintenance and Condition Monitoring Definitions

Condition Monitoring: The continuous or periodic monitoring and diagnosis of equipment and components in order to forecast failures.

Maintenance: Work performed to retain facility, system, or equipment capacity/availability in a safe and reliable condition to retain function.

Predictive Maintenance: See Condition Monitoring.

Preventive Maintenance: Work performed at set intervals (calendar, run time, cycles, etc.) to prevent failure. Interval-based preventive maintenance is often the appropriate approach to maintenance for systems that have defined, age-related wear patterns. Condition monitoring is a subset of preventive maintenance because the inspection is interval-based.

Proactive Maintenance: Those work items that are performed to avoid functional degradation. In the operation and maintenance portion of a system's life cycle, this could include establishing refurbishment standards,

Combined Use of Various Techniques to Determine Equipment Condition

Visual	Ultrasonic Noise	Infrared Thermography	Motor Circuit Evaluation	Wear Particles	Lubricant Condition	Temperature	Vibration
							Vibration
						Temperature	•
					Lubricant Condition		
				Wear Particles	•	•	•
			Motor Circuit Evaluation				•
		Infrared Thermography	•			•	•
	Ultrasonic Noise	•	•	•		•	•
Visual		•		•		•	

Figure 16.5

analyzing failures to identify trends or root causes, replacing obsolete components prior to failure, and other actions that seek out failure modes and identify solutions to prevent them.

Repair: The work needed to restore function. Repair can range from minor adjustment to major overhaul.

Run-to-Failure: Often called *reactive maintenance* or *breakdown maintenance*, work is performed only when deterioration in a machine's condition causes a functional failure. Run-to-Failure is typically associated with a high percentage of the total work being unplanned maintenance, high replacement part inventories, and the inefficient use of maintenance personnel. However, Run-to-Failure is acceptable for noncritical, low-cost equipment.

Conclusion

Preventive and Predictive Maintenance are important components to consider in developing a successful long-term facility maintenance strategy. The merits and specific implementation methods of each must be considered in the context of each particular facility and owner organization. Clearly, these programs can play a major role in controlling costs and improving the quality of facility operations.

Chapter 17

General Maintenance Estimating

Phillip R. Waier, P.E.

General Maintenance (GM) is defined as custodial maintenance or cleaning. This is the easiest work for the facility manager to accomplish because it can be planned or identified in advance. A notable feature of general maintenance is that, if necessary, it can usually be deferred without adversely affecting productivity.

Common Recurring Maintenance Work Activities

Grounds Maintenance

Lawns

Lawn and field mowing
Trimming and edging
Fertilizer application
Insect and rodent control
Leaf raking and removal

Plants, Trees, and Shrubs

Weed removal or control
Annual plantings
Trimming, pruning
Insect and rodent control
Fertilizer application
General flower bed maintenance

Roads, Parking Lots, Walkways

Street sweeping, cleaning
Painting of traffic striping
Sign maintenance
Curb maintenance, painting

Miscellaneous, Site Work

Litter collection
Trash collection and disposal
Painting fences, small buildings
Exterior lighting (bulb replacement)
Cleaning out storm drains

Building Exteriors

Walls
Painting
Cleaning, mildew, mold, fungus removal
Caulking renewal

Windows
Window washing
Storm window/screen installation/removal

Doors
Cleaning, washing
Storm/screen door changeouts
Lubricating hinges, locks

Roof
Gutter and downspout cleaning
Roof drain cleaning
Trash removal off flat roofs

Building Interiors

Floors
Vacuuming and sweeping
Washing and waxing
Stripping wax
Mopping
Carpet steam cleaning

Walls, Ceilings
Dusting
Cleaning, spot cleaning
Touch-up painting

Doors, Windows
Cleaning, washing
Drape, shade, curtain cleaning
Touch-up painting

Public Spaces
Replacing paper towels
Replacing toilet paper
Replacing soap—cake, liquid, powder
Emptying trash receptacles
Emptying/cleaning public ash trays

Offices, Shops, Workplace
Waste collection
General dusting
Vacuuming
Sweeping
Washing and waxing floors

Requirements for the Estimate

GM is the most common element of facilities maintenance to be subcontracted, because it typically has the least impact on the organization's mission. Many of the tasks are seasonal, such as landscaping and snow removal. These tasks readily lend themselves to subcontracting, because it is not economical for an organization to keep permanent staff on the payroll who will be underutilized for a significant part of the year. (See Chapter 3, "Outsourcing," for more on this issue). Similarly, building cleaning is frequently subcontracted to avoid the high cost of permanent employees' wages and benefits. Whether or not a function is contracted out, it is necessary to have some idea of the time and dollars associated with the maintenance.

This chapter addresses the information and steps required to prepare a general maintenance estimate, including:

- Incurred costs
- Worker productivity data
- Site visits
- Identification and frequency of required tasks
- Pricing of tasks and indirect expenses

Incurred Costs

GM estimates start with a thorough understanding of all expenses incurred. Generally, the expenses are categorized as *direct* and *indirect*. The direct expenses are related to performing work at the specific location, including any labor, supervision, equipment, cleaning supplies, and uniforms. The indirect expenses are related to the contractor's office and other administrative costs. Indirect expenses exist whether the cleaning services are contracted out or performed by an in-house staff. Indirect expenses include administrative salaries, office supplies, telephones, and office space. Figure 17.1 lists typical direct and indirect expenses.

Productivity

Once the direct and indirect expenses are identified, production rates are the greatest variable. *Production rate* is the time required to complete a specific task under normal conditions. It is usually measured in hours per task or hours per 1,000 S.F. Cleaning organizations can either develop their

---

Typical Direct and Indirect Expenses

Direct Expenses	Indirect Expenses
• Labor wages, including:	• Home office salaries:
• FICA	• Owner
• Federal and state unemployment taxes	• Sales staff
• Workers' compensation	• Accounting staff
• Health and welfare benefits	• Professional fees
• Site supervision	• Accountant
• Cleaning materials	• Attorney
• Equipment, including repair	• Dues and publications
• Vehicle expense	• Bad debts
• Uniforms	• Entertainment
• Paper products	• Travel and lodging
	• Office expenses
	• Telephone
	• Office supplies
	• Utilities
	• Transportation
	• Taxes and insurance

Figure 17.1

own rates or apply standard rates. (*Facilities Maintenance & Repair Cost Data* provides productivity plus material, labor, and equipment information for over 350 common maintenance tasks. The presentation and use of this data will be discussed later in this chapter.) The most common sources of productivity information are as follows:

- The facility's own historical data and experience
- *Facilities Maintenance & Repair Cost Data*
- The Building Service Contractors' Association International
- International Sanitary Supply Association
  (See the Appendix for information on these and other associations.)

Your own individual time standards are developed by listing all the tasks necessary to clean a specific building. Then measure the total area to be cleaned in a task. Next observe the length of time required for different people to perform the same task. Ideally, this procedure should be applied to several buildings with similar characteristics. For example, say 5 workers each performed the task of wet mopping and rinsing an 800 S.F. lobby area in times of 32, 33, 35, 38, and 36 minutes. Calculate the production rate. The average of the 5 times is 34.8 minutes, or 34.8/60 = .58 hours. The productivity rate is .58 hours/800 S.F. = .000725 labor hours per S.F. or .725 labor hrs./thousand S.F. (M.S.F.).

Note: It is usually desirable to express times in decimals of an hour; for example, 20 min. = .33 hrs.

After the productivity standards are developed, they should be checked periodically. New equipment or improved cleaning supplies may improve productivity.

The Site Visit

The next step in producing a general maintenance estimate is the site visit. Work should never be estimated in an existing facility without the benefit of an inspection. Ideally, the owner or facility manager should be present to point out areas of special concern and answer specific questions. The contractor or in-house maintenance manager should obtain a floor plan of the building, including furniture arrangement. This plan will refresh your memory and will save you from having to measure areas in the building before calculating costs. Ideally, the facility manager or owner should also provide a scope of work with proposed frequencies for each task. The scope of work or specification has a great impact on the estimate because it identifies what tasks are to be completed and at what frequency. Not every surface is cleaned every day. For example, when vacuuming is required, the main travel areas and spot areas are usually done daily; the secondary travel areas are done weekly; and all areas are done monthly.

During the site visit, particular attention should be paid to the following building characteristics. Each of these may have an impact on your final estimate.

- *Traffic into the building:* Observe the amount of traffic into the building and note the number of visitors. A large number of outside visitors increases the cleaning requirements.
- *Number of building occupants (density):* The number of desks and other furnishings in a building will affect production.
- *Existing level of cleaning:* Does the owner want to meet or exceed the existing level of cleaning?
- *Division of floor space:* Partitions and cubicles are additional obstructions that can result in reduced productivity.

- *Types and areas of different floor surfaces:* Different floors require different maintenance. The floor plan should provide adequate dimensions to quantify each finish.
- *Restroom finishes and number of fixtures:* Restroom cleaning is usually priced by the fixture, so the number has a direct impact. Wall and floor finishes also affect time; consider the impact of tiled vs. painted vs. wallpapered walls.
- *Smoking or nonsmoking building:* Nonsmoking buildings are easier to maintain. Smoking adds time not only to clean ashtrays, but also to vacuum and clean vents and light fixtures.

After these observations have been made, compare the working conditions and building characteristics at this location against those where your productivity standards were calculated. There may be a good match, in which case the standard productivity can be directly applied. It is more likely that there will be a variance from known conditions, which is where the true art of estimating enters the picture. You will have to determine that this building is more or less difficult to clean than your standard, and that it is more or less difficult by a quantifiable degree, such as 10%, or 1.5 labor-hours for a specific task. This is a judgment based on your experience.

Identifying Specific Tasks

When preparing an estimate it is always desirable to use an estimating form. The headings from two Means forms are shown in Figures 17.2 and 17.3. These forms provide space to list the tasks by area. The site visit and scope of work form the basis for this list. It is not uncommon for additional items to be added to the list as they occur, such as additional time required to enter a secure area of the building.

The standard frequencies used in cleaning are daily, weekly, monthly, quarterly, semiannually, and annually. Tasks required daily are performed 250 times per year.

5 days/week × 52 weeks = 260 days less 10 legal holidays = 250.

The Quantity column on the estimating form is used to address the frequency.

Pricing All Tasks and Adding Indirect Expenses

After all the tasks have been entered onto the estimating form and the frequencies added, it is time to begin pricing. A unit price estimate provides costs for material, labor, and equipment associated with each task. The sum of all the items provides all the direct costs associated with the work. Now the indirect costs must be added. These are usually expressed as a percentage of direct costs. The final number to be added is profit, which is expressed as a percentage of total costs. This number usually varies from 5 to 15%, depending on competition.

Using Means Cost Data

Means cost data is used by cleaning contractors, facility managers, and owners. The data is used to estimate cleaning times, compare and assess estimates submitted by cleaning companies, and budget for in-house staff. The unit price data shown in Figure 17.4 lists the crew, daily output, and labor-hours to perform the task. Bare material, labor, and equipment costs are also listed. The last two columns provide a total cost including indirect expenses for an in-house staff and total with overhead and profit for an outside contractor.

Each line item on the unit price page is identified by a unique ten-digit line number adjacent to the description. The description is followed by three

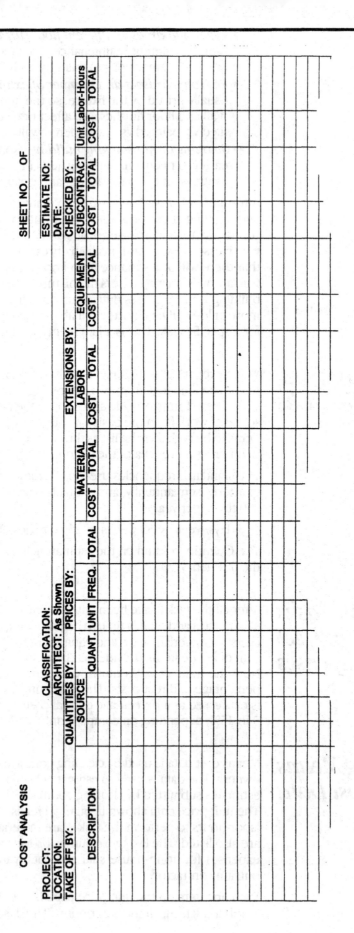

Figure 17.2

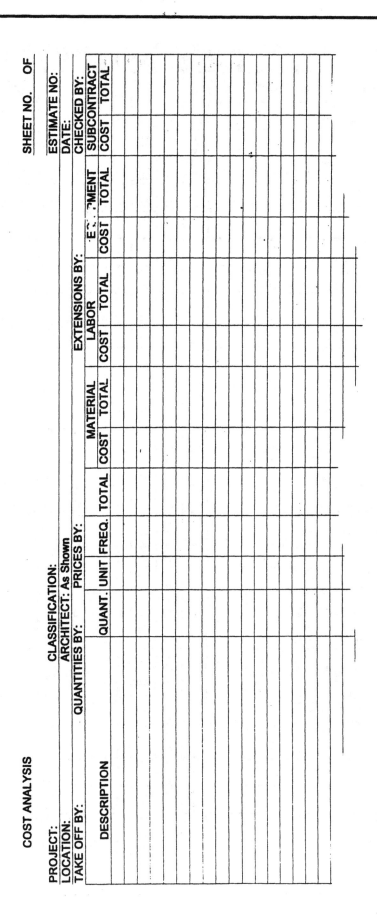

Figure 17.3

018 400	Exterior General Maint.	Crew	Daily Output	Labor-Hours	Unit	Mat.	Labor	Equip.	Total	Total In-House	Total Incl O&P	
410	**0010 LAWN CARE**											**410**
	0020 Mowing lawns, power mower, 18"-22"	A-15	80	.100	M.S.F.		1.49	.39	1.88	2.24	2.74	
	0030 22"-30"		120	.067			.99	.26	1.25	1.49	1.82	
	0040 30"-32"		140	.057			.85	.22	1.07	1.28	1.56	
	0050 Self propelled or riding mower, 36"-44"	A-16	300	.027			.40	.29	.69	.78	.93	
	0060 48"-58"		480	.017			.25	.18	.43	.49	.58	
	0070 Tractor, 3 gang reel, 7' cut		930	.009			.13	.09	.22	.25	.30	
	0080 5 gang reel, 12' cut		1,200	.007			.10	.07	.17	.19	.23	
	0090 Edge trimming with weed whacker	A-15	5,760	.001	L.F.		.02	.01	.03	.03	.04	
420	**0010 YARD WASTE DISPOSAL**											**420**
	0030 On site dump or compost heap	2 Clam	24	.667	C.Y.		9.90		9.90	12.30	15.35	
	0040 Into 6 CY dump truck, 2 mile haul to compost heap	A-17	125	.064			.95	1.36	2.31	2.54	2.97	
	0050 4 mile haul to compost heap, fees not included	"	85	.094			1.40	1.99	3.39	3.73	4.36	
430	**0010 RAKING**											**430**
	0020 Raking, leaves, by hand	1 Clam	7.50	1.067	M.S.F.		15.85		15.85	19.65	24.50	
	0030 Power blower	A-15	45	.178			2.64	.70	3.34	3.98	4.87	
	0040 Grass clippings, by hand	1 Clam	7.50	1.067			15.85		15.85	19.65	24.50	
440	**0010 EDGING**											**440**
	0020 Hand edging, at walks	1 Clam	16	.500	C.L.F.		7.45		7.45	9.20	11.50	
	0030 At planting, mulch or stone beds	"	7	1.143			16.95		16.95	21	26.50	
	0040 Power edging, at walks	A-15	88	.091			1.35	.36	1.71	2.03	2.49	
	0050 At planting, mulch or stone beds	"	24	.333			4.95	1.31	6.26	7.45	9.10	
450	**0010 LAWN RENOVATION**											**450**
	0020 Lawn renovations, aerating, 18" walk behind cultivator	A-15	95	.084	M.S.F.		1.25	.33	1.58	1.88	2.30	
	0030 48" tractor drawn cultivator	A-18	750	.011			.16	.21	.37	.41	.48	
	0040 72" tractor drawn cultivator	"	1,100	.007			.11	.15	.26	.28	.33	
	0050 Fertilizing, dry granular, 4#/M.S.F., drop spreader	1 Clam	24	.333		1.65	4.95		6.60	7.95	9.75	
	0060 Rotary spreader	"	140	.057		1.65	.85		2.50	2.87	3.38	
	0070 Tractor drawn 8' spreader	A-18	500	.016		1.65	.24	.32	2.21	2.43	2.78	
	0080 Tractor drawn 12' spreader	"	800	.010		1.65	.15	.20	2	2.20	2.51	
	0090 Overseeding, utility mix, 7#/M.S.F., drop spreader	1 Clam	10	.800		10.45	11.90		22.35	26.50	31.50	
	0100 Tractor drawn spreader	A-18	52	.154		10.55	2.28	3.09	15.92	17.55	20	
	0110 Watering, 1" of water, applied by hand	1 Clam	21	.381			5.65		5.65	7.05	8.75	
	0120 Soaker hoses	"	82	.098			1.45		1.45	1.80	2.25	
460	**0010 FLOWER, SHRUB & TREE CARE**											**460**
	0020 Flower or shrub beds, bark mulch, 3" deep hand spreader	1 Clam	100	.080	S.Y.	1.60	1.19		2.79	3.24	3.84	
	0030 Peat moss, 1" deep hand spreader		900	.009		.85	.13		.98	1.10	1.27	
	0040 Wood chips, 2" deep hand spreader		220	.036		.90	.54		1.44	1.66	1.96	
	0050 Cleaning		1	8	M.S.F.		119		119	148	184	
	0060 Fertilizing, dry granular, 3 #/C.S.F.		85	.094		12.50	1.40		13.90	15.50	17.80	
	0070 Weeding, mulched bed		20	.400			5.95		5.95	7.40	9.20	
	0080 Unmulched bed		8	1			14.85		14.85	18.45	23	
	0090 Trees, pruning from ground, 1-1/2" caliper		84	.095	Ea.		1.41		1.41	1.76	2.19	
	0100 2" caliper		70	.114			1.70		1.70	2.11	2.63	
	0110 2-1/2" caliper		50	.160			2.38		2.38	2.95	3.69	
	0120 3" caliper		30	.267			3.96		3.96	4.92	6.15	
	0130 4" caliper	2 Clam	21	.762			11.30		11.30	14.05	17.55	
	0140 6" caliper		12	1.333			19.80		19.80	24.50	30.50	
	0150 9" caliper		7.50	2.133			31.50		31.50	39.50	49	
	0160 12" caliper		6.50	2.462			36.50		36.50	45.50	56.50	
	0170 Fertilize, slow release tablets	1 Clam	100	.080		.19	1.19		1.38	1.68	2.08	
	0180 Pest control, spray		24	.333		11	4.95		15.95	18.25	21.50	
	0190 Systemic		48	.167		11	2.48		13.48	15.15	17.60	
	0200 Watering, under 1-1/2" caliper		34	.235			3.49		3.49	4.34	5.40	

Figure 17.4

columns; *crew*, *daily output,* and *labor-hours.* These columns are interrelated so you can always calculate the third if you know any two. *Labor-Hours* is the most important column when preparing an estimate. Labor-hours are the number of hours required to perform one unit of work. The unit may be *M.S.F.* (1,000 S.F.) or *Ea.* (each). The *Bare Cost* is the cost of Material, Labor, and Equipment without overhead and profit. The costs shown are National Averages. These costs should be adjusted to your particular location by the City Cost Indexes discussed in Chapter 14.

The last two columns, *Total In-House* and *Total Incl. O&P,* are calculated by adding markups to the bare costs. In-house labor markups include Workers' Compensation plus *Average Fixed Overhead* (Federal & State Unemployment + Social Security + Builder's Risk + Public Liability Insurance). The National Average In-house total markup is 24.2% for the Common Maintenance Laborer *(Clam).* Note: Markup for in-house office overhead has not been included. A percentage may be added to match your organization's operating costs. Means has found that this can range from 0% to more than 25%, depending on the organization and its accounting methodology. *Total Incl. O&P* includes a 25% markup on materials, 10% on equipment plus labor markups. Contractor's labor markups include Workers' Compensation + Average Fixed Overhead (Federal & State Unemployment + Social Security + Builder's Risk + Public Liability Insurance) + Overhead (office) + Profit. See Figure 17.5, How to Use the General Maintenance Pages, and Figure 17.6, Installing Contractor's Overhead and Profit, for additional explanations of markups and wage calculations.

Sample Estimate

The best way to explain estimating is through an example. Figure 17.7 is the ground floor plan for a six-story, 60,000 S.F. office building located in a suburban office park. Figure 17.8 is a sample specification received from the owner.

The items on this specification have been listed on the standard estimating form, Figure 17.9, according to the frequency. The quantities and frequencies were entered, and the total annual quantity was calculated. For example, .20 M.S.F. (thousand square feet) of entryway mats are vacuumed daily: .20 M.S.F/day × 250 days/year = 50 M.S.F./year. Means line number and unit bare material, labor, and equipment cost for each task were entered. (Alternatively, the contractor's or facility manager's own data could be used to calculate these costs.) Finally, the total material, labor, equipment, and labor-hours were calculated by multiplying the number of units times the unit cost. The total quantity of entrance mat vacuuming is 50 M.S.F./year. The calculation is as follows:

Material	=	0.00	
Labor	=	50 M.S.F. × $14.85/M.S.F.	= $742.50
Equipment	=	50 M.S.F. × $3.13/M.S.F.	= $156.00
Labor-hours	=	50 M.S.F. × 1 hour/M.S.F.	= 50 hours

The remainder of the estimate is extended similarly. The summation of each column is the bare material, labor, and equipment cost for the year. Now that the totals of the bare costs have been calculated, it is necessary to add the labor markup plus indirect costs. These are usually expressed as a percentage of the bare costs. The in-house estimate would add 0% to the material and equipment and 24.2% (Workers' Comp. + F.I.C.A. + Federal & State Unemployment + Insurance) to the labor, plus a markup for office overhead as appropriate. The contractor would add 25% to bare material cost, 55.2% to labor, and 10% to equipment as shown on the estimate.

Conclusion

General maintenance estimating is similar to any other estimating function. The most important element is to identify all the tasks and allocate labor-hours to accomplish the work. The next most important step is to apply an hourly wage rate. Means used a bare labor rate of $14.85/hour, then marked up in-house labor by 24.2% and contractor's labor by 55.2%. Your labor rate may be different; if so, the total bare costs may be adjusted, either by using the City Cost Index, or proportionate to your actual labor costs. For example, if your wage rate for maintenance labor is $7.50/hour, the labor on the sample estimate would be calculated as follows:

$$\frac{\$7.50}{\$14.85} \times 97{,}784 = \$49{,}381$$

After you start the work, it is important to periodically check your estimate against the times and expenses reported from the site. Using this information, you can reevaluate your labor-hours or crew productivity. This ensures that you know the financial status of every job and always have the most up-to-date information to use in future estimates.

How to Use the General Maintenance Pages

The following is a detailed explanation of a sample entry in the General Maintenance Section. Next to each bold number that follows is the item being described with the appropriate component of the sample entry following in parenthesis.

1 Division Number/Title (018/General Maintenance)

Use the General Maintenance Section Table of Contents to locate specific items. This section is classified according to the CSI MasterFormat.

2 Line Numbers (018 450 0020)

Each General Maintenance unit price line item has been assigned a unique 10-digit code based on the 5-digit CSI MasterFormat classification.

MasterFormat Mediumscope
MasterFormat Division

018 400
018 450 0020

Means Subdivision
Means Major Classification
Means Individual Line Number

3 Description (Lawn renovations, etc.)

Each line item is described in detail. Sub-items and additional sizes are indented beneath the appropriate line items. The first line or two after the main item (in boldface) may contain descriptive information that pertains to all line items beneath this boldface listing.

018 | General Maintenance

		018 400	Exterior General Maint.	Crew	Daily Output	Labor-Hours	Unit	Mat.	Labor	Equip.	Total	Total In-House	Total Incl O&P	
440	0010	**EDGING**												440
	0020	Hand edging at walks	1 Clam	16	.500	C.L.F.		7.45		7.45	9.20	11.50	
	0030	Edging, mulch or stone beds		7	1.143			16.95		16.95		26.50	
	0040	Power edging at walks		88	.091			1.35	.36	1.71		2.49	
	0050	edging, mulch or stone beds		24	.333			4.95	1.31	6.26		9.10	
450	0010	**LAWN RENOVATION**												450
	0020	Lawn renovations, aerating, 18" walk behind cultivator	A-15	95	.084	M.S.F.		1.25	.33	1.58	1.88	2.30	
	0030	48" tractor drawn cultivator	A-18	750	.011				.21	.37	.41	.48	
	0040	72" tractor drawn cultivator	"	1,1	.007			.11	.15	.26	.28		
	0050	Fertilizing, dry granular, 4#/M.S.F., drop spreader	1 Clam	2				1.65	4.95		6.60	7.95		
	0060	Rotary spreader	"	1	.007			1.65			2.50	2.87	3.38	
	0070	Tractor drawn 8' spreader	A-18	500	.016			1.65	.24	.32	2.21	2.43	2.78	
	0080	Tractor drawn 12' spreader	"	800	.010			1.65	.15	.20	2	2.20	2.51	
	0090	Overseeding, utility mix, 7#/M.S.F., drop spreader	1 Clam	10	.800			10.45	11.90		22.35	26.50	31.50	
	0100	Tractor drawn spreader	A-18	52	.154			10.55	2.28	3.09	15.92	17.55	20	
	0110	Watering, 1" of water, applied by hand	1 Clam	21	.381				5.65		5.65	7.05	8.75	
	0120	Soaker hoses	"	82	.098				1.45		1.45	1.80	2.25	
460	0010	**FLOWER, SHRUB & TREE CARE**												

III – 2

Figure 17.5

4 Crew (A-15)

The "Crew" column designates the typical trade or crew used to install the item. If an installation can be accomplished by one trade and requires no power equipment, that trade and the number of workers are listed (for example, "2 Clam"). If an installation requires a composite crew, a crew code designation is listed (for example, "A-15"). You'll find full details on all composite crews in the Crew Listings.

• For a complete list of all trades utilized in this book and their abbreviations, see the last page of the book.

Crews

Crew No.	Bare Costs		In House Costs		Incl. Subs O&P		Cost Per Labor-Hour		
Crew A-15	Hr.	Daily	Hr.	Daily	Hr.	Daily	Bare Costs	In House	Incl. O&P
1 Maintenance Laborer	$14.85	$118.80	$18.45	$147.60	$23.05	$184.40	$14.85	$18.45	$23.05
1 Lawn Mower, small		31.40		31.40		34.55	3.93	3.93	4.32
8 M.H., Daily Totals		$150.20		$179.00		$218.95	$18.78	$22.38	$27.37

5 Productivity: Daily Output (95)/ Labor-Hours (.084)

The "Daily Output" represents the typical number of units the designated crew will install in a normal 8-hour day. To find out the number of days the given crew would require to complete the installation, divide your quantity by the daily output. For example:

Quantity	÷	Daily Output	=	Duration
200 M.S.F.	÷	95 M.S.F./ Crew Day	=	2.1 Crew Days

The "Labor-Hours" figure represents the number of labor-hours required to install one unit of work. To find out the number of labor-hours required for your particular task, multiply the quantity of the item times the number of labor-hours shown. For example:

Quantity	x	Productivity Rate	=	Duration
200 M.S.F.	x	.084 Labor-Hours/ LF	=	16.8 Labor-Hours

6 Unit (M.S.F.)

The abbreviated designation indicates the unit of measure upon which the price, production, and crew are based (M.S.F. = Thousand Square Feet). For a complete listing of abbreviations refer to the Abbreviations Listing in the Reference Section of this book.

7 Bare Costs:

Mat. (Bare Material Cost) (0.00)

This figure for the unit material cost for the line item is the "bare" material cost with no overhead and profit allowances included. *Costs shown reflect national average material prices for the current year and generally include delivery to the facility. Small purchases may require an additional delivery charge. No sales taxes are included.*

Labor (1.25)

The unit labor cost is derived by multiplying bare labor-hour costs for Crew A-15 by labor-hour units. In this case, the bare labor-hour cost is found in the Crew Section under A-15. (If a trade is listed, the hourly labor cost – the wage rate – is found on the last page of the book.)

Labor-Hour Cost Crew A-15	x	Labor-Hour Units	=	Labor
$14.85	x	.084	=	$1.25

Equip. (Equipment) (.33)

Equipment costs for each crew are listed in the description of each crew. The unit equipment cost is derived by multiplying the bare equipment hourly cost by the labor-hour units.

Equipment Cost Crew A-15	x	Labor-Hour Units	=	Equip.
$3.93	x	.084	=	$0.33

Total (1.58)

The total of the bare costs is the arithmetic total of the three previous columns: mat., labor, and equip.

Material	+	Labor	+	Equip.	=	Total
$0.00	+	$1.25	+	$0.33	=	$1.58

Figure 17.5 (cont.)

8 Total In-House Costs (1.88)

"Total In-House Costs" include suggested mark-ups to bare costs for the direct overhead requirements of in-house maintenance staff. The figure in this column is the sum of three components: the bare material cost plus 10% (for purchasing and handling small quantities); the bare labor cost plus workers' compensation and average fixed overhead (per the labor rate table on the last page of the book or, if a crew is listed, from the crew listings); and the bare equipment cost.

Material is Bare Material Cost + 10% = $0.00 + $0.00	=	$0.00
Labor for Crew A-15 = Labor-Hour Cost ($18.45) x Labor-Hour Units (.084)	=	$1.55
Equip. is Bare Equip. Cost	=	$0.33
Total	=	$1.88

9 Total Costs Including O & P (2.30)

"Total Costs Including O&P" include suggested mark-ups to bare costs for an outside subcontractor's overhead and profit requirements. The figure in this column is the sum of three components: the bare material cost plus 25% for profit; the bare labor cost plus total overhead and profit (per the labor rate table on the last page of the book or, if a crew is listed, from the crew listings); and the bare equipment cost plus 10% for profit.

Material is Bare Material Cost + 25% = $0.00 + $0.00	=	$0.00
Labor for Crew A-15 = Labor-Hour Cost ($23.05) x Labor-Hour Units (.084)	=	$1.94
Equip. is Bare Equip. Cost + 10% = $0.33 + $0.03	=	$0.36
Total	=	$2.30

III – 4

Figure 17.5 (cont.)

In-house & Contractor's Overhead & Profit

Below are the average percentage mark-ups applied to base labor rates to calculate a typical in-house direct labor cost or contractor's billing rate.

In-House Labor Rate = Hourly Base Rate + Workers' Compensation + Avg. Fixed Overhead.

Contractor's Labor = Hourly Base Rate + Workers' Compensation + Avg. Fixed Overhead + Overhead + Profit.

Column A: Labor rates are based on union wages averaged for 30 major U.S. cities. Base rates including fringe benefits are listed hourly. These figures are the sum of the wage rate and employer-paid fringe benefits such as vacation pay, employer-paid health and welfare costs, pension costs, plus appropriate training and industry advancement funds costs.

Column B: Workers' Compensation rates are the national average of state rates established for each trade.

Column C: Column C lists average fixed overhead figures for all trades. Included are Federal and State Unemployment costs set at 7.3%; Social Security Taxes (FICA) set at 7.65%; Builder's Risk Insurance costs set at 0.34%; and Public Liability costs set at 1.55%. All the percentages except those for Social Security Taxes vary from state to state as well as from company to company.

Columns D and E: Percentage mark-ups in Columns D and E are applied to the mark-ups of columns B and C to calculate the contractor's billing

rate. They are based on the presumption that the contractor has annual billing of $500,000 and up. The overhead percentages may increase with smaller annual billing. The overhead percentages may vary greatly among contractors depending on a number of factors, such as the contractor's annual volume, logistical support costs, and staff requirements. The figures for overhead and profit will also vary depending on the type of work, the job location and prevailing economic conditions.

Column F: Column F lists the total in-house markup of Columns B plus C.

Column G: Column G is Column A (hourly base labor rate) multiplied by the percentage in Column F (in-house overhead percentage).

Column H: Column H is the total of Column A (hourly base labor rate) plus Column G (in-house overhead).

Column I: Column I lists the total contractor's mark-up of Columns B, C, D, and E.

Column J: Column J is Column A (hourly base labor rate) multiplied by the percentage in Column I (contractor's O&P).

Column K: Column K is total of Column A (hourly base labor rate) plus Column J (total contractor's O&P).

Abbr.	Trade	A Base Rate Incl. Fringes Hourly	B Workers' Comp. Ins.	C Average Fixed Over-head	D Over-head	E Profit	F Rate with In-House Mark-Up %	G Amount	H Hourly	I Rate with Contractor's O & P %	J Amount	K Hourly
Skwk	Skilled Workers Average (35 trades)	.$25.95	20.2%	16.8%	16%	15%	37.0	$ 9.60	$35.55	68.0	$17.65	$43.60
	Helpers Average (5 trades)	19.25	21.4				38.2	7.35	26.60	69.2	13.30	32.55
	Foreman Average, Inside ($.50 over trade)	26.45	20.2				37.0	9.80	36.25	68.0	18.00	44.45
	Foreman Average, Outside ($2.00 over trade)	27.95	20.2				37.0	10.35	38.30	68.0	19.00	46.95
Clab	Common Building Laborers	19.80	21.9				38.7	7.65	27.45	69.7	13.80	33.60
Clam	Common Maintenance Laborer	14.85	7.4				24.2	3.60	18.45	55.2	8.20	23.05
Asbe	Asbestos Workers	28.55	19.7				36.5	10.40	38.95	67.5	19.25	47.80
Boil	Boilermakers	30.05	17.7				34.5	10.35	40.40	65.5	19.70	49.75
Bric	Bricklayers	25.90	19.4				36.2	9.40		67.2	17.40	43.30
Brhe	Bricklayer Helpers	20.00	19.4				36					33.45
Carp	Carpenters	25.20	21.9									
Cefi			12.8									

Figure 17.6

282

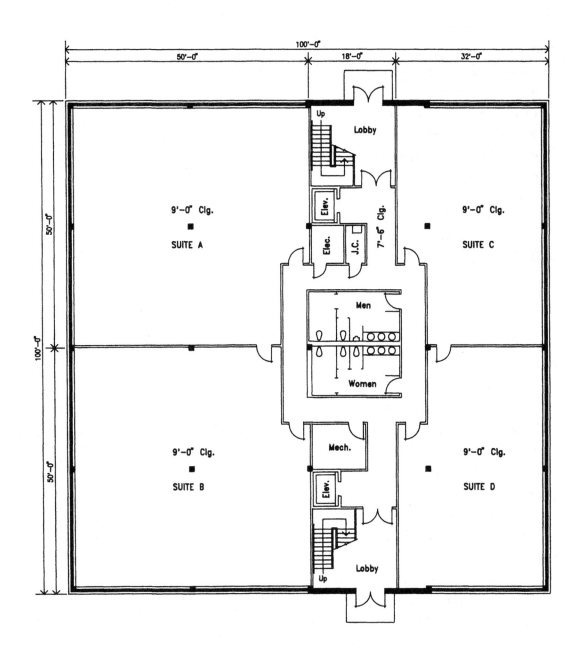

FLOOR PLAN
Scale: 1/16" = 1'-0"

Figure 17.7

Cleaning Specifications for a Typical Office Building

Contractor shall provide all labor, equipment, and material necessary to clean a 6-story, 60,000 S.F. building located at 156 Main Street, Anytown, USA. Note: paper towels, toilet tissue, and hand soap are furnished by owner.

Daily Work

Vacuum entrance mat carpeting

Vacuum traffic areas

Dust mop all resilient floors

Wash glass at entrance doors

Clean and polish drinking fountains

Clean elevators

Clean restroom fixtures, mop floor, empty receptacles, replenish supplies (paper and soap)

Empty all wastepaper containers

Weekly

Vacuum all areas

Damp mop and spray buff all resilient floors

Sweep stairwells

Damp wipe restroom walls and partitions

Monthly

Damp mop stairwells

Machine scrub restroom floors

Shampoo elevator cab carpet

Dust venetian blinds

Scrub and wax resilient floors

Quarterly

Vacuum all diffusers

Wash ceramic tile restroom walls

Semiannual

Strip and refinish resilient tile floors

Shampoo all carpeting

Annual

Clean all fluorescent light fixtures

Figure 17.8

COST ANALYSIS

PROJECT: General Maintenance Estimate CLASSIFICATION: 6 Story Office Building, 60,000 S.F. Total Floor Area
LOCATION: Anytown, USA ARCHITECT: As Shown ESTIMATE NO: 1701
TAKE OFF BY: PRW QUANTITIES BY: PRW PRICES BY: Means EXTENSIONS BY: TJA DATE: 1996 CHECKED BY: PRW

DESCRIPTION	REFERENCE	QUANT.	UNIT	FREQ. /YEAR	TOTAL UNITS	MATERIAL COST	MATERIAL TOTAL	LABOR COST	LABOR TOTAL	EQUIP. COST	EQUIP. TOTAL	SUBCONTRACT COST	SUBCONTRACT TOTAL	LABOR-HOURS /UNIT	LABOR-HOURS TOTAL
Daily															
Vacuuming, entry mats	FM&R 018 570 0750	0.20	M.S.F.	250.00	50.00			14.85	742.50	3.13	156.50			1.000	50.0
Vacuuming, traffic areas	FM&R 018 510 0090	11.25	M.S.F.	250.00	2,812.50			2.26	6,356.25	0.48	1,350.00			0.152	427.5
Dust mop floors, 24" mop head	FM&R 018 530 0150	18.00	M.S.F.	250.00	4,500.00			3.49	15,705.00					0.235	1,057.5
General cleaning, glass doors	FM&R 018 570 0390	8.00	Ea.	250.00	2,000.00			0.71	1,420.00					0.048	96.0
General cleaning, water fountains	FM&R 018 570 0490	6.00	Ea.	250.00	1,500.00			0.25	375.00					0.017	25.5
Clean carpeted elevators	FM&R 018 570 0080	2.00	Ea.	250.00	500.00			2.48	1,240.00	0.52	260.00			0.167	83.5
Empty waste baskets	FM&R 018 570 0290	400.00	Ea.	250.00	100,000.00			0.12	12,000.00					0.008	800.0
Restroom, fixture cleaning, toilets	FM&R 018 570 0110	30.00	Ea.	250.00	7,500.00			0.66	4,950.00					0.044	330.0
Restroom, fixture cleaning, urinals	FM&R 018 570 0120	6.00	Ea.	250.00	1,500.00			0.62	930.00					0.042	63.0
Restroom, fixture cleaning, sinks	FM&R 018 570 0130	36.00	Ea.	250.00	9,000.00			0.58	5,220.00					0.039	351.0
Restroom, floor, damp mop, 24"	FM&R 018 530 0060	2.40	M.S.F.	250.00	600.00			3.39	2,034.00					0.229	137.4
Refill soap dispenser	FM&R 018 570 0210	24.00	Ea.	250.00	6,000.00			0.19	1,140.00					0.013	78.0
Refill sanitary napkin dispenser	FM&R 018 570 0220	6.00	Ea.	250.00	1,500.00			0.31	465.00					0.021	31.5
Refill paper towel roll dispenser	FM&R 018 570 0240	12.00	Ea.	250.00	3,000.00			0.35	1,050.00					0.024	72.0
Refill toilet tissue dispenser	FM&R 018 570 0250	30.00	Ea.	250.00	7,500.00			0.08	600.00					0.006	45.0
Refill sanitary seat dispenser	FM&R 018 570 0260	30.00	Ea.	250.00	7,500.00			0.12	900.00					0.008	60.0
Weekly															
Vacuum carpet, 16" upright, all areas	FM&R 018 510 0100	36.00	M.S.F.	52.00	1,872.00			4.75	8,892.00	1.00	1,872.00			0.320	599.0
Sweep stairs and landings	FM&R 018 570 0020	12.00	Floor	52.00	624.00			1.49	929.76					0.100	62.4
Washing walls, hand, vinyl surface	FM&R 018 540 0080	5.00	M.S.F.	52.00	260.00			36.50	9,490.00					2.462	640.1
Restrooms, partition cleaning	FM&R 018 570 0270	30.00	Ea.	52.00	1,560.00			1.24	1,934.40					0.083	129.5
Damp mop, 24oz., resilient tile floor.	FM&R 018 530 0060	18.00	M.S.F.	52.00	936.00			3.39	3,173.04					0.229	214.3
Spray buffing, 350 RPM, 24" dia.	FM&R 018 530 0910	18.00	M.S.F.	52.00	936.00			4.07	3,809.52	0.86	804.96			0.274	256.5
Monthly															
Damp mop stairs and landings	FM&R 018 570 0040	12.00	Floor	12.00	144.00			2.64	380.16					0.178	25.6
Scrubbing, 350 RPM, 24" dia., resilient	FM&R 018 530 1060	2.28	M.S.F.	12.00	27.36			3.22	88.10	0.68	18.60			0.217	5.9
Finish application, mop, resilient floor	FM&R 018 530 1160	18.00	M.S.F.	12.00	216.00			8.50	1,836.00					0.571	123.3
Damp wipe, venetian blinds	FM&R 018 560 0070	600.00	Ea.	12.00	7,200.00			0.74	5,328.00					0.050	360.0
Carpet cleaning, portable extractor	FM&R 018 520 0030	0.14	M.S.F.	12.00	1.73			14.85	25.66	0.89	1.54			1.000	1.7
Cleaning supplies	FM&R	1.00	L.S.				150.00								
Quarterly															
Vacuum overhead air vents	FM&R 018 570 0760	600.00	Ea.	4.00	2,400.00			0.25	600.00	0.05	120.00			0.017	40.8
Restrooms, ceramic wall cleaning	FM&R 018 570 0280	5000.00	S.F.	4.00	20,000.00			0.05	1,000.00					0.003	60.0
Semiannual															
Carpet cleaning, portable extractor	FM&R 018 520 0030	36.00	M.S.F.	2.00	72.00			14.85	1,069.20	0.89	64.08			1.000	72.0
Stripping, 350 RPM, 24" dia., resil. floor	FM&R 018 530 1130	18.00	M.S.F.	2.00	36.00			11.30	406.80	2.38	85.68			0.762	27.4
Wet pick-up, 20" opening attachment	FM&R 018 530 1150	18.00	M.S.F.	2.00	36.00			4.93	177.48	1.04	37.44			0.332	12.0
Finish application, mop, resilient floor	FM&R 018 530 1160	18.00	M.S.F.	2.00	36.00			8.50	306.00					0.571	20.6
Annual															
Fluorescent fixture, clean	FM&R 018 160 2518	600.00	Ea.	1.00	600.00			5.35	3,210.00					0.182	109.2
TOTAL BARE COSTS							150.00		97,783.87		4,770.80				6,468.3
Markups: Mat 25%, Labor 55.2%, Equip. 10%							37.50		53,976.70		477.08				
TOTALS							187.50		151,760.57		5,247.88				

Figure 17.9

Chapter 18

Value Engineering: Continued Maintenance vs. Capital Improvements

William H. Rowe III, AIA, P.E.

Hypothetically, it is possible to construct a building so well that it will never have any maintenance costs. Such a building would be impossible to afford, however, because it would need to be built perfectly. At the other extreme, it is hypothetically possible to spend no money on a building, which would result in an environment that must be constantly "maintained" by an endless stream of materials and labor. While this "building" costs nothing to construct, it too is expensive because of the impossibly high maintenance costs.

As a general rule, systems that are more durable cost more. A concrete building will last longer than a wood shack, and is initially more expensive. In value engineering, the critical question is, "What level of durability is needed?" Hospitals, town halls, and telephone company buildings fill crucial functions that require a high level of construction durability. On the other hand, retail stores, restaurants, and speculative office buildings have a shorter economic life, with a more frequent turnover and more substantial renovation. Consequently, less durable building construction is acceptable for this type of building.

Durability can sometimes be a disadvantage if components are used that are not flexible. In certain types of buildings that are renovated frequently, such as hospitals and speculative office buildings, it is generally better that the systems are easy to remove and replace. In fact, "walls on wheels" are often the objective.

When major repairs or renovations are considered, it is important to ask the question, "How committed will the owner be to proper maintenance?" There are significant pressures to maintain market-driven facilities such as shopping malls, restaurants, private offices, and theaters. Failure to do so can result in loss of revenue to competitors who provide a cleaner, more comfortable, and pleasant environment. There are, however, users with a history of poor maintenance, and for them overall cost effectiveness is usually accomplished by choosing better quality components at the outset, thereby minimizing the damage that will result from a less than ideal maintenance policy. Municipalities, schools, churches, and nonprofit groups, for example, do not have the market-driven pressures related to upkeep, and are often subject to "once in a lifetime" funding opportunities to build or substantially renovate their buildings.

All buildings have expenses that are distributed between the first (acquisition or construction) cost and the annual (maintenance, repair, and improvement) costs. Once the first cost choices are made, the consequential maintenance costs are, to a great extent, established. The value engineering process analyzes the combined effect of initial and long-term costs to determine this aspect of a building's financial success over its useful life.

What is Value Engineering?

Value engineering for the facility manager is simply a rational method of making cost-efficient choices between alternatives when upgrading, maintaining, and making system replacements. By analyzing both the program needs and the design capabilities of a facility, it is possible to determine how to maximize overall performance while minimizing costs. Most building systems wear out in time; many deteriorate aggressively. Maintenance of roofs, windows, mechanical systems, carpeting, and paint finishes requires more attention and more money each year they age. Value engineering compares the annual costs for a particular system with the cost of replacement alternatives.

Common Examples of Value Engineering

For certain building systems that are likely to wear out, the annual operating costs increase each year. In determining the annual budget, it is necessary to decide whether the escalating annual operating costs for a particular system (or decreasing building revenue) justify replacement, in that replacement would result in overall lower costs (or increased revenue). This makes the case for a capital improvement to that system.

Some choices do not affect cost directly but may result in a change for the better. For example, building colors are generally thought of as an aesthetic judgment only. However, a poor choice of color may have a negative impact on the usability of the space, be difficult to maintain, or affect the attitudes of both building users and maintenance staff, and hence the care or maintenance associated with the facility. In value engineering, these factors are identified so that choices can be made as effectively as possible. The following is a discussion of some of the factors to consider in analyzing particular systems.

Lighting

Maintenance of lighting is an annual expense. The costs of relamping, the energy cost to operate the lights, and the additional costs to cool the building from lighting loads are all significant. Electric companies play an active role in encouraging efficient use of electric power. In many areas, the electric company will pay all or part of the cost for a building to be completely retrofitted with high-efficiency flourescent fixtures. In this case, not only is there no cost for the capital improvement, but the following year's costs for both the lighting and the related HVAC cooling costs will be lower.

Boilers

Making the choice between maintaining existing boilers and replacing them is one of the easier value engineering analyses to undertake. Older boilers are less efficient than newer models and require more care. In establishing costs of boiler operations, annual repair and maintenance and annual fuel costs must be considered. Maintenance and fuel costs are easily obtainable, and the comparative costs of boiler replacement, as well as the fuel and maintenance costs for a new boiler, are relatively simple to calculate.

Reference books as well as sales representatives, consulting engineers, maintenance staff, purchasing departments, and operations histories can all provide useful information regarding boiler efficiency. Boilers over 20 years old usually have operating efficiencies below 75% and new boilers have operating efficiencies above 90% — thus it is not difficult to calculate the relative merits and payback associated with replacement of an older boiler.

An additional benefit of boiler replacement may be an increase in useable floor area for other programmatic functions, since newer boilers are considerably smaller than older boilers.

When contemplating removal of an existing boiler, the costs of demolition and related asbestos removal, if any, must be considered as a part of the replacement cost. In value engineering boiler replacement, the options of dual fuel boilers, modular boilers, and burner replacement only should be considered.

Roofs

Over time, roofs wear and leak. Joints separate, membranes become worn or punctured, and flashings corrode. Each year, the building's repair history can illustrate the costs of patching. The following factors should be considered when evaluating the costs of maintaining a deteriorating roof:

- Annual repairs to fix leaks
- Water damage
- Plaster repair
- Carpet replacement
- Replacement of equipment damaged by water
- Loss of rent (or below-market rent) because of tenant lawsuits or perception of a poor quality building
- Increased energy costs resulting from a poorly insulated roof

The expected annual operating costs should be established using the above list as a guideline. Then, the costs for replacing the roof should be obtained for more than one alternative. For example, if the existing roof is tar and gravel, alternatives may include a synthetic rubber roof, ballasted and mechanically fastened. The estimated cost of the new roof can be compared with the present worth of the annualized repairs for the current roof to determine how effective a replacement will be.

Windows

In older buildings, window replacements are often considered. The principal direct and indirect costs associated with window repair and replacement are as follows:

- Maintenance costs for painting, sash repair, and caulking
- Energy losses resulting from single-thickness glass, leaks at cracks, or poorly insulated window sections
- Water damage to carpets, paint, and plaster from leaking windows

When the above costs are analyzed and compared with alternatives such as storm windows, interior glazing, new double-insulated glass, and a variety of frame types including aluminum and vinyl-coated wood, some rational choices can be made. Unfortunately, it is often difficult to justify the cost of full replacement.

Discussions concerning window replacement often involve tenants and users of the space who are dissatisfied with windows that are inoperable and/or drafty. The value engineering process is complicated by the fact that it is difficult to assign costs to the implications of dissatisfaction. Clearly, unhappy tenants and users will cause management to waste time on complaints and may result in difficulty in collecting or raising rents. While it is difficult to establish hard costs, these human relations issues are often the determining factors in window replacement decisions.

Carpet

The choice to replace carpet is usually made when the existing carpet is so worn that it is a hazard or so unsightly that it is unacceptable to management. In some instances, carpet may harbor fungi or other organisms that are determined to be a health hazard. The carpet may also be upgraded as part of a general building or tenant space improvement.

In selecting new flooring, alternate materials may be appropriate in certain locations such as lobbies, toilets, laboratories, and offices. When carpet is selected, the weave, density, type of fabric, pile height, and related specifications will have a bearing on its cost and life span, depending on the application. Generally, for high traffic areas a low-pile, high-density carpet will wear best.

Control Systems

Perhaps the best opportunity for value engineering is a review of the building's energy control systems. Simple, effective controls can often save considerable money, far beyond their implementation cost. Setback thermostats, outdoor reset temperature devices, occupancy sensors, and economizer systems are all comparatively inexpensive and can be dramatically effective in saving money when properly used. For large facilities with large maintenance departments, properly designed direct digital control (DDC) systems with remote station terminals can permit facility managers to diagnose and often cure many problems without dispatching staff to buildings. The savings in time to travel to and around a site can justify the costs of installing the more elaborate DDC systems.

Key Considerations of the Value Engineering Team

Facilities usually employ several key players who make major budget decisions: the owner, the users (or customers or tenants), the building operators, the designers (architects and engineers), and the contractors. When working as a team or with a specialist in value engineering, they are expected to challenge each other cooperatively to reach a consensus on the goals of the facility and the choices to be implemented. This is done during the preparation for the annual budget and continues after it.

The value engineering team usually works together in a workshop fashion. By meeting regularly and giving each other assignments to review the feasibility of alternative methods and generate their associated costs, the value engineering team can establish a broad overview of successful alternatives.

The goal of the team is to reach decisions with the greatest acceptance at the least cost. Since the owner must live with the decisions, he or she usually takes the lead in establishing how acceptable one alternative is compared to another. The value engineering team can provide the relative costs. When a particular alternative is clearly preferred by the owner and costs less than other options, the choice is obvious and the work pays off. When the owner chooses a more expensive alternative, it may be necessary to determine whether that alternative will pay for itself over time and thus be justified.

When considering capital improvements, it is helpful to divide the improvements into the following categories, which will assist management in understanding the importance of and establish the priorities for recommended improvements:

- Life safety code requirements
- ADA requirements
- Repairs to currently inoperative equipment
- Upgrades to existing systems to enhance operation
- Programming of new, changed, or additional functions into the facility

Once the costs for each type of capital improvement are broken down, it is possible to reach better decisions. When a facility undergoes a major upgrade, the results of the value engineering process will also trigger discussion on the overall building suitability, as well as consideration of alternatives such as building a brand new facility, moving to a rented facility, deferring certain expenses, or moving ahead as planned with selected modifications developed by the team.

Value Engineering and Constructability

One key element in a value engineering study is the review of the project for constructability. This involves a determination of how easily the project can be built. Are streets too narrow for delivery of certain parts? Do excavations require unreasonable amounts of sheeting, shoring, and bracing? Do the project delivery dates coincide with the availability of manufactured equipment that is to be delivered to the site? Do asbestos or confined spaces limit the workforce that can be assigned to execute the work? In evaluating the constructability of the project, the actual means and methods of construction will be examined as a reality check on how suitable the project is to build.

Additional Considerations in Justifying Capital Expenditures

To evaluate the condition and program needs of a facility, a facility audit or assessment is performed. This audit results in an analysis of the current condition of the facility, the objectives for the facility in the future, and the actions that are necessary to make the changes to reach the goal. (See the Introduction to Part 2 for more on the facility audit.)

There are many valid reasons for capital expenditures that do not involve an analysis of maintenance and repair costs: older facilities faced with obsolescence, ceilings that are too low for necessary mechanical systems, the presence of hazardous materials such as asbestos or PCB's, uneven floors that make access for the disabled difficult, poor building controls, and insufficient outdoor air.

In addition, many buildings are being used differently. Offices, for example, are being affected by the growth of home computer use, which has resulted in many more part-time workers, increased need for computer wiring, and sharing of office work space by more than one worker. The introduction of programmatic elements like day care, restaurants, commuter vans, locker rooms and showers has also led to capital improvements. Overall, the very nature of buildings and how they are used is changing.

The single most effective part of value engineering is knowing what work is to be undertaken in the first place. In some instances, facility managers or operators tend to "create" work that maintains staff because their fees are based on a percentage of the work undertaken. Such procedures do not contribute to improving the outcome of a facility. Providing extra emergency generators, upgrading facilities to code even when not required, and redoing entire sections of electrical or mechanical systems without first determining the cost benefit of such work have resulted in excessive expenditures that, when discovered, have in some cases resulted in dissolving the entire facility department and outsourcing the maintenance function.

Value Engineering and Facility Assessment

A facility assessment reviews an existing facility to determine how well it is currently functioning and what deficiencies exist. In particular, facilities are examined to see what work is needed regarding codes, ADA, repairs, upgrades, and programming. Following are questions to ask in conducting the facility audit.

- **Codes:** What are the life safety code deficiencies? Items such as legal means of egress, panic hardware, fire alarms, smoke evacuation, fire-rated walls, door closers, and sprinkler systems are commonly reviewed.
- **ADA:** What must be done to meet ADA requirements? Access to public spaces is most important. Major categories of work include ramps, toilets, signage, and door hardware.
- **Repairs:** How well do the systems operate? What systems are broken or need upgrade? What is the expected useful life of the existing systems? Audit checklist forms are available in *Means Facilities Maintenance & Repair Cost Data.* The entire process is addressed in *The Facility Manager's Reference*, by Harvey H. Kaiser, also published by R.S. Means Co., Inc. The basic idea is to assess the condition and cost to repair the systems.
- **Upgrade:** What areas should be improved beyond code or repair of breakdowns? Are there items, such as finishes or HVAC control systems, that should be redone or improved in conjunction with the required improvements? Many of these issues involve assessment and review of the engineering systems of a facility.
- **Programming:** What programmatic changes are likely to occur that will trigger adaptation of the building for new functions? In determining these changes the following are helpful to review:
  - *Quantity of Space.* Does the facility have enough space for the projected needs of the users? By performing a detailed space-by-space analysis of each area or department and interviewing both staff and management where necessary, a picture of the growth needs for the facility can be obtained.
  - *Quality of Space.* Are the finish materials too worn? Does the facility have a successful positive working ambiance?
  - *Spatial Relationships.* Are the spaces and departments dispersed throughout the facility in a way that they relate well together? Are the departments that work together close to one another—are the "adjacencies" correct?
  - *Space Efficiency.* Are individual layouts efficient and is space well utilized? Are the fixtures and furnishings functional and adequate?

The above categories are further analyzed by trade as follows:

- Expenditures for repairs and/or testing of *Foundations, Substructures, and Superstructures* are not common for existing buildings. Such expenditures do occur when new loadings such as mechanical equipment are planned. In choosing a structural system, consider the effect on building height. A thick structure or sandwich means more enclosure wall material, and greater story-to-story height. One foot thicker construction means loss of one entire story in ten—or a 10% loss in revenue or use—with no significant decrease in overall project cost.

- Expenditures for *Exterior Closure and Roofing* include roofing, garage work, windows, doors and siding materials.
- Expenditures for *Interior Construction* include improvements to common areas such as lobbies and corridors, common toilets, elevator lobbies, and building insulation. This category can also include tenant improvement items and interior and/or exterior signage.
- Expenditures for *Conveying Systems* include modernization, repairs and mandated improvements to elevators, escalators, lifts, cabs, and/or controls.
- Expenditures for *Plumbing* include toilet upgrades and/or plumbing system repairs.
- Expenditures for *HVAC* include replacement of major and/or minor equipment. Consideration is given to first or installed cost vs. annual cost, comparing annual energy use and maintenance costs for the new vs. existing system.
- Expenditures for *Fire Protection* include sprinkler work. Retroactive fire protection legislation has been enacted in many states for high-rise buildings. These requirements would also include repair and upgrades of fire protection equipment. Another consideration is the added benefit of insurance premium reductions if improvements such as fire-rated walls or sprinklers are made.
- Expenditures for *Electrical* work include generation, distribution, lighting, security, and/or fire alarms. Issues to consider are the use of emergency generators, incandescent vs. fluorescent lighting, copper vs. aluminum wire, and reserve capacity.
- *Landscaping* expenditures include plazas, walks, paths, roads, parking, and planting.
- *Miscellaneous* expenditures include items such as asbestos removal, accessibility renovations required by the ADA, and other expenditures not directly tied or limited to one of the above listed systems.

Capital improvements for property should be reviewed in some detail under the above categories. The individual items in each category should be placed in a table that summarizes the capital expenditures in these categories. An example is shown in Figure 18.1. The Capital Improvements Summary is initiated when an improvement is budgeted, and maintained until the work is completed. The first two columns in the table show the budgeted and actual capital expenditures for the year. The third column indicates the variance between the actual and budgeted amounts. The fourth and fifth columns show the property manager's budgets for the projected fiscal year.

Tenant improvements and leasing commissions should be included as separate line items in the table because they can represent a large proportion of the capital budget. However, the close relationship between actual tenant improvement expenses and lease negotiations should be carefully understood in making a meaningful analysis.

Effective Choices
In value engineering, choices often must be made in the larger context of construction cost, constructability, phasing, and future renovation and

operating costs. If a brick facade is being considered, it has two disadvantages: it is expensive to build and it may take considerable time as compared to an EIFS (Exterior Insulation Finish System). On the other hand, the building may require a brick facade to remain consistent with the neighborhood or the organization's desired image.

Many opportunities exist for value engineering in the building's mechanical, electrical, and plumbing systems. These systems involve the greatest share of maintenance costs and offer the greatest number of options. Buildings can be heated with steam, hot water, and air. In each of these systems, the type of pipe and duct that can be used involves many additional choices. Steam can use Schedule 40 or Schedule 80 pipe, welded or threaded. Hot water pipe can be steel or copper, which can be welded, threaded, or soldered. Air systems can use aluminum, galvanized steel, or a host of preformed duct materials made of fiberglas, gypsum wallboard, and aluminum coatings. These can be evaluated against the expected maintenance to be provided and the resulting operating and maintenance costs.

Phasing

Another value engineering related consideration in planning repairs is determining how major projects should be phased. One important question is whether or not the building's occupants will move out of their affected space during the work. If they stay, construction time is likely to be extended and the presence of construction activity will be a disruption to daily activities. The extra costs and inconveniences to the contractor may be such that it is worth relocating the activities elsewhere for the construction period.

In many projects, the work may proceed in distinct phases, for example: exterior work, then interior, or boiler plant replacement followed by distribution piping and interior painting. When such divisions are possible, there may be an advantage to dividing the project into phases and starting them at different times. Inevitably there are some coordination problems,

	Current Year Capital Expenditures			Next Year Capital Expenditures (Proposed)	
	Budget from MAL*	Actual	Variance	Budget from MAL*	Variance
Property A Building Imp.					
Tenant Imp.					
Leasing Comm.					
Total Capital Exp.					

*MAL refers to Management and Leasing reports prepared by the individual property managers.

Figure 18.1

but these are often offset by the considerable savings in overall project time or reduction in construction cost because of early or seasonal starts.

Presenting a Request for Capital Improvements

Facility and plant maintenance managers are responsible for ensuring that a facility operates efficiently within the budget limitations of the owner's organization. Maintaining that efficiency sometimes requires going beyond normal annual maintenance, to making a case for capital improvements.

If an emergency system replacement is needed due to an unpredictable breakdown of equipment or other damage, it must be attended to directly. Non-emergency situations call for an organized, well-documented presentation to senior management. The following issues must be addressed:

- The system's maintenance history, including cost and frequency (presumably escalating) of maintenance, as well as downtime (in the case of equipment).
- The projected increase in efficiency that will be realized by the recommended replacement system. This information might include reduced maintenance and operating/energy costs, increased employee productivity and/or morale, or an improved image in the community. New equipment may offer tremendous advantages in terms of technological developments that did not exist in the original equipment. These documented points represent the "payback" behind the proposed replacement or improvement.

Conclusion

The useful life and eventual replacement of a system should, of course, be considered when it is first installed, as part of a facility's long-range maintenance plan. This approach, together with ongoing maintenance and repair documentation over the system's years in service, provides the best opportunity for a positive outcome when a capital improvement appropriation request must be made.

Chapter 19

Deferred Maintenance

James E. Armstrong

All aspects of maintenance are necessary to extend the life of equipment and facilities. The purpose of this chapter is to put order to the chaos and to provide some tools to manage one of the most difficult aspects of facilities management: deferred maintenance. The key to this system is a collective awareness of the condition of equipment and facilities, and prioritization of those items in a collective system.

Most facility managers do not actively defer maintenance. We know that someday it will get done. In the ideal world, things don't break, because you know the life span of an item and it is replaced or maintained prior to failure. In the real world we manage the best we can with what we have for resources.

Deferred maintenance is an approach we all practice in our personal as well as professional lives. Just take a minute to think of all those issues or problems that you will attend to someday. Well, today is the beginning of someday! Almost every facility has some sort of deferred maintenance — from ductwork that needs cleaning, to paint that is peeling. How many times have you passed by an item and said, "I need to take care of that?" To maintain your sanity as a facility manager, you develop a special sort of tunnel vision, deferring items that need attention, yet not necessarily this instant. The first step is making the commitment to attend to these items, and to document those that have been deferred. In the process, a true understanding of the facility's condition emerges. The deferred list is essential to the management of limited financial resources, both today and into the future.

The discussion of deferred maintenance in this chapter will utilize the following terms.

Repair Maintenance: Work that is performed to put equipment back in service after a failure, to extend the life of equipment, or to make its operation more efficient. Examples include replacing valve packing on a leaky valve, replacing a burned out lightbulb, or repairing the broken wheel on the boss' chair. Think of how many times you have been called because it is "hot" or "cold" in someone's office; providing a response to such complaints also falls into the Repair Maintenance category.

Planned Preventive Maintenance: Service or replacement of equipment components for the purpose of extending its life or making its operation more efficient. Examples include cleaning air registers and replacing air filters, group re-lamping of light fixtures, lubricating the bearing of a pump or blower, or conducting a furniture replacement program.

Predictive Maintenance: An inspection process designed to estimate the condition of equipment and possibly predict the failure rate of equipment or its components. Examples include monitoring bearing vibration, or sampling engine oil to check for acidity and metal content. Other examples include testing gauges on a bench to ensure true readings, or monitoring wear of bearings.

The Facility Audit

Understanding the starting point for "getting around to" deferred maintenance is key to an effective system. We are now entering the stage of identifying all known and potential problems. This requires a thorough audit of the equipment, buildings, and grounds that are the responsibility of the maintenance department. Note both the condition and any potential deficiencies. Look for items that could develop into future repair, preventive or predictive maintenance issues — or define the item as a deferred maintenance item. The key is thinking creatively of potential problems, both now and in the future. The audit process involves four steps:

1. Break the facility down into categories that make auditing most effective. You know the needs of the operation, so develop a breakdown that is simple and appropriate. Some categories might be office space or buildings, public areas, mechanical rooms, roofs, and manufacturing. Audit checklist forms are available in *Means Facilities Maintenance & Repair Cost Data.*

2. Develop audit teams comprised of people who are most familiar with the particular areas, as well as capable of finding potential maintenance issues. Try to get people who work in the area on the team. This is not a time-consuming process if you listen to or involve the people whose work space you are surveying. Everyone has ideas (as well as complaints), and these ideas become a tool to fully understand the maintenance needs of the facility. (Remember when *you* had an idea and no one would listen?) Sample teams might consist of:
   - For the office space: 1 custodian, 1 secretary, 1 carpenter
   - For a mechanical space: 1 mechanic and 1 electrician

3. Audit the facility, making sure that the staff has no false hopes that everything will get an immediate commitment, but also that nothing is too small to write down. Look carefully at the equipment and buildings when auditing your facility, noting any unusual or excessive equipment noise, potential tripping hazards, questionable air flows, creaks in floors, etc. One positive side effect of the surveys is that staff become more aware of their work area. They learn, for example, the location of their fire exits (staff can now find their way out in an emergency), electrical outlet and extension cord locations and numbers (did you ever get a work order for new outlets or extension cords when all along there was one behind the file cabinet?), and bookcase capacity and its security (consider the liability if one falls).

4. Assemble a list of all items that require attention. Some of the items on the list could be repair maintenance, and some preventive maintenance; categorize them accordingly. There will be odd items like a light switch whose purpose no one can identify, and a lot of repair work orders may be generated that will

keep you busy. It will become apparent that some previously unidentified things need to be done on a regular basis (planned maintenance). It's important not to become overwhelmed. Assemble the list of items that you know you and your staff cannot get to within the next few months. This is your deferred maintenance list.

Figure 19.1 is an excerpt from a Sample Facilities Audit.

Note: Audits should be done regularly, or at a minimum every two years.

Sample Audit Issues
General Building Condition

- Roof: Note any leaks, tears, cupping, and/or bubbling.
- Flashing: Note condition on the perimeter and rises. Also, note expansion joints, drip edge, and wear patterns.
- Sidewall: Brick, metal, clapboards, etc. Note transitions to rake boards or wall penetrations such as windows and doors. Record any spalling or rot conditions.
- Building Trim: On rake boards or window trim, look for rot or dents. Also, note water flow patterns and/or ventilation issues.
- Windows: Whether or not windows are insulated, look for caulking deterioration, water flow, cracks, and rot.
- Exterior Doors: Check for threshold rot or water tracking. Also check lock security, and ensure accessibility of fire exits. Note the integrity of the doors and exit hardware.
- Structural: Look for sill rot, beam fire-coating integrity, floor loading, accessibility issues, and sagging floors or roof lines.

Office Conditions

- Carpet: Look for wear patterns, tears, stains, and tripping hazards.
- Walls: Check for dents or holes, water stains, bowing and spalling.
- Interior Doors: Evaluate lock security and ensure that fire exits are accessible. Note the integrity of the doors and hardware.
- Office Partitions: Check wall security; e.g., make sure there is no excessive movement.
- Electrical: Note the number and condition of outlets and light switches.
- Telephone: Find out if the system is fulfilling requirements.
- HVAC: Make sure the system is maintaining a comfortable environment.
- Ceiling: Check for water stains, plaster cracks, and holes.
- Associated Wall Trim and Treatments: Check condition of the cove base, curtains, blinds, etc.
- Office Furniture: Check the door slides, wheels on chairs, and drawer locks.
- Security: Find out if there are problems with room and building accessibility and/or occupant issues.
- Bathrooms: Check to see that facilities are adequate and investigate any cleanliness, odor, and/or ventilation problems, as well as faucet leaks and compliance with ADA requirements.
- Hallways and Corridors: Maintain four-foot-wide egress; look for excessive storage of items that obstruct passageways.
- Entrances: Check to see that entrances are unobstructed and comfortable.

Excerpt from a Sample Facilities Audit

a	C	Project Description	COST	Prio.	Remarks
	PRO01	Roof renovations (Sports Science Building)	$35,000.00	2	
	PRO03	Repair ventilation to restrooms	$1,000.00	2	
	PRO04	Asbestos abatement	$8,500.00	3	
	PRO05	Rebuild pneumatic system	$8,000.00	3	For bldg.heat control
	PRO06	Rebuild six univent in nursing Lab.	$3,000.00	3	
	PRO07	ADA Compliance	$68,000.00	3	see ADA report under separate cover
Campus wide					
	PRO09	Paving of roadways and parking lots	$80,000.00	2	Recommendation is to perform this over a
	PRO15	Purchase Roto-Tiller	$5,000.00	3	
	PRO16	Misc. small drainage projects	$10,000.00	3	
	PRO17	QUAD area drainage	$10,000.00	3	
	PRO18	Renovate quad lawn	$20,000.00	3	
	PRO19	Renovate Burpee Quad lawn	$7,000.00	3	
	PRO20	Post &rail installation (campus wide)	$10,000.00	3	
Residence Halls-GEN;					
	PRO21	Repair water main shutoffs	$24,000.00	2	
	PRO22	Replace thermostatic control valves	$37,500.00	2	Approx.500 valves: Payback as soon as 2 YRS. IN PROGRESS,ON GOING
SMIT HALL					
	PRO23	Repair roof,flashing,ice guard,slates Chimney and Chimney caps	**********	2	
	PRO24	ASBESTOS ABATEMENT	$3,500.00	2	
	PRO25	ADA Compliance	**********	2	SEE ADA REPORT UNDER SEPARATE COVER
	PRO26	Painting & Patching-Restrooms	$3,000.00	4	
	PRO27	Reinsulate pipes &Breeching in boiler room	$10,000.00	3	
HALL					
	PRO28	Repair roof flashing	$2,250.00	2	
	PRO29	ADA COMPLIANCE	$54,100.00	3	SEE ADA REPORT UNDER SEPARATE COVER
	PRO30	Recarpet building	$6,000.00	4	
	PRO31	Upgrade kitchen	$4,500.00	4	Ventilate, new sink,cabinets, drawers Counters, Ceiling, & Repaint.
AUSTIN HALL:					
	PRO32	ASBESTOS ABATEMENT /	$7,000.00	3	Stairway&Room Floor tile replacement
	PRO33	Upgrade hallway&room lighting	$3,500.00	3	INCREASE LUMENS & EFFICIENCY
	PRO34	Replace floor covering in STUDENT Rooms	$11,000.00	3	
JONES HOUSE					
	PRO35	ADA COMPLIANCE	$8,000.00	3	SEE ADA REPORT UNDER SEPARATE COVER
LIBRARY					
	PRO36	Bldg.-Wide Ventilation	$35,000.00	2	Air circulation only. Does not include A/C
	PRO37	FOUNDATION LEAKS/LANDSCAPING	$10,000.00	2	
	PRO38	Paint Exterior	$23,500.00	3	Entire Exterior of BLDG.
	PRO39	ADA COMPLIANCE	$11,000.00	3	SEE ADA REPORT UNDER SEPARATE COVER
GREENE SCI.CTR.					
	PRO40	Repair Ventilation	$25,000.00	2	Includes Replacement of one H&V coil
	PRO41	ADA COMPLIANCE	$55,400.00	2	SEE ADA REPORT UNDER SEPARATE COVER

Projections   February 16, 1997

Project Description	COST
SAFETY/COMPLIANCE	
Asbestos Removal	$75,000.00
Repair of Fire Escapes	$65,000.00
Oil Tank Repair (Bulk Tanks)	$775,000.00
Outside Lighting	$8,000.00
Update Fire & Sprinkler Systems	$252,000.00
Removal of PCB transformers & relocate	$375,000.00
Install Stairtreads in Stairwells	$22,500.00
Fire Rated Doors	$465,000.00
Locking Systems Campus-Wide	$161,875.00
ADA Compliance Campus-Wide	$926,300.00
Safety Totals	$3,125,675.00

Project Description	COST
Recruitment / Retention	
Update Heating / Ventilation / Controls	$402,500.00
Carpeting Repairs / Replacement	$104,550.00
Painting / Patching / Interior / Exterior	$120,555.00
Residential Kitchens	$76,000.00
Lighting Repairs / Upgrades	$86,000.00
Bathroom Renovations	$692,000.00
Recruitment / Retention Totals	$1,481,605.00
Deferred / Other Maint	
Roofing / Structual / Brick Repointing	$1,350,000.00
Window Replacement	$643,358.00
Insulation Buildings / Piping	$76,480.00
Grounds Repair	$152,550.00
Plumbing Repairs	$85,000.00
Deferred / Other Totals	$2,307,388.00
Grand Total	$6,914,648.00

Figure 19.1

Excerpt from a Sample Facilities Audit (cont.)

Project #		Project Description	COST	Prio.	Remarks
BROWN HALL					
PRO42		Install stairtreads in stairwells	$2,500.00	1	SHOULD BE DONE IMMEDIATELY
PRO43		Waterproofing /Structural Repairs	$12,500.00	2	Brick and Cement work to S.E. & S.W. exteriors
PRO44		Roof flashing & reshingling	$8,000.00	2	Partial repairs only required
PRO45		Asbestos Abatement -Basement	$4,500.00	2	Patching & light removal for temporary compliance
PRO46		ADA COMPLIANCE	$34,200.00	2	SEE ADA REPORT UNDER SEPARATE COVER
PRO47		Modify heat distribution	$15,000.00	3	Divide Building into (3) Zones
PRO48		Recarpet hallways	$8,000.00	4	
MORRIS HOUSE					
PRO49		Repair stairs leading to records room	$1,000.00	1	PERSONNEL HAZARD
PRO51		Reinsulate under building	$9,000.00	2	
PRO52		Repair window in attic	$800.00	2	Leaking into office below
PRO53		Seal windows from wind	$2,500.00	2	
PRO54		Repair roof & reshingle	$22,291.00	2	Ensure flashing along main bldg. is replace
PRO55		Modify heat distribution	$4,500.00	3	Divide Bldg. into 3 or 4 Zones.
PRO56		Chimney repairs	$1,000.00	3	Minor repairs needed
PRO57		ADA COMPLIANCE	$17,000.00	3	SEE ADA REPORT UNDER SEPARATE COVER
CARETAKER'S					
PRO58		Replace gutters and downspouts	$3,500.00	3	
PRO59		ADA COMPLIANCE	$12,200.00	3	SEE ADA REPORT UNDER SEPARATE COVER
WILFORD HALL					
PRO60		ADA COMPLIANCE	$34,600.00	3	SEE ADA REPORT UNDER SEPARATE COVER
FOX CENTER					
PRO61		ASBESTOS ABATEMENT	$3,000.00	1	
PRO62		Reinsulate pipes in boiler room	$6,800.00	2	
PRO63		ADA COMPLIANCE	$62,100.00	2	SEE ADA REPORT UNDER SEPARATE COVER
PRO64		Masonry, chimney, & roof repair	$6,500.00	3	
PRO65		Repair 2 unit heaters	$1,500.00	3	Kitchen & Dining room
BREWSTER HALL					
PRO66		Install Fire Sprinkler System	$65,000.00	1	
PRO67		Insulate hot water tank (500 gal. cap)	$1,600.00	2	
PRO68		Chimney repair	$2,500.00	2	
PRO69		Paint exterior	$8,500.00	3	Wood surfaces
PRO70		ADA COMPLIANCE	$46,000.00	3	SEE ADA REPORT UNDER SEPARATE COVER
PRO71		Replace carpeting in hallways	$12,000.00	4	In progress
Project #		Project Description	Estimated Co	Prio.	Remarks
KENDALL CENTER					
PRO72		ASBESTOS ABATEMENT -(IMMEDIATE)	$8,000.00	1	Overhead (AHU'S) -Stage left & Stage Right
PRO73		ADA COMPLIANCE	$58,200.00	2	SEE ADA REPORT UNDER SEPARATE COVER
PRO74		Repair air handlers & controls	$38,000.00	3	Sawyer Center Theatre
PRO75		Landscaping /Drainage	$7,500.00	3	
PRO76		DECOMMISSION DUMBWAITER	$5,000.00	3	Permanently Sealed Off
PRO77		ASBESTOS ABATEMENT-ELECTIVE	$5,000.00	3	Various locations throughout building
PRICE HALL					
PRO78		ASBESTOS ABATEMENT	$2,500.00	2	
PRO79		Upgrade kitchen,sink,counters,cabinets	$3,500.00	4	
PRO80		ADA COMPLIANCE	$62,000.00	3	SEE ADA REPORT UNDER SEPARATE COVER
BELLOWS CENTER					
PRO81		ASBESTOS ABATEMENT	$2,500.00	2	NON-BOILER ROOM
PRO82		ADA COMPLIANCE	$11,100.00	2	SEE ADA REPORT UNDER SEPARATE COVER
PRO83		Reshingle roof Reflash	$7,000.00	3	
PRO84		Replace Boiler	$35,000.00	3	BOILER ROOM ASBESTOS ABATEMENT
PRO85		Paint Exterior	$3,500.00	3	Wood surfaces

Figure 19.1 (cont.)

Excerpt from a Sample Facilities Audit (cont.)

CHASE HALL

PR086	ASBESTOS ABATEMENT	$5,000.00	2	
PR087	ADA COMPLIANCE	$50,000.00	2	SEE ADA REPORT UNDER SEPARATE COVER

Willis Hall

PR088	Repair foundation leaks	$3,500.00	2	Window, grading, waterproofing
PR089	Repair overhang	$4,000.00	2	Renovate front entrance
PR090	Improve lighting in all bedrooms	$3,500.00	3	Increase lumens &efficiency
PR091	ADA COMPLIANCE	$50,100.00	3	SEE ADA REPORT UNDER SEPARATE COVER

ACADEMY BUILDING

PR092	Repaint walls/ceilings	$6,500.00	3	Strip,stain & refinish floors & cabinets first floor
PR093	Patch & repaint walls & ceilings	$2,500.00	3	SECOND FLOOR
PR094	ADA COMPLIANCE	$15,800.00	3	SEE ADA REPORT UNDER SEPARATE COVER

HOPEWELL CENTER

PR095	Walkway drainage	$6,000.00	2	
PR096	ADA COMPLIANCE	$21,000.00	3	SEE ADA REPORT UNDER SEPARATE COVER

MAINTENANCE BLDG.

PR097	ADA COMPLIANCE	$3,800.00	3	SEE ADA REPORT UNDER SEPARATE COVER
PR098	ADA COMPLIANCE	$17,100.00	3	SEE ADA REPORT UNDER SEPARATE COVER

HOMESTEAD

PR099	ADA COMPLIANCE	$12,600.00	3	SEE ADA REPORT UNDER SEPARATE COVER

PRESIDENT'S HOME

PR100	ADA COMPLIANCE	$23,100.00	3	SEE ADA REPORT UNDER SEPARATE COVER
PR101	ADA COMPLIANCE	$14,700.00	4	SEE ADA REPORT UNDER SEPARATE COVER

GAME FIELD

PR102	ADA COMPLIANCE	$3,300.00	3	SEE ADA REPORT UNDER SEPARATE COVER

SOFTBALL FIELD

PR103	ADA COMPLIANCE	$7,300.00	3	SEE ADA REPORT UNDER SEPARATE COVER

TENNIS COURTS

PR104	ADA COMPLIANCE	$1,500.00	3	SEE ADA REPORT UNDER SEPARATE COVER

Figure 19.1 (cont.)

Excerpt from a Sample Facilities Audit (cont.)

CAPITAL BUDGET 96/97

ITEM # 1

SAFTY/COMPLIANCE			REMARKS
PR105	BEST FIRE SPRINKLERS	$50,000.00	
PR106	ASBESTOS REMOVAL	$25,000.00	IN PROGRESS
PR107	PIPE REINSULATION	$25,000.00	
PR108	INSPECT/REPAIR FIRE ESCAPES	$30,000.00	BADLY NEEDED
PR109	TREE PRUNING (SAFTY ONLY)	$10,000.00	ON GOING
PR110	OIL TANK REPAIR (BULK)		BADLY NEEDED
PR111	OUTSIDE LIGHTING	$3,000.00	
PR112	Update Colgate Fire & SPRINKLER Sys.	$4,477.00	

ITEM # 2

Deferred Maintenance		
PR113	BAIRD ROOF	$10,000.00
PR114	SPORTS SCIENCE BOILER	$35,000.00
PR115	BAIRD BOILER	$6,000.00
PR116	KITCHEN DISH MECH/CONVEYOR	••••••••••
PR117	MASONRY REPAIR	$25,000.00
PR118	COLGATE ROOF & SKYLIGHT REPAIR	••••••••••
PR119	MISC. PAVING	$25,000.00
PR120	SEAMAN'S ROOF	$23,000.00
PR121	HARRINGTON ROOF	$10,000.00
PR122	NEW DORM SHUTTERS	$13,000.00
PR123	SPORTS SCIENCE BOILER FEED	$6,000.00
PR124	TENNIS COURT REPAIR	$5,500.00
PR125	TENNIS COURT FENCE REPAIR	$3,200.00
PR126	SAWYER CTR. C.B. GAS CONVERSION	$12,150.00
PR127	LIBRARY COMPUTER (A.C.)	$10,905.00
PR128	HOGAN MULTI-PURPOSE RM (A.C.)	$17,500.00
PR129	HOGAN CTR DRAINAGE FRONT ENT.	$6,000.00
PR130	HOMESTEAD ARCHIVE STRUCTURE	$87,166.00
PR131	LIBRARY EXTERIOR PAINTING	$23,500.00

ITEM # 3

Customer-Sensitivity		
PR132	BURPEE BATHROOM RENOVATION	••••••••••
PR133	ABBEY BATHROOM RENOVATION	••••••••••
PR134	BEST BATHROOM RENOVATION	••••••••••
PR135	COLGATE BATH ROOM 2&3 FLOOR	$50,000.00
PR136	WALKWAY NEW DORM	$4,391.00

ITEM # 4

OTHER PRIORITIES		
PR137	GROUNDS TREE PRUNING	$20,000.00
PR138	TENNIS COURT LIGHTING	$30,000.00
PR139	CARPET BEST DORM ROOMS	$25,000.00
PR140	LIBRARY MECH. REPAIRS	$50,000.00
PR141	Update Bldg. Heat (Zone's)	$90,000.00

Figure 19.1 (cont.)

Mechanical Spaces

- Floor Drains: Make sure they work and are within code.
- Ductwork: Look for insulation tears, large dents, accessibility for service, and condensate drains. Also check inside for cleanliness and flow issues.
- Motors: Evaluate noise, belt guards, and temperatures. (Feel with the back of your hand; if there is a short, your reflex is to close your hand; if you cannot hold your hand on the motor, it's running too hot.)
- Pumps: Look for leaks and excessive noise.
- Piping: Check the condition of insulation; look for leaks, excessive sweating, asbestos, and/or wall insulation.
- Electrical: Check the condition of panels or conduit, covers on panels, and clearance (OSHA requires 36" clearance in front of all electrical distribution panels).
- Pneumatics: Note the condition of the system and look for leaks and/or moisture.

Warehouse Space or Storage

- Egress: Are aisles maintained with 32" egress?
- Electrical Panels: Ensure accessibility.
- Floor Loading: Make sure there are established parameters for floor loading.

Classrooms

- Fire Safety Issues: Look for items hanging from sprinklers, excessive hangings of any kind from the ceiling, storage of paper products near a heat source, and blocking of secondary exits.
- Electrical Outlets: Look at loading issues, GFI applications near sinks, and an adequate number of outlets for new or changing technology.

Auditoriums

- Fire Safety Issues: Make sure fire exits are not blocked, and curtain is flame-proofed. Look for flammables being stored.
- Patron Seating: Check the condition of seats for tears and other damage, and to be sure they are operable.

Grounds

- Parking Lots: Look at lots both when they are busiest and when they are empty. This will give you perspective on traffic patterns and parking issues. When the lot is empty, look for ponding or drainage issues, the pavement condition, and clarity of parking lines.
- Overgrowth of Plant Materials: Look for blocked air intakes, blocked exits, etc., as well as building envelope problems resulting from prolonged moisture exposure from plant shading.

Manufacturing Spaces

For manufacturing facilities, deferred maintenance must be managed with particular care, since a well-maintained facility is key to production, and thereby, to the organization's competitive advantage. Deferring scheduled maintenance is a risky proposition. Most of the items discovered in the building audit may, therefore, prompt attention, and there may not be many candidates for deferred maintenance.

- Air or Steam Line Configuration: Look for low points that are not drained or separators. Check all line ends for drip legs, etc.

- Flow Review: Follow parts or process flow with the manufacturing staff and investigate ways to make the process less involved. Ensure that traffic patterns, such as painted lines on the floor, have a minimum of 48″ clearance and are well-defined.

Priority Matrix

Once the audit is completed and reviewed by the facilities department, it is time to prioritize the list. This will help everyone involved in the decision-making process fully understand the extent of deferred maintenance within the facility. This can be a discouraging part of the process in that you and your maintenance staff will see just how much has been deferred. Keep in mind that it is also helpful for the administration to understand the facility's condition and what you are up against.

The examples in the following section show how each prioritizing team establishes which category has precedence and/or other categories that are needed.

By using the matrix (shown in Figure 19.2), facility administration and operations can mutually agree on the priorities of deferred maintenance. Operations include the maintenance and/or facilities departments. Administration includes the plant managers, department managers, and finance departments. This structure may vary somewhat by facility.

Operations Review

The Maintenance/Facilities team should prioritize all of the noted items shown in Figure 19.1. The following is a suggested format, although again, each facilities organization has its own needs, and may require different categories.

Priority Matrix					
	Urgent	Important	Necessary	Would Be Nice	Other
Life Safety					
Regulating Requirement					
Safe Environment					
Known Requirement					
Equipment Life Cycle					
Energy Efficiency					
Other					

Figure 19.2

- *Life Safety:* Items that have the immediate potential for harm to personnel or visitors. Some examples are exposed live wires or loose or missing stairs.
- *Regulating Requirements:* Items that represent violation of an OSHA, EPA, AHERA, NFPA or other regulation. The facility may be fined or closed because of noncompliance. Examples include a lack of exit signs, or exit doors with the wrong swing.
- *Safe Environment:* Items that have the potential for harm to personnel or visitors who are not familiar with the operation. Examples include light switches that are ten feet inside of a room, or a motor that has two starting stations.
- *Known Requirements:* Items that represent a possible failure to comply, yet do not require rapid response for compliance. Examples include some ADA issues, posting signs and maintaining 36" clear space in front of electrical panels.
- *Equipment Life Cycle:* Items that require life cycle maintenance. An example is EPDM roofing, which has a normal life span of 20 years, and an actual life, on your facility, of 22 years, although it does not currently leak. These items need to be considered for maintenance planning.
- *Energy Efficiency:* Existing fixtures or equipment that are in good operating condition, yet have the potential for energy savings through maintenance. For example, when lightbulbs are replaced, use energy-efficient bulbs or fixtures. When motors burn out, make sure the replacement motor is the correct size, and not necessarily the size normally used.
- *Other:* Any item that does not fit in the above list, but still must be considered; for example, customer-related issues that require a high priority.

Administrative Review

The administration team should prioritize all of the noted items. The following is a suggested format; actual lists may vary.

- *Urgent:* These items can adversely affect the operation of the facility if they are not taken care of soon. They may relate to customer liability or adverse business relations. An example is a fuel oil tank that has leaked and has the potential of spreading onto abutting property.
- *Political Impact:* Depending on the organization, some items take precedence based on who needs them done, or what effects they may have on the surrounding community. Aesthetic considerations may fit into this category.
- *Important:* These items have the potential for liability or adverse repercussions. For example, staff housed mainly in one part of the building have been repeatedly sick.
- *Necessary:* These items affect the way staff, visitors, or the community views the facility. An example is sidewalks in front of a building that have settled and are tripping hazards. Aesthetics can be involved in this category as well.
- *Would Be Nice:* These items could possibly come under capital improvements, yet may be considered by the organization to be deferred maintenance. For example, the shrubs at the main entrance may be overgrown and need replacement.
- *Other:* Any item that does not fit in the above categories, but still must be considered.

Once the matrix is completed, both you as the facilities manager and your administration will have a mutually agreed upon set of priorities. You must also realize that the deferred maintenance matrix is a fluid document that is constantly changing as jobs are completed and as new problems arise. If your operation is cost-driven, you may want to use the priority matrix described in Chapter 23, "Health Care Facilities."

Chapter 20

Reserve Funds

Charles J. Stuart, CPM

All facilities require some type of a capital plan or reserve. The components of a facility that are exposed to weather, climate, and user wear and tear have a higher degree of need for maintenance and repair than structural elements. Components that maintain income and market values may also have a priority.

The perfect reserve sets aside funds equal to the rate of attrition of the components, and schedules component replacement just prior to the onset of either functional or economic obsolescence. The formula for such a concept creates a reserve equal to the cost of a day's use of the facility. Any type of reserve plan should maximize the useful life of each component, while retaining accumulated reserve savings for the greatest holding/investment period.

Facility managers without a capital reserve plan tend to operate on a crisis-driven basis, do not employ a strategy for cash flow management or reinvestment, and actually build functional obsolescence into a cash flow stream with crisis situations. They need to be aware of the constant movement their facilities make toward obsolescence, and plan accordingly.

The History of Capital Reserve Funds

Capital reserves have evolved from the fortunate sites with enough cash flow to allow for proactive reinvestment, to a necessity required by either law or competitive market conditions. In the early days of income-producing real estate and facilities, the need to reserve was observed in shopping centers and office buildings that faced new competition entering the market. For existing older sites, however, functional obsolescence could occur without warning. Managers realized that to remain competitive, capital improvements would constantly be required. Reserve funding and capital planning soon became an everyday function of the management process.

Unprepared, aging properties were forced downward in class structure, resulting in a loss of rental income. These damaging cyclic events forced crisis management, presenting little choice but to defer maintenance functions, build-outs, and other improvements. Periods of decline were usually addressed by intense management workouts backed by new mortgages. The process usually resulted in a losing investor, personal or corporate bankruptcy, a property stigmatized with a poor reputation, and an overall negative impact on the marketplace through increased vacancy and turnover rates.

During the most recent recessionary period of 1989 to 1993, the lack of prior funding for reserves in conventional property types resulted in major financial losses for mortgage holders to whom property was returned through the foreclosure process. The downward spiral of a troubled property resulted in an inability to protect or reinvest. Mortgage lenders found themselves holding property worth substantially less than the loan balance, not only in terms of market value but in actual physical reconstruction value. Lenders of all types have since realized the need to protect future physical values, and require reserve funding as part of the mortgage agreement. The following chart is a list of government loan programs and the reserve status required of each agency. These loans are typically used in all types of properties.

- **HUD 223(f):** Established at closing and paid monthly, commencing with amortization.
- **HUD 221(d)(4):** Deposited monthly commencing with amortization, and based on a percentage of structure costs.
- **HUD 223(a)(7):** Established with original mortgage; annual deposits continue for the life of the loan; additional deposits may be required.
- **HUD 232:** Deposited monthly, commencing with amortization and based on a percentage of structure costs.
- **Farmer's Home Administration (FmHA):** Required to some extent, dependent on Agency review of the investment. Open in terms.
- **Freddie Mac Conventional Cash Program:** Replacement reserve escrows typically required.
- **Fannie Mae Prior Approval Product Line:** Depends on structure but not automatically required for transactions of under ten years.
- **Fannie Mae Delegated Underwriting and Servicing:** Dependent on pricing tier, but not automatically required of properties under ten years of age.

Some of these requirements have been known to discourage a number of developers or investors, as heavy reserve payments limit cash flow or dividend distribution. On the other hand, a facility manager whose first exposure to reserves involves one of these mortgages is somewhat lucky, especially if enough margin can be found to create a cash flow. The facility is then bound by program parameters and should concentrate on meeting the obligation. An important aspect may be to have specific disclosure in writing as to the application of interest income, which, over time, will become substantial. The type of expense that would qualify as a reserve expense or reimbursement should be clearly stated. A process should be arranged for submitting a draw against reserves, and a qualified accountant consulted as to the tax implications.

Virtually every type of lender has a reserve policy; from the neighborhood bank to HUD, lenders insist on a program that will ensure physical reinvestment and protection. Understanding reserve methods will help the owner or manager address this responsibility.

Establishing a Reserve

The proactive owner/manager is searching for the methodology that will address the capital needs of the property. There are three main ingredients in planning a reserve, each of which involves various functions. The first is determining the current capital need, the second is analyzing current values, and the last is timing expenses and income, or the capital cash flow program.

Determining Current Capital Need

The first process is to identify inventory by nomenclature and quantity. Think of the inventory as a master listing of all improvements built on raw land. The analysis format that is suggested in the following section sorts out long-term, low-capital-need items. Begin with utilities owned by the subject property (subsurface systems and others), followed by grounds improvements as a whole, asphaltic and concrete improvements, etc. The building envelope would be next, including interior improvements, decoration, mechanical systems, and life support. The table of contents for this inventory list may look like the one shown in Figure 20.1.

Each major topic listed in the table of contents may contain five to ten components, or even more. The main goal is to identify all of the property components, concentrating on user-type items such as decor or roofing. Once the master list of inventory is created, a quantitative total for each inventory item is determined.

Inventory List Table of Contents	
Site Improvements	
Grounds retainage	6
Asphalt surfaces	7
Site misc.	8
Building Envelope	
Roofing	9
Exterior siding & deck allowance	10
Doors, windows, & misc.	11
Interior Improvements	
Carpeting & decor	12
Interior misc.	13
Mechanical Systems	
Fire pump, elevators	14
Lighting, intercom, mail, emergency systems	15
Contingency Report	16
Subsurface waste and water systems, encroachments and easements, electrical systems, structural allowance (areas not available for inspection or evaluation of condition).	

Figure 20.1

Analyzing Current Values and Conditions

Once the inventory is identified and quantified, the condition analysis may begin. This analysis can be performed by analysts at a variety of proficiency levels, from novice to expert engineer. It can involve service vendors specializing in certain components, general contractors with several areas of expertise, a series of expert engineers' opinions, or a combination of several types blended to reach a conclusion.

Government authorities, ownership via trusts, non-profit organizations, and others managed by committees usually search for a company specializing in condition needs studies with reserve analysis conclusions. Many firms employ engineers of various disciplines, and some use an architect as well. Use of this type of specialized firm eliminates self-interests associated with contractors and vendors, and encourages honest conclusions through supported analysis. Unless otherwise specified, the typical firm specializing in condition surveys and reserve studies is not conducting an engineering study. To eliminate any questions relative to functions and responsibility, that point should be reviewed with the firm prior to granting the assignment.

The typical study will be based on non-invasive observations. For example, an analyst would not normally remove asphalt shingles to view substrate conditions.

In creating a capital needs summary and a capital budget, many of the procedures and long-term projections are similar to an annual operating budget. Although many line items in a facility's annual operating budget will perform very close to projections, usually some line items will not. Like an operating budget, a capital plan will require verification and updates on a regular basis—anywhere from one to three years, depending on the conditions found in the initial study. Those conditions could include environment, product or component quality, degrees of maintenance and preventive maintenance, and resulting wear and tear. One of the most likely influences could be economic conditions; remember, when income streams are reduced or expenses are disproportionate, proactive approaches to life extensions tend to be reduced or abandoned.

In a reserve study on an aged property for which no reserve has been established, the years of catch-up make the capital plan appear unrealistic, with the recommended funding levels unattainable. This type of property must view the capital need based on priority. In this situation, protecting the income stream would be of paramount importance.

Figure 20.2 identifies the criteria that should be used in examining a component of inventory to establish current value and conditions. This proforma can be adapted to your specific specialty or use. The analyst determines the current capital need for the inventory through the use of the proforma. The proforma becomes a tool for specific attention and detail to each component, a function considered a luxury during the hectic management day.

Timing of Expenses and Income

There are two values associated with reserve planning: the typical price at which a component can reasonably be replaced, and the value of time.

Pricing Replacement of Components

Pricing for product replacement can be achieved through several methods, including R.S. Means' publications, statistical pricing, and market-driven competitive pricing such as contractors' estimates. Although contractors' estimates provide an accurate, realistic cost, they must be updated if the work is postponed because of the competitive nature of the market and swings in the economy. Statistical pricing endures because the information on which it is based is collected over a period of time. Whichever method is used, component replacement costs should include the entire function, including disposal costs, installation costs with related services, etc. For example, let's examine the total cost associated with replacement of two roof-top chillers:

Suggested Analysis Proforma	
Category	**Site Improvements**
Component	**Asphalt Roads and Parking**
Original life	Standard life expectancy of the component.
Chronological age	Actual number of years in service.
Effective age	Chronological age, plus or minus the level of maintenance proficiency and practices; condition represented.
Life years	Estimated remaining life years.
Deferred maintenance	Maintenance proficiency and maintenance services required.
Obsolescence	Comments on whether any component has reached, or is soon to reach, the economic or functional end of useful life.
Capital costs over 5 years	Highlights of dedicated expense during the next five-year period.
Rehabilitation/life extension	Comments on measures to overcome deferred maintenance or decrease effective age. Examine the payback.
Alternatives	Comments on advancements in technology of a component, or suggestions of another product use.
Comments and descriptions	Comments on observations, maintenance levels, preventive maintenance criteria, specifications, or general enhancement suggestions. Allowed application of reserve funds may also be covered. Examine the pros and cons of rehab and life extension costs.
Inventory and cost data	The quantity of the components and the costs of capital repairs or replacements, or both, in current year dollar values.
Dedicated reserves	Highlights of capital repair and replacement expenses during the next five-year period. This category may be dependent on existing funds.
Segregated reserves	Description of the annual funding based on replacement cost divided by the number of remaining useful life years. For example, the charted expense is for ten-year-old asphalt needing capital enhancement in year three, and a $65,000 replacement in year 16 is demonstrated below.

Chart capital costs of replacements with function year, to be brought into the reserve schedule:

Dedicated expense	1996	1997	1998	1999	2000
Add apron overlay:	0	0	5,500	0	0

Segregated reserves

New wearing course:	1996 through 2010: $4,333/year ($4,333/yr. x 15 yrs. = $65,000)
Next cycle:	2012 through 2015: $2,600/year ($2,600/yr. x 25 yrs. = $65,000)

Figure 20.2

Disconnect plumbing and electric	$ 970
Trucking & disposal	2,000
4 hours helicopter services	6,000
2 chillers (purchase price)	143,000
Connect plumbing and electric	1,500
Permits, fees, engineering costs	3,500
	$156,970

Certain situations may incur higher costs, such as the removal of equipment railings or roofing systems. The object is to examine the entire job as the cost unit. In the case of the above exhibit, the facility may be short by some $14,000 if it had reserved for only the chiller component cost.

At this point in the capital planning process, we have completed a well-rounded condition analysis, or condition survey. From this point, the information evolves into the reserve planning process and the timing of income and expenses for the reserve schedule.

The Value of Time

The value of time is the single greatest influence on a reserve analysis/reserve budget. The useful remaining life of the component is the divisor of the component replacement costs. The more extended a useful life period, the more time a component has to achieve funding. Accordingly, the greater amount of time will allow for lower annual contributions.

Let's look at time in detail when applied to a reserve schedule. First, abandon all thoughts of "straight-life" use periods. This type of projection creates unrealistic results. The real world of product or component use depends on "effective age" as opposed to "chronological age." This can be a negative or positive factor. It is relative to the maintenance and preventive maintenance a component receives over its life. This is where the well-maintained property is rewarded for its proficiency and proactive management. Likewise, if a component has been neglected through extended deferred maintenance, effective age will show the accelerated use life. The following is an example of effective age.

Roadway Asphalt, in Years of Remaining Life		
	Straight Life	**Effective Age**
Original use life	25	25
Chronological age	15	15
Effective age	**	8
Useful life remaining	10	17

Assume the roadway asphalt has received seal coating and crack filling programs during the use period. While these maintenance functions are likely to be classified as operating expenses, they have a direct payback to capital need through the use of effective age. In this scenario, the component was recognized to have attained a 28% increase in useful remaining life. The straight-life method does not recognize diligence in maintenance practices, and is likely to raise the question, "Why conduct preventive maintenance when replacement timing is somewhat pre-determined?"

The Value of Time Extended

Effective age can be lowered at any time. As long as a component continues to function as intended, it can probably be enhanced in some way to extend its useful life. A simple math analysis can determine if the cost of enhancement will achieve longer life and make economic sense. Using the exhibit above, assume the asphalt had not yet been enhanced; its effective age is equal to the chronological age of 15 years. Also assume there are 200 yards of asphalt that will require a reserve for an overlay in ten years:

200 yds. × \$7.00 per square yard = \$1,400 in total replacement cost, creating an annual contribution of \$140.00.

Say the manager is planning a crack filling and seal coating program for the inventory at a cost of \$300, with a resulting decrease in effective age to eight years. Annual contribution is lowered to \$82, but the real value of the enhancement becomes \$986 over the useful life [(\$140-\$82) × 17 yrs.] The manager will probably be pleased with the results of the maintenance program and repeat the enhancement costs one or two more times during the use period. Even with a break-even cost analysis, the property benefits from the maintained appearance in a number of ways, one of which is always market value; the property demonstrates prudent management; and it creates extended use and funding periods. Finally, when replacement is imminent, it is controlled by the manager. Component replacement is not a surprise, and certainly not a crisis.

A note should be added regarding the concept of effective age. For many components, it is possible to continually add to value enhancement. Life remaining can become almost perpetual. On the other hand, if effective age indicates an age greater than chronological age, or accelerated use, then useful life remaining is shown at the accelerated rate. For example, say the asphalt's effective age was indicated to be 18 years as opposed to 15 years in chronological age. With an anticipated life span of 25 years, the remaining useful life is not 7 years, but is an accelerated rate of only 5 years.

Support Tools for Capital Planning

The capital plan up to this point consists of inventory, condition analysis, component costs, and timing. Before we move ahead into building the cash flow program that will support the capital need and reserve program, we will examine several tools that will help maintain current information.

- The maintenance delivery or work order request system is an important support tool for the capital plan review. Each component of inventory should have a record within the delivery system that specifies dates, failures, required service, and associated costs. This not only helps for review during plan updates, but may indicate that a component is draining services beyond its worth.
- Regular physical plant inspections can benefit from a structured proforma, at least for condition analysis. Consider swapping property inspections with friends in the business or other employees in the company.
- When sending requests for proposals to contractors for repairs, always ask for the replacement price as an alternative and invite them to submit a revision every two years. The assignment will eventually be given out when the time is appropriate.
- Trade publications are usually a good source for new products coming onto the market. Review the material, searching for alternatives to or modernization of components.
- Review statistics from organizations such as the Institute of Real Estate Management (IREM) or Building Owners and Managers Association (BOMA). Both have extensive experience exchange services that will give budget variations by line items, as well as reserve funding information. All of the systems are based on comparable use and size, so information is instantaneous. (See the Appendix for more information on associations.)

Deciding Which Items Merit Reserves

There is a fine line in determining the items for which to establish a reserve. Priority should be given to components that support current use or operations, and those that protect or enhance property value. These components usually include everything you see when you look at the property, and everything you experience when you use the property, such as heating, lighting, and conveyance. The items that are left over, or low priority, are typically foundations and other structural items. This may be misleading, especially for the property with a history of structural trouble — service history and demand generally determine whether a component is to receive funding, and to what degree.

Some analysts may suggest that it is not necessary to reserve for a building component while more than 20 years remain of its predicted life. However, this approach tends to place an unreasonably heavy financial burden on those who pay into the reserve fund for the 20 year period, to cover these big ticket items. Spreading the payments out over 30 years makes them more affordable and equitable to those who use the building over any portion of that time period. By totaling the reserve funds needed to cover all major building components, one can calculate the needed annual, monthly, and even daily cost of extending the life of the building, thereby establishing fair payments for users.

Some analysts discount the need to reserve for grounds improvements or landscaping. While a natural growth landscape may be satisfied with a simple line item in the operating budget for pruning or removal, improved plantings require their own reserve coverage. In all types of property uses there will be interaction with residents, tenants, employees, or vendors. This will include foot or vehicle traffic with resulting damage, the need to plow snow and/or treat with chemicals and sand, etc. in many locations. There is also a natural attrition with plantings; some may not be planted properly or well suited to their surroundings. Often insurance coverage does not include landscaping. At least a viable percentage of the original improvement cost should be reserved.

Building the Cash Flow Program for Reserves

The process of building cash flow for reserves is best served in two parts. The first part builds the capital need on an annual basis, with the analyst choosing the term of five, ten, fifteen, or twenty years. The longer the term, the more distorted the projection is in runout because of factors such as inflation or actual use. A term of ten years is suggested for the first-time analysis for managers who are performing their own capital plan. Working with this time period provides a greater likelihood that they will see the outcome of many of their projections, and will achieve experience in planning.

The annual capital need document should be produced on a spreadsheet. In a topic column to the far left, open a row for each component. In each of the corresponding years to the right of the component title, insert the constant increment resulting from component cost divided by remaining life. For example, if a roof system has a replacement cost of $30,000, with six years of life remaining, insert $5,000 for each of the six columns (years). In the year of the actual replacement, leave the column with a zero, because it does not have a capital need in that year. With the example above, the seventh year would be $0. In the eighth year, however, reserves for cyclic replacement starts. In this case, for planning purposes, the $30,000 roof system receives a new 25-year life, or $1,200 per year, inserted into the columns for the ninth and tenth years of a ten-year plan. The next step is to total both the yearly columns and the actual expenses. Figure 20.3 clarifies these functions.

There are several items to highlight in Figure 20.3. First and foremost is the addition of a line item titled "contingency." For most properties that are allowed to set aside certain income as non-taxable for reserves, the IRS will not recognize slush funds or some percentage-based allowance as a buffer. The contingency should be supported for areas not available for evaluation, such as foundation cracks, waste, pipe failure, correction of inherent defects, etc. Be specific as to its use.

Although the dollar values of annual contribution change proportionately, this scenario should be used without regard to property age. It would appear amazingly proactive if an operator applied this concept on the first day of a new property's use, but that is never the case. Accordingly, almost all properties are "used" when the plan is created. Any real estate entity, at any point in its existence, qualifies to start with day one of a long-term plan.

Annual Capital Need Worksheet										
Plan Year	1	2	3	4	5	6	7	8	9	10
Component	1996	1997	1998	1999	2000	2001	2002	2003	2004	2005
Landscape	$ 3,000	$ 3,000	$ 3,000	$ 3,000	$ 3,000	$ 3,000	$ 3,000	$ 3,000	$ 3,000	$ 3,000
Asphalt	2,500	2,500	2,500	2,500	2,500	2,500	2,500	0	750	750
Site Lighting	525	525	525	525	525	525	525	525	525	525
Surface Water	100	100	100	100	100	100	100	100	100	100
Fence and Misc.	460	460	460	460	0	120	120	120	120	120
Roofing	5,000	5,000	5,000	5,000	5,000	5,000	0	1,200	1,200	1,200
Thermal Siding	750	750	750	750	750	750	750	750	750	750
Decks	250	250	250	250	250	250	250	250	250	250
Windows	105	105	105	105	105	105	105	105	105	105
Doors	50	50	50	50	50	50	50	50	50	50
Carpeting	425	425	425	425	425	425	425	425	0	160
Decorations	200	200	200	200	200	200	200	200	200	200
Furnishings	150	150	150	150	150	150	150	150	150	150
Elevators	3,100	3,100	3,100	3,100	3,100	3,100	3,100	3,100	3,100	3,100
Heating	500	500	500	500	500	500	500	500	500	500
Ventilating	330	330	330	330	330	330	330	330	330	330
Air Conditioning	645	645	645	645	645	645	645	645	645	645
Hot Water	610	610	610	610	610	610	610	610	610	610
Contingency	1,500	1,500	1,500	1,500	1,500	1,500	1,500	1,500	1,500	1,500
Indicated Need	$20,200	$20,200	$20,200	$20,200	$19,740	$19,860	$14,860	$13,560	$13,885	$14,045
	1996	1997	1998	1999	2000	2001	2002	2003	2004	2005
Actual Expense	0	0	0	0	$1,840	0	$30,000	$17,500	$3,400	0

All projections are shown in current dollar value.

Figure 20.3

This exhibit is simplified for understanding. It does not suggest that a property should establish 19 savings accounts and deposit money each month. It does not suggest that each line item is inflexible and cannot be used to fund a shortfall in another line item. There are some property uses where this would have far-reaching tax implications. This is only a tool to illustrate the capital need, over time, with supported analysis.

The Cash Flow Program

There are a number of methods by which to take the capital need and create a cash flow model. Strict segregating by line item is not a suggested method. This creates a larger accounting load, and a possible issue of safety in handling numerous accounts. Further, inter-account loans may be difficult to properly structure, especially under some IRS codes. Some forms of ownership may be 50% or 60% funded, and have shortfalls absorbed by rent increases or pass-throughs to tenants.

Most conventionally financed properties, no matter what their nature, do not have a mechanism to shelter money from taxes for future expenses. Ironically, the real estate entity without any type of safety net or government-backed program must rely on the owner's skill and determination, and is not allowed to shelter capital reserves. These sites might conduct a capital needs worksheet as support for a new mortgage or existing mortgage modification. This leaves about two-thirds of the entire real estate stock in non-profits, institutional, government-backed, and government programs. Residential single-family dwellings are not included.

As a primer and exercise, try to create a reserve schedule for a single-family residence. Include the roof, painting (which is not allowed in some applications), furnace, parking, etc. You will quickly see how conventional properties depend on a healthy income stream, just as you do for your residence, to address obsolescence. The single-family house, on a square-foot basis, is one of the most expensive entities to reserve for.

The Liability Method

Returning to the cash flow method, the author's favorite process is called the "liability method," aptly coined by the American Institute of Certified Public Accountants (AICPA). This method applies any previously accumulated funds to the current capital need, and all future funding levels to a rate equal to the daily attrition use of the property. The formula itself sets the stage for fairness. If every building's owners embraced this concept, there would be no further need for the wrecking ball. While perpetuation of existing stock may not be quite that proficient, let's take a look at the liability formula in practice.

The concept of this method is:

> Starting balance + annual contribution − annual capital expense
> = annual cash balance adjusted with any interest earned during
> the year

The adjusted balance is compared to the reserve required, or liability/exposure, indicating any shortfall. A spreadsheet proforma would appear as shown in Figure 20.4.

Referring to the Capital Needs Worksheet in Figure 20.3, note the row entitled "Indicated Need." In the first year of the schedule, the indicated need is $20,200. In year one of the plan, that is the liability to satisfy. In year two, the liability becomes the first and second year's indicated need— $20,200 + $20,200—for a total liability at the end of the second year of $40,400. The adjusted cash balance in the reserve account should meet or exceed the liability amount to achieve full coverage.

There is also a mechanism for reducing the reserve liability by showing the reinvestment back into the property, or the capital expenses performed as a credit. See the "Indicated Need" and "Actual Expense" lines in Figure 20.3.

The first six years total $120,400 of indicated capital need. In the fifth year, however, the property received $1,840 of a capital expense, which is used to reduce the liability to $118,560. The reserve required can also have an inflation factor added, compounding a rate over a period of years.

There are some disadvantages to using this method:

- If a component will require replacement in a relatively short time, say two to four years, the reserve rate is distorted during this period. We consider immediate capital expenses as a "dedicated" expense that should be addressed through existing reserve funds or operating accounts. If existing reserve funds are used, return to the capital needs worksheet, inserting the predicted expense into the total expense, and also onto the cash flow chart, reducing cash accordingly.
- Distortion is a problem. If solid analysis and quality input are used in the initial plan, then the projections might be reliable for three to five years, at which time a review or update should be performed. If the plan is reviewed on a regular basis, such as when operating budgets are reviewed and formulated, then the projections have an improved chance of remaining viable.
- Interest income and inflation becomes unpredictable over a long term.

Figure 20.5 is an actual cash flow chart from a recent reserve report.

Cash Flow Chart of Income and Expenses					
Starting Balance or Balance Forward ($)	Annual Contribution + ($)	Annual expenses < – >$	Adjusted Cash Balance w/Interest = ($+)	Reserve Required	Shortfall
1996					
1997					
1998					
1999					
2000					
2001					
2002					
2003					
2004					
2005					

Figure 20.4

The methodology and scope of reserve planning used in some property types can vary greatly. As mentioned earlier, government-backed or insured mortgages may come with an established set of guidelines. The prudent facility manager will become proficient in the methods described herein to find a negotiating point that will change a burdensome, arbitrarily established reserve rate into a rate supported through analysis. The difference may be the margin needed for financial survival. In this situation, the facility manager would be expected to share the reserve study with the agency.

Properties with conventional mortgages can use the reserve as an operating tool, and possibly share the results with a lender to show condition analysis and proactive planning. The analysis is not related to the entity tax

Cash Flow Program Start Date 1/1/96 Annual Contributions and Expenses							
Year	Balance Forward	Annual Funding	Actual Expense	Shortfall	Balance w/Interest	Reserve Required	Funding Shortfall
1996	$ 46,318	$ 23,500	$ 13,600	$0	$ 57,905	$ 23,219	$0
1997	57,905	23,500	14,617	0	68,791	32,838	0
1998	68,791	23,500	12,555	0	82,128	41,440	0
1999	82,128	23,500	9,109	0	99,415	52,104	0
2000	99,415	23,500	750	0	125,830	66,214	0
2001	125,830	23,500	0	0	153,810	88,683	0
2002	153,810	23,500	0	0	182,629	111,902	0
2003	182,629	23,500	4,500	0	207,678	135,121	0
2004	207,678	23,500	12,980	0	224,744	153,840	0
2005	224,744	23,500	0	0	255,691	164,079	0
2006	255,691	23,500	0	0	287,567	187,298	0
2007	287,567	23,500	64,350	0	254,118	210,517	0
2008	254,118	23,500	0	0	285,947	169,386	0
2009	285,947	23,500	1,050	0	317,649	192,605	0
2010	317,649	23,500	2,241	0	349,075	214,774	0
2011	349,075	23,500	93,105	0	287,854	235,752	0
2012	287,854	23,500	0	0	320,695	165,866	0
2013	320,695	23,500	0	0	354,521	189,085	0
2014	354,521	23,500	47,895	0	340,029	212,304	0
2015	340,029	23,500	0	0	374,435	187,628	0
Cycle End Totals		$470,000	$276,752	$0	$374,435	$187,628	$0

Interest is compounded at 3% per year. 2% is allocated for inflation of expenses. The inflation rate may vary depending on the current interest rates. This scenario is good for a limited time of approx. 3 years. All projections require regular updates.

Figure 20.5

position, while conventional properties lack the ability to shelter money for future expenses.

Non-profit organizations, such as condominium associations, institutions, and some trust forms, have a direct obligation to meet certain Internal Revenue Service codes with regard to handling, or qualifying for, reserves. (See IRS Code Section 528, specifically rulings 75-370 and 75-371, and The American Institute of CPAs' *Common Interest Realty Guidelines* for additional information.) Some of these property types must meet state laws for reserves, such as condominiums in Florida, California, Massachusetts, and a growing number of others. The trend is to require trustees and managers to at least know the liability amount, and to develop a plan to address it. In many states, trustees can charge member owners with special assessments beyond normal maintenance dues. However, the typical homeowners' association with a history of special assessments develops a stigma in the marketplace, lowering its viability. In a property type with a large or transient group of owners, the goal should be to set a proportionate rate of reserve.

Government-insured mortgages, typically from HUD or other agencies, may also participate in programs such as subsidized housing. Most government-related projects fall under Title IV or Title VI Federal Preservation Law. Program participants may also conduct extensive reserve studies such as the HUD Comprehensive Needs Assessment, which also includes social and fiscal issues.

Conclusion

Capital planning and reserves have increased dramatically in popularity during recent years; however, it is only the start of a growing trend. Operators, owners, lenders, and various authorities are becoming increasingly aware of the need to create a capital plan to safeguard and nurture the future of their investments.

Part 4

Special Maintenance Considerations by Building Type

Part 4

Introduction

The chapters in this part of the book are intended to point out some of the unique maintenance requirements of different building types. The authors have provided guidelines on the overall approach typically taken for each, including which building systems to focus on, and the unique challenges in areas such as funding and administration. For example, while the major building component concerns in apartments and office buildings may be mechanical and electrical, hotel maintenance managers may be equally concerned with general maintenance of interior finishes, landscaping and conveying systems. Manufacturing facilities will naturally focus on superior equipment maintenance, and hospital maintenance will require an in-depth knowledge of plumbing and HVAC systems, including the vital intricacies of medical gas piping and maintenance of healthy air quality at the correct temperatures. Those who maintain historical buildings must be more concerned than others about their choice of cleaning methods and materials, and have the challenge of locating aesthetically suitable materials for repairs and compliance with new codes or laws, like the ADA. University maintenance departments will require expertise in just about every area, since this type of facility typically incorporates many different building types. Knowing what systems to focus on, readers can return to Part 2 to find out more about the actual inspection, maintenance, and repair requirements for those systems.

There are also differences among building types in safety requirements, scheduling and access restrictions, and other factors that affect maintenance planning and management. In different ways, the activities that take place in manufacturing plants and hospitals are both governed by strict and numerous safety rules and codes. Access restrictions are common among different facility types, for a variety of reasons, including aesthetic, security, safety, and other considerations. Scheduling maintenance is frequently a challenge to varying degrees, whether because a manufacturing plant is on-line 24 hours a day, or because retail or hotel facilities are striving to minimize inconvenience to their customers. The chapters in Part 4 include many suggestions for addressing these issues.

Funding sources and methods and the influence of the community are other factors that differ from public to private institutions, and from educational facilities to private retail or office buildings. Successful planning of maintenance and major repairs is tremendously affected by financial restrictions and opportunities, as is evident in the different approaches put forth in these chapters. Chapter 23, "Health Care Facilities," includes a sample budget.

The specialized guidance in Part 4, together with the cost estimating techniques in Part 3, and the coverage of overall financial issues and building components in Parts 1 and 2, paves the way for the budgeting process outlined in the last part of the book.

Chapter 21

Apartment Buildings and Condominiums

William H. Rowe III, AIA, PE

Public vs. Private

Housing has a wide range of use and ownership, from single-family dwellings to large multi-family, mixed-used developments in both the private and public sectors. While some housing is generally considered all private (such as a single-family dwelling) or all public (such as an elderly housing complex), in reality most projects are intertwined. The "private" residence receives federal and state income tax benefits on the interest and property taxes paid, and the "public" housing project is paid for by taxes collected from private citizens and businesses. The biggest difference between public and private housing is that while they can cost the same to build, the rents that are collected in public housing are much lower than those in private housing.

Private Housing

A significant factor in the operations and maintenance of housing stock is whether the owner is public or private. Private managers have more control than public owners, can take more risks, and generally control the best (luxury condominiums) and the worst (slum tenements) of housing. Private owners can use the marketplace to increase revenue and can decide what portion of the revenue to apply to the upkeep and maintenance of the facility. Anecdotal evidence suggests that it is best to live in a building that is also the owner's principal residence. The success of condominiums, cooperatives, and private communities seems to show that private ownership provides well-run, attractive, and desirable housing for those who can afford it.

Availability of Funds/ Financing Mechanisms for Private Housing

For facility managers, the three basic rules of real estate (location, location and location!) determine more than anything else the value of a facility, and consequently the monies and effort that can be expected for its upkeep. In well-run housing, the annual costs are generally predictable and recoverable in market rents. The building owner (or owners) often hires a management company to collect rents (or fees), to conduct a preventive maintenance program, and to respond to emergencies as they occur. In owner-operated facilities such as condominiums, the Board of Trustees meets annually to plan for capital improvements and to assess the costs associated with the desired improvements. The facility manager usually presents information concerning the need for such improvements, and the costs associated with the improvements. When a shortfall is expected, private owners can usually obtain a bank loan for the funds necessary to make repairs, as long as there is enough equity remaining in the building to cover the loan. This is generally possible because new buildings have few problems requiring loans for capital improvements, and older buildings have usually had several years to pay off a portion of the debt service, leaving the remainder for a new loan. In private housing in run-down areas, or areas where rent control is in effect, the rents are too low to cover the costs of upkeep, the neighborhood eventually deteriorates, and the assessed value becomes too low to cover any debt.

Management of private housing is a stressful occupation. At the high end of housing where tenants are paying considerable rent, the tenants are likely to be fussy, and their contact with the property manager is usually for repairs that are distressing them. Also, this type of housing has the greatest number of contact persons — every tenant. Property managers need to develop a strong, friendly, service-oriented approach to their customer relationships, since people are most demanding when it comes to issues involving their residence; thus considerable effort is required to maintain a residential property.

Public Housing

Public housing programs are run by the Federal Government, states, cities and towns. Public housing is generally provided for low-income citizens who are not able to afford private housing or who have been excluded from it. While there are fair housing laws that prohibit discrimination in a variety of areas, because low-income families are not able to afford moderate or high rents, they are effectively excluded from them.

Housing is among the more expensive building types to construct. It makes little difference if the housing is for the rich or the poor. There is no substantial difference between the basic cost of building low-income housing and the cost of building luxury housing. It is true that large sums can be spent on finishes and larger room sizes, but attempts to lower the cost of low-income housing usually results in cheaper building materials, and significantly increased maintenance costs. Rich or poor, dwelling units operate 24 hours a day, and have several systems that are costly to build such as bathrooms, kitchens, bedrooms and living rooms, as well as the other systems common to all buildings such as corridors, elevators, heating and electrical systems.

Availability of Funds/Financing Mechanisms for Public Housing

While public housing does make housing available to those who might not otherwise afford it, it does not have the ability to be economically self-sustaining — operating revenues must be sought, often in competition with other valuable government programs. In the evolving climate of reductions in government programs, public housing will continue to be underfunded compared to the actual costs of running the facilities. In many communities, improvements to public housing require special legislation, or annual appropriations to supplement the income from low-rent units, to fund the operating budget.

Public Funding

Housing is funded by the public in several ways. Public housing is funded annually with supplemental appropriations that are added to collected rents. Currently, federal programs provide low-interest loans to build certain types of low-income housing. There are also programs that pay landlords the portion of rent that cannot be paid by low-income tenants. This is an effort to dismantle the image of those public housing projects that foster crime and are a poor environment for children. Other programs, those that have turned the management of public housing over to the tenants, have also been successful. In neighborhoods that have deteriorated, the government often initiates incentive programs to revitalize the area. Such efforts include programs that only the government can provide, such as better public transportation, improved review and approval processes, seed money to foster increased investment in the neighborhood, or low-interest loans for critical improvements.

General Maintenance and Repair Strategies for Housing

Following is a description of the items most commonly requiring attention in the management of buildings for housing.

Exterior Maintenance

Roofs: Roofs in housing generally last the full extent of their predicted useful life since there are typically few changes in the roofscape over the life of the building. Most roof work is focused on dealing with unpredicted leaks. In cold climates, *sloped shingle roofs* often experience ice dams resulting from improper gutters and poor flashing. Repair of the chimney or flashing at roof penetrations and transitions is often required; this can be a signal to examine the roof for its overall condition and for possible replacement.

Flat roofs are most common in multi-family residences. These buildings often have mechanical equipment on the roof. Because of the vibration or regular maintenance of the equipment, these roofs require more frequent repairs. Condensate lines that drip on the roof, poor location or clogging of roof drains, and flashing at parapet walls are common causes of frequent repairs. When replacing flat roofs, consideration should be given to providing an insulated roof. Poor ventilation can cause serious problems of roof rot. Water moisture that is trapped under roof decks can cause rotting, resulting in structural damage as well as the need for a full roof replacement. This problem is made worse by the use of fire-treated plywood, which, while good for minimizing fire damage at roofs, actually is less stable than preservative pressure-treated wood and delaminates in poorly vented environments.

Enclosure walls: Wood facades require periodic painting, and must be maintained so that the wood is 6″ above the ground. Over time, as plants grow and water causes soil to migrate near foundation walls, periodic maintenance is necessary to avoid wood rot at the sills.

Masonry: Masonry generally requires little annual maintenance until the mortar needs repointing. However, repointing is a significant cost to consider. Masonry cracks are usually not extensive, but should be repaired to avoid accelerated damage, particularly where freeze-thaw cycles are common.

Windows: Window repair is usually done in conjunction with exterior painting. Repair of broken windows, window washing, and weatherstripping should be done annually. When windows are painted, the following miscellaneous repairs are usually completed.
- Replacement of deteriorated caulking and putty work
- Replacement of cracked glass
- Replacement of sash cords and hardware
- Replacement of deteriorated wood trim

Window replacement is usually expensive and not done unless tied to other issues such as extreme discomfort from leakage, severe rot, or the need to reconfigure the windows for air conditioners or other changes in the building.

Doors: Hardware for doors in residential buildings often requires rekeying when tenants leave and can be a source of maintenance for minor repair.

Waterproofing: Below-grade leaks occur in basement areas. Sump pumps or foundation drains can be added to control moisture. Often, the source of moisture is an inadequate storm water system or poor exterior grading that does not properly conduct rainwater away from the building.

Exterior painting: Exterior painting of trim and wood siding is to be expected. Proper priming, coupled with touch-ups between major paint jobs, can minimize painting costs and extend the life of wood siding.

Exterior ramps, walks, paths and signage: These items require annual review and patching for wear and tear, frost heaves, salt damage, and weathering. Some items may require improvement to meet ADA requirements.

Landscaping: Annual maintenance is expected for restriping of parking areas, maintenance of roads and parking areas, cleanout of site manholes, maintenance of lawns (mowing) and planting areas, and snow removal.

Utilities: Generally, residential buildings do not experience significant costs resulting from maintenance of utilities. However, trash removal and lawn sprinklers may be considered in this category.

Interior Maintenance

Life safety: Residential buildings generally do not require unusual costs for code upgrades for life safety. There are expected annual expenses to maintain fire alarm, elevator, HVAC, and fire protection systems.

Upgrades for the disabled: Modifications for the disabled in existing housing for private residences are not common since much of private housing is exempt from ADA requirements. Entrances, doors, thresholds, hardware, toilets, elevators, and signage in common areas used by the public may require some upgrade as annual repairs and improvements are made.

Upgrades to improve building performance: Each year some allowance should be made for building upgrade. Examples of typical maintenance include the following:

- HVAC and energy management systems.
- Lighting upgrades for improved efficiency, such as replacing incandescent lighting fixtures with fluorescent ones in public and common areas.
- Installation of improved security, CATV, or intercom systems.

Laundry rooms — appliances: Laundry rooms are often provided in residential buildings. When the laundry equipment is located within the individual apartment unit, the maintenance of the equipment can be made the responsibility of the tenant. When laundry rooms are located in common areas, maintenance is performed by the building manager. Where coin-operated machines are used, the laundry is usually a revenue source for the landlord. Operation of laundries, including maintenance of machines, can be provided by service vendors who keep a percentage of the coin receipts.

Water leaks from equipment: In buildings where individual water heaters or laundries are installed, it is possible for the water heaters to corrode or the washing machines to flood, causing water damage to units below. Careful maintenance of such equipment is necessary to minimize damage. When possible, floor drains installed throughout the building can provide some measure of protection, although such systems are expensive to retrofit in existing buildings.

Trash rooms: Weekly trash collection, as well as trash room washdown and cleanup, are a regular part of apartment maintenance. Common costs associated with trash rooms include rodent control, protection of sprinker systems from freezing in cold weather, maintenance to avoid fire hazards, and maintenance of special equipment such as compactor systems.

Tenant changeover: At the conclusion of rental leases, it is customary for the landlord to provide some improvement to the apartment — typically apartment cleaning and painting. In some cases, floor sanding and refinishing or new carpet are provided.

Conclusion

Apartment buildings and condominiums are actively maintained properties, since tenants who reside there naturally care about where they live. Owner-occupied buildings, such as condominiums, have the most repair and maintenance work. Much maintenance work in apartment buildings is custodial in nature and consists of generally predictable repairs, usually performed by staff with general "handyman" skills.

Chapter 22

Educational Facilities

Pieter van der Have and James Armstrong

Section I: Higher Education

Elsewhere in this publication, the interested reader has ample opportunity to become familiar with many of the issues that affect facilities operations and management. One might argue, with some accuracy, that a building is a building. After all, once one looks past the glittery stuff that makes an architect an artist, building structures serve one generic purpose. *They house stuff!*

That *stuff* might be people, things, processes, or functions. Essentially, buildings are boxes designed to allow a certain function to be achieved within their walls, contingent on certain specifications and parameters. Thus one might question the validity of the proclamation offered by many in the higher education facilities business that, "Our needs are different!" In fact, those individuals will also most often object to any attempt to lump their problems in with those identified for facilities designed to handle grades K – 12, or with facilities that house manufacturing or warehousing operations. The following text is an attempt to identify certain tangibles which may clarify some of those differences to the flexible reader.

What Is Higher Education?

Purists involved in higher education will insist that they are associated with and committed to providing a learning environment and an opportunity for growth to the intellectually curious. Such a facility caters to those who want to go beyond a twelfth-grade education, even if they do not intend to acquire a degree. It offers an environment where knowledge is shared, challenged, and enhanced. It offers associate degrees, bachelor degrees, masters degrees, and doctorates of many different flavors and specialties. Because there is such a wide range of needs to be served for the potential student, institutions tend to specialize in various ways. Nonetheless, institutions of higher learning exist to satisfy a thirst for self-improvement by anyone who can meet some basic established standards and satisfy certain criteria.

How Is Higher Education Funded?

Publicly funded institutions receive a portion of their base through taxes, with a substantial supplement from tuition payments. They are frequently further supported by endowments developed through extremely aggressive fund drives orchestrated by paid staff and alumni of the individual institution. Such endowments may reach eight or nine figures in size. At some institutions, such as typical two-year or liberal arts colleges, tuition may account for only a fraction of the institution's total operating budget, with most of the budget generally supported by taxes (local, state, federal, or a combination thereof). At other institutions, especially research and medical campuses, taxes and tuition make up less of the total budget, and a large portion of the operating budget may come from research grants and other revenue streams. Because these institutions are always in part supported by tax revenue streams, they frequently fall under tight control of the state legislature and/or other designated bodies. Furthermore, as a result of that funding philosophy, their accounting methods prevent them from depreciating their capital facilities, making them different from private industry and from private institutions of higher education. In some not-so-subtle ways, publicly funded institutions operate in a less flexible and more closely scrutinized environment.

Private institutions, in a sense, follow the same general profile as that described for publicly funded schools, except that the taxpayers fall out of the picture, and another group of stakeholders takes their place. These institutions may be church-supported, corporation-sponsored, or privately-owned and totally self-supporting. Students at some private institutions may be asked to pay virtually no tuition, whereas others may have to fork over very large amounts, depending on the institution's philosophy, and frequently the students' own financial situations. Facilities professionals at private institutions will often quietly admit they feel just as restrained in the way they are allowed to operate as do their peers in the public sector.

Private institutions, as a group, offer the same types of programs and degrees as are offered by the public institutions. There are a number of very successful private research institutions. One finds prestigious programs in both groups, and each can point to highly successful alumni. Depending on the day one asks, there is about an 8:1 ratio between public and private institutions (excluding highly specialized colleges, such as beauty colleges, etc.)

One can generally say that higher education has been around for many centuries as a formalized part of society. Since its inception, its method of meeting the challenge has changed very slowly. Today, its method is under challenge constantly, largely because (like the health-care industry) it is in extreme danger of virtually pricing itself out of business — or at least out of reach of many would-be students. Since the late 1980s, the rate of increase of the cost of higher education has far outstripped the Consumer Price Index, by as much as 80%! This phenomenon makes students and the public still more aware of other perceived shortcomings with higher education. Insensitivity to the needs of the student is one tender spot. The perceived workload of its faculty also provides an area for criticism. The cost of running the plant is another, and a measurable one.

What Is the Physical Environment in Which Higher Education Operates?

Most campuses are more complex than a small city. Although there is no such thing as an "average" campus, it may be useful to create one, for the purpose of illustration:

The campus has over 200 buildings (one-third of which are more than 30 years old), totaling 6 million gross square feet, over 20,000 students, and half again that many staff and faculty. At every class change, most occupants converge on the rest rooms, hallways, pedestrian walks, vending machines, etc., creating a virtual 15-minute human traffic jam, while seriously taxing affected systems. The campus has several sites within the community, which are not all contiguous. This campus has over 700 acres of landscaping to maintain, runs its own bus system, and has more than 100 elevators. It has miles of roadways and sidewalks to clean and clear of snow or other impediments.

This campus has a centralized chiller and/or heating and power plant. Its staff operates the campus utility distribution system, including gas, water, communications, sewer, storm sewer, distilled water, and electrical, and generates much of its own electricity, steam/heat, and chilled water. The campus has most of its buildings connected to a centralized energy management system. Since the campus is the home of much significant research, its staff has responsibility for hundreds of fume hoods, as well as the most sophisticated research support equipment currently available, ranging from autoclaves to MRI units, to elaborate radiology equipment, with new and advanced equipment being added daily (challenging the level of training required by staff).

The campus has a parking system, a law enforcement system, a food service operation and a bookstore, as well as a configuration of housing facilities for both individual students and those with partners and families. With that many people on campus, there is likely a state-of-the-art card reader (swipe or proximity) system in place, allowing and restricting access. All of this investment in physical, human, and capital resources has to be supported, operated, and maintained. Most frequently, such responsibilities are planted squarely on the desk of the chief facilities officer, even if the cost for such support is reimbursed by different revenue-generating entities.

The campus provides a football stadium and other facilities for athletic events. It has two or more theaters, frequented by the public. It operates a fleet of several hundred vehicles for use by researchers, administrators, and staff. It has a component of heavy equipment (such as snow plows, backhoes, etc.).

Even when the institution happens to be "smaller" than the average campus described above, the responsibilities do not dilute. Quite the opposite — facilities managers at the smaller institutions have a wider range of responsibilities and often greater impact on the institution than do their counterparts at the larger campuses. There are also more of them. Of the 3,500 accredited, degree-granting institutions in the United States today, nearly two-thirds of them are "small," but only in terms of student FTE (full-time equivalents). They are certainly not small in terms of responsibilities and accountabilities for the physical campus. They just have less staff to help them do it.

No city public works department has to work with so many components and systems. The only type of installation or site that closely approximates a university campus in complexity is a military base. Even there, the function is more singular.

Contrasting K – 12 and Higher Education

Programs offered in grades K – 12 tend to follow a fairly narrow track and cater to a rather well-defined market within controlled age groups. Course work, calendars, and curricula are reasonably well-defined. The students (frequently still called "pupils") know they are expected to be in class (although they may not always choose to do much while occupying space).

There is a moderate degree of parental support in nudging the kids toward success. Most taxpayers do not question the need for the system, although they may question the way it is being applied. The pupils are relatively inexperienced with (and may not care much about) the physical environment in which they find themselves. Consequently, they do not ordinarily prove to be a critical group of building users. Stereotypically speaking, they spend more time looking at and interacting with each other than they do at the brick and mortar that surrounds them.

As a rule, high schools do not operate twelve months out of the year, nor do many of them offer full use of the facilities in the evening and weekend hours. Thus, the demands on the facility are less intense than those of universities. The level of sophistication demanded of K − 12 facilities is increasing, to where accommodations have to be made for specialized equipment, such as computer labs. They have not yet developed a demand for highly specialized space, such as research facilities.

In contrast, institutions of higher learning offer a much wider range of programs, with more focused specialization, to virtually all types of individuals, from all walks of life, ranging in age from teens to octogenarians. Today's college student is not always the kid who moves on right after high school graduation. He or she is more and more what used to be called "nontraditional," in other words, individuals who have been out of the educational system for a number of years. As a result, they tend to have higher expectations of what they get for their investment (tuition and/or taxes). Apart from their high (and sometimes frustrated) expectations of faculty, today's college students expect the campus' physical environment to provide convenience, proximity, and physical comfort. Many students spend evenings, weekends and summers on campus. At research and medical institutions, staff is on site 24 hours every day of the year.

The campus provides the focal point for its host community's cultural life. Aside from sponsoring high school athletic championships, many campuses play host to theater productions, entertainment, rock concerts, and political debates, among other activities. This provides the neighbors (often very vocal stakeholders as a group) with frequent access to the campus' facilities, placing them in a prime position to see where their dollars are going.

Most campuses offer housing on-site. Even though this is a revenue-producing venture for most campuses, and expected to pay its own way, there is a ripple effect from this constant presence of segments of the student body. For example, security becomes a much more pressing issue. Frequently, a food service function has to be provided every day of the school year, including weekends. Parking and/or at least access to some sort of public/mass transportation has to be made available. The net result, of course, is that the level of wear and tear on the physical facility increases.

One cannot forget the ever-present influential alumnus, the one who has reached financial success, who supports substantial endowments or funds whole buildings, and who is not timid about identifying standards for the facilities. This person has the president's home phone number!

A challenge shared by K − 12 facilities and higher education is that they have to rely on an outside funding source, often the same one. As described elsewhere in this publication, there are a number of code issues which both have to heed. Because the two entities often compete for already inadequate funding (especially among publicly funded institutions), both groups often go wanting for funds to resolve critical problems. Since the funding bodies are not frequently inclined to increase the base allocation, the funding that is provided often remains barely adequate to address daily needs. In recent years standards have been increasingly compromised as a result of funding cuts. Funding to take care of ADA, OSHA, EPA and other mandated corrections is frequently extracted from other existing funding sources, thereby cutting them short.

Contrasting Higher Education with the Private Sector

Obviously, one variation between educational facilities and a manufacturing facility is that the measurement of success is much more elusive in the first. There is no profit measurement. Higher education can measure how many students are accepted, or how many it retains or graduates. It has not yet identified a method of measurement to determine how successfully these students have reached their potential.

From a facilities perspective, the differences are more obvious. For instance, stewards of successful enterprises in the private sector consistently question the cost of new construction for new buildings on campus, and the subsequent cost of operations and maintenance. They may not be sensitive to some of the differences in the programmatic functions of those buildings. For example:

- In the private sector, buildings are designed to last a specific number of years, which could be 20, 40, or 50. They are depreciated that way on their financial statements to the stockholders. In higher education, the expectation is that buildings will last forever.

- In the private sector, buildings are designed to serve a specific purpose, often very singular in nature. In higher education, individual buildings often serve a variety of purposes, from teaching, to administration, to research.

- In the private sector, buildings frequently are not designed for open public access in the way that educational facilities are; therefore, code compliance reaches a different magnitude.

- In the private sector, buildings are often designed to be built at a low first-cost, anticipating that future revenue streams will allow for upgrades and replacement as they are needed. In education, priority must be placed on getting the best product possible up-front, with the knowledge that there is little likelihood that additional funding will be forthcoming at a later time. (One is more likely to find donors for a new building or a new wing, than to find a benefactor for a new chiller or a better roof—since you cannot put their name on these essential, but "mundane" systems.)

- In the private sector, the occupancy level and intensity of usage (refer back to the class change description) are often not as high as they are in higher education. Building occupants in a campus environment (who often consider the building as *theirs*) also have widely varied expectations.

- With some exceptions, few individuals in the private sector are familiar with the construction, operations, and maintenance of a sophisticated research facility. It is understandably difficult for them to appreciate why a new research facility on a university campus should cost three times as much per gross square foot to construct as it does to build a commercial building anywhere else. Most do not deal with or understand the impact of vibration characteristics, latent noise transmission, and other factors on research activity, nor do they appreciate the need for mandated flexibility in communication technology – all in the same building. Finally, ADA compliance in a "public" research facility is notably cost intensive.

It is fair to acknowledge that higher education administrators (facilities managers included) could historically have been more visionary in administering use of facilities, budgets, O&M (operations and maintenance) programs, facility designs, and quality of construction, among other issues. The well once appeared bottomless, but it was inevitable that many stakeholders would begin to insist on evidence of accountability and responsibility. Understandably, an increasing number of institutions have recently tested the private sector. This is in reaction to the demand that the process could surely be done more cheaply and equally effectively through the use of private contractors, rather than relying exclusively on in-house staff.

Operation and Maintenance Issues in Higher Education

There is no magic about the operation and maintenance of a higher education facility. Nevertheless, there are a number of characteristics, perhaps somewhat unique to the higher education environment, which must always be considered. It matters not whether the institution of concern is privately or publicly funded. A nonexclusive list might include:

- Institutional vision, missions, strategic plans
- Class schedules and special events
- Established standards, long-range development plans, and perceived precedents
- Impact on/by new construction and/or remodeling
- Impact on students, staff, faculty who have impairments or disabilities
- Perception of response time, quality of work, and cost to the institution
- A feeling of ownership of buildings by their occupants
- Special needs of certain faculty, administrators, and researchers
- Perception by the community
- Communication with all pertinent stakeholders
- Morale of staff (including management)
- Resistance to change vs. willingness to be entrepreneurial
- Consistent application of fairness, not necessarily sameness
- Importance of human resource skills
- Objective evaluation of importance of system or component to the mission of the facility and its function
- Contrasting of O&M costs to those of perceived peers, and to the private sector

Full textbooks could be and have been written on most of these topics. For the purpose of this section, a few thoughts will be shared on the most critical items on the preceding list. Also, for the purpose of these discussions, it is important to agree that the term "facilities staff" may include all those individuals whose primary function is to operate, maintain, clean, and extend the life expectancy of the physical facilities on the campus and its capital assets, regardless of the source of funding. This may well include contractors' representatives who have been hired to accomplish certain tasks or functions defined under the facilities mission.

Who Has the Vision?

Among higher education facilities officers, it is often privately admitted that most of their institutions do not yet have a strategic plan, or a vision. They will also acknowledge that it is the existence of that kind of foresight and insight that should drive the O&M plan. With a strategic plan in place, questions such as which buildings should be razed, leased out, remodeled, or built can be much more effectively addressed. The good news, fortunately, is that an increasing number of institutions are in the process of operating a long-range development plan (which may still function independently of any academic plan), and which will help to identify the relative priority of facilities and infrastructures.

Why Are Standards Necessary?

Too often, the facilities organization helps drive the academic/institutional strategies. Therefore, the level at which the facilities are maintained will have a direct impact on those strategies, no matter how subtle. The *standards program* in place for each critical O&M function must be carefully designed with cooperative, collective input and review. Subsequent revisions must be in line with institutional values and political realities. There is definitely an advantage to involving key building occupants as well as front-line staff in the definition of most standards programs. Questions such as, "How often can I expect my office walls to be painted?" or "Who is responsible for cleaning off the top of my desk?" need answers, and those answers must be public knowledge. The more the building occupants know and understand, the more supportive they will be. The more the front-liners understand and support the standards program, the better they will be able to represent the facilities organization. On the flip side, as more work is done on demand only, expectations will rise, and discontent will increase (even among maintenance staff).

Whose Needs Take Priority?

Staff employed in facilities organizations in higher education often joke, in poor taste, "This would not be a bad place to work if it was not for the students!" Realistically, all the effort spent by facilities staff has to be orchestrated in consideration of the long-term welfare of the student population. Furthermore, occasionally work must be scheduled around individual classes, functions, or special events. This may lead to shift work, overtime, or adjusted work weeks. Minimal disruption of ongoing processes is an essential objective. In some cases, this may be something as apparently insignificant as changing the timing of automatic sprinklers. Other times, it may involve rescheduling an overhaul of a frequently used rest room facility, or the replacement of an entry door or floor covering.

Increasingly, the education environment has learned to be sensitive to individuals with physical or mental impairments that previously may have prohibited them from gaining access to a higher education. (See Chapter 6, "Codes and Regulations".) Obviously, while new buildings are being planned and designed for construction, these considerations take a front-row seat and can have a significant effect on budget. Some estimates suggest those requirements may increase a new construction project cost as much as 15% , depending on the type of facility. In the O&M side of the house, constant vigilance also has to be paid to those codes and acts. It is no longer possible to barricade off a walk or a hallway without due consideration for the individual who may not be able to get to his or her destination via an alternative route that is not accessible to disabled people. For that matter, he/she may not even be able to see the barricade. Facilities' staff has to be constantly aware of how campus occupants get around, and must be able to offer opportunities for safe passage to wherever they are going. The knowledge that these facilities are often used virtually around the clock, and around the calendar, only further challenges one's creativity. *It is valuable to remember that programmatic and reasonable access is the key!* The campus is not obliged to make every floor of every building accessible, as long as all courses and programs offered can be accessed! The courts will, of course, further define these issues for us.

Why Does the "Squeaky Wheel" Cost Extra?

At times, a request may arrive at a scheduling desk for replacement of carpeting in a frequently used area – for example, the main library. Before such replacement is scheduled, however, it is critical to identify the short- and long-term plans for that space. If funding is in place for complete remodeling of that space, it is wise to work with the occupant to identify an intermediate or temporary solution. Too frequently, the maintenance side of a facilities organization does costly repair, replacement, or restoration work, only to have it torn out as part of a remodeling or upgrade project a short while later.

It is true that such negotiations have to take into consideration the special needs of key faculty and researchers (both real needs and ego needs), and may provide opportunity for delicate negotiations, and an abundance of flexibility. However, simply ignoring the request will not make it go away. *A side note: A standards program for the replacement of floor coverings in general, carpeting in specific, is essential. With budgets as they are, carpeting should not be replaced until it has become a safety issue.* With today's fibers, and if the right design specifications are applied, carpet does not easily wear out. *It uglies out!* That in itself may not be a valid criterion for replacement.

Why Be Entrepreneurial?

Equipment fails all the time. The best possible preventive maintenance program (which every facilities organization, whether higher education-related or otherwise, must have in place) will not prevent failure. If done correctly, a preventive maintenance program will defer failure, and likely decrease the extent of the failure or damage. When a system or component fails, a rational decision must be made regarding its repair or replacement. In a higher education environment, especially where the institution may provide a home for extensive research and/or a medical school, the failure of even a minor component can have catastrophic impact on a research project, or a patient's well-being. Objectively, the decision maker (often the front-liner) must evaluate the relative importance of the item to the facility or the function it serves. Even when it is less expensive on the surface to send an item out for repair, it may be to the benefit of the institution to spend the extra money to buy a new one, if it minimizes downtime of a critical element. There are also situations where the reverse argument can be utilized. *Another side note: There is increasing evidence that over-maintaining certain systems can be more destructive than not servicing them at all. Close adherence to manufacturers' instructions is therefore always appropriate. Comparing notes with others in the same business also offers guidance on such decisions.*

A well-trained and empowered staff can make these kinds of decisions on the spot. It is an unfortunate irony that in a higher education environment, maintenance staff are often seen as the least trained and enabled to make key, on-the-spot decisions. Massaging organizational philosophies and hierarchies to allow continuous growth (in the form of training and education of staff), is highly essential. Close communication with the building occupants or system users is once again an obvious component of the decision-making process.

Who Are the Customers?

The campus community as a whole is well aware of the support staff. Many of the campus stakeholders are aware of the challenges facing the facilities organization, and are supportive. However, there is justification to paying constant vigilance to response time to emergency calls and routine calls, recognizing this is a two-edged sword. Too quick a response can create as much criticism as too casual a response. Sending two or more individuals when one could have done the job (a frequent perception, sometimes unjustified, of university maintenance staff by their customers), can create fodder for criticism. Since the customers are often not aware of safety or other requirements, effective communication again can only help avoid conflict. Remember the customer, then support the customer's needs!

All faculty and staff are aware of their own wages. Students are painfully aware of what they have to pay for tuition. Thus the perception that technicians' time is not effectively utilized can be injurious to the organization, and therefore to the institution. When this perception pertains to in-house staff, and becomes widespread, there will be increasing pressure to privatize or outsource – to hire a contractor to do what has been done by in-house staff until now.

The decision to privatize is often based on perceptions, and not necessarily on realities. That decision is often made at levels furthest removed from the facilities operation itself. Once again, a valid and pertinent standards program, when applied in conjunction with appraisals and customer satisfaction surveys, will help the facilities manager and the campus community to stay in touch with reality. Similar attention has to be paid to quality of workmanship, call-backs, and perceived costs, as opposed to real costs. A top notch cost accounting program and job costing system must exist to help the facilities manager lead his/her organization effectively. There are many fine software packages available that operate in a PC environment designed to help the facilities manager cross this bridge in a moderately inexpensive way.

Why Mingle?

Facilities professionals need to be part of the campus community. They need to know and be known by the other leaders on campus (and off campus), to meet with them, and to communicate openly with them. Working in isolation from the general campus is counterproductive, especially in light of the fact that for most institutions, the facilities budget is the largest single portion (under a single department or supervisory head) of the institution's operating budget, varying all the way from 2% to over 10%, depending on the type and size of the institution. The other administrators and leaders associated in any way with the institution are aware of this ratio. They have a right to demand assurances that this investment is handled professionally and competently.

Who Makes It Happen?

Many institutions' facilities organizations work in a union environment. The facilities' manager must be aware of the environment, and must know how to work cooperatively with union representatives (and there may be a number of unions expecting representation). Where there is no overt union involvement, facilities' leaders have to be aware of and sensitive to staff issues. There is little doubt that the rest of the campus, which operates much like a small town, may know of morale problems in the facilities organization before the facilities manager learns of it. Bad morale affects productivity and quality of workmanship, effectiveness suffers, and the institution is no longer served. In today's society, the effective leader must be comfortable with human relations skills (and it seems, legal skills), without having to rely constantly on the Human Resources department.

Why Use Cost as a Measurement of Success?

The obvious answer is, "Because it is there." Yet, many facilities managers would be reluctant to have his/her cost of doing business compared with those of (perceived) peer institutions, or even the private sector. Arguably, there are no two institutions exactly alike anywhere in the United States. They have unique academic profiles, unique programs, unique climates, unique labor situations, unique benefit packages, highly individualized purchased energy relationships and distribution systems, and unique warehousing/vendor environments. There may be institutions that appear similar, but no two are alike! Thus, comparing costs is very much like comparing apples and oranges.

One approach with which facilities officers are experimenting is to use *Current Replacement Value (CRV)* as a basis for benchmarking. Conceivably, such ratios will minimize regional differences, and will definitely allow any institution to measure itself against itself. It will also allow an institution to look at factors such as deferred maintenance backlog and code compliance backlog, and contrast those ratios against those of other institutions. CRV is a figure that is readily available at most institutions through the Risk Management office, although care must be exercised to get the real value, not the one adjusted/reported for insurance purposes.

What Is the Higher Education Facilities Officer's Most Critical Challenge?

Finally, there is the big one: Deferred Maintenance. This is generally defined as "the accumulated backlog of maintenance projects that are too large in scope (cost) to be covered by the annual maintenance allocation." Current estimates place the national backlog at over $60 billion. Many institutions have a backlog of $10 or more per gross square foot in identified deferred maintenance projects. With the current shift of emphasis from brick and mortar to technology and fiber, long-distance learning and the virtual university, this backlog is not likely to receive serious attention at many institutions. As stated before, it is not easy to find donors to resolve deferred maintenance issues. For this reason, institutions are turning to alternative financing options, including third-party financing for retrofits (often energy-related), which are paid back out of savings or out of the annual operating budget. In some cases, buildings are closing down.

What Makes the Higher Education Environment So Unique?

It can arguably be said that much of the preceding is not totally unique to the higher education environment. However, what is unique is that all of it applies to higher education. It is doubtful that all of it applies as clearly to any other sector of the market. As described earlier in this chapter, a university campus is unique in function and operation, and needs to be serviced as such. These unique characteristics have a direct and obvious impact on the cost of doing business, and will continue to do so until or unless the general vision for higher education changes significantly.

Summary

It is appropriate to re-emphasize that there is no standard approach to O&M that will fit all institutions of higher learning. However, it is possible to list these basic truths regarding facilities costs in higher education:

- The cost breakdown supporting higher education facilities has changed over the years.
- Until 1973, utility/energy costs constituted about one-fourth to one-fifth of the total facilities budget. Today, those costs often exceed the other O&M costs.
- Utility costs vary widely across the country, and even within the same region or state. It is impossible to make cost comparisons on purchased utilities. It is possible to make comparisons on units of consumption or production.
- Of the costs not assignable to purchased utilities (labor including benefits, materials, equipment, vehicles, etc.), where the ratio used to be about 4:1 (labor to utilities), it now approaches 3:1.
- As real labor costs have increased, and as total budgets have effectively decreased, funds available for non-labor costs have simply shrunk.
- The benefit package offered to higher education staff has generally been quite generous over the years. Recent surveys indicate that benefit packages can equal anywhere from 25% of wages to over 50%. Some institutions are starting to pull back some of the benefits, not without its corresponding price on the morale of long-time employees.
- Publicly funded institutions are finding it increasingly difficult to obtain funding from their legislatures, or other sources to operate and maintain newly constructed facilities.
- Combined with the compounding problem of deferred maintenance, the challenge to continue providing acceptable levels of support increases in its complexity.
- Custodial costs and grounds' costs are the most flexible. Standards programs can be adjusted and, while they have an impact on the aesthetics of the campus, they can be reduced fairly significantly without impacting safety or the life of a building.
- Decisions negatively impacting the physical appearance of the campus can turn away potential students, researchers, and special events — all of which have an impact on budget.
- Environmental health and safety issues are important and expensive.
- Most campuses could operate more efficiently if space were more effectively utilized and distributed, especially where evening classes and summer school are offered. Many might find they have surplus space.

Higher education is in a state of turmoil. It is undergoing a period of inspection, both from within and without, as evidenced by the popularity of benchmarking studies and privatizing or outsourcing analyses. Much of this turmoil is based on alleged or perceived shortcomings in the way these institutions conduct business. Facilities officers certainly do not find themselves exempted from questions related to such issues. Since maintenance, operations, and utility costs make up the largest single portion of an institution's budget, it follows that this is an area where many look for the most significant improvements. Privatization, or utilization of off-campus personnel on a short-term basis, is seen by many as at least a partial solution to this crisis. Unfortunately, ongoing changes in the methods of presenting learning opportunities to students leaves almost all institutions with a tremendous and unfunded backlog of deferred maintenance projects. Facilities officers find they have had to become more open, more communicative, and more flexible, and display an increasing willingness to take risks. They know that continuous change, even in their management and leadership style, is the foundation for success in facilities management!

Section II: Elementary, Middle and High Schools

General Maintenance and Repair Strategies

The key to maintaining both your sanity and your educational facility is to constantly remind yourself that the mission is education, not facilities. Maintaining a building is an overhead cost. On average 85–90% of the school department budget is salaries. The educational background of most school and university administrators is in education, and they may have little operations experience. Successful maintenance of educational facilities means providing comfortable surroundings that support an effective learning environment.

Funding for education is usually from steady funding sources — tax dollars, endowments, tuition or other private funding; yet both public and private schools have been tightening their belts since the early eighties. Budgets and planned expenditures have been established in the prior year and, in most school districts, are subject to voter approval.

Most public and private schools have parent organizations and/or public meetings where the public can address issues of concern. In some states the only direct voice or vote taxpayers have concerning the control of their tax dollars is with the school district budget. Public and parental scrutiny is natural, since it involves an environment that may affect their children's welfare; but it makes decision-making difficult.

Contracts for services and purchases are also open to scrutiny, and must be carefully documented. In many operations bidding, or obtaining at least three quotes, is required for all purchases. The terms for payment can be negotiated during the bidding process, but usually not afterwards. Most operations run on net 30-day payment terms, with some going as far as net 45 days.

Key Building Elements
Existing Deferred Maintenance Conditions

The majority of public schools were built within three general time periods:

1. During the thirties, as part of the government work projects (WPA).
2. The Baby Boom, which started when the GIs came home from World War II to start families, and continued through the fifties and early sixties. This caused educational institutions to expand during the late fifties until the early seventies.
3. The Second Baby Boom, resulting in part from the coming of age of the baby boomers who were having children. This caused schools to expand from the late eighties to the present.

Because many schools were built in the fifties and sixties many school districts have had to start dealing with major deferred maintenance items. After twenty years of a school's operation, many major maintenance issues occur, such as roofs, heating oil tanks, brickwork, boilers, etc. During the fifties and sixties, energy was relatively cheap, so the construction during this period was not energy-conscious. Uninsulated roofs, oversized boilers, aluminum single-pane windows, asbestos insulation, and inefficient lighting have presented a challenge to cost-conscious facility managers over the past twenty years. ADA has been an issue because many school buildings sprawl over a large area or have more than one story. In areas of rolling hills it was easier to make the building multi-level, which now makes ADA

344

compliance for the building difficult. Plumbing construction in the fifties and sixties used galvanized pipe for drains, and these eventually rust and fail. The preferred flooring for schools was hard-surface, and because vinyl asbestos tile was a good product for the cost at that time, many schools have 9″ x 9″ VAT.

For schools that were built during the twenties and thirties, maintenance of historical integrity has become a costly issue. These buildings tend to have a lot of character, yet are often suffering from a lack of proper and protective maintenance. The repairs that are needed are costly, and many institutions are looking at cheaper alternatives. Other issues on older school sites are ground contamination from leaking fuel oil tanks, and asbestos and lead paint removal.

Schools have some fairly unique maintenance and repair issues that differ from those of other facilities. One relates to the demographics of the students and another is the issues that arise from parent organizations or school groups.

Elementary Schools (K–5)

At this point in a child's development, the parents are naturally protective of their children, and parent organizations tend to be very strong. Many parents make time to be involved with school activities, and schools have the greatest scrutiny, as well as possibly the most positive parent interaction.

Physical aspects of elementary schools include smaller furniture (desks, chairs and counters). Plumbing may include lower toilets, water fountains and sinks. Some of the more noted maintenance issues are tripping hazards and sticking doors that are noticed more by little feet and hands. Elementary schools tend to have a lot of carpeted area, primarily in the lower grades when students often sit on the floor. Grounds issues, such as fence maintenance, removal of dead branches and repair of playground equipment, are prevalent, because the students spend time outside for recess.

Middle Schools (6–8)

At this age students tend to be the most demanding and the most destructive. This can be the age when students are exploring and trying new things on their own. The biggest critics of a middle school facility are both parents and students.

Physical aspects of middle schools include traditional classrooms, yet more diversity in the labs and grounds. Some of the labs are for computer science, chemistry, home economics, wood and metal working. The grounds have athletic fields for competition, such as baseball, football and soccer or field hockey. The mechanical systems tend to be more consistent with general construction guidelines regarding fixture placement or counter size. Electrical diversity (e.g., more outlets and capacity) becomes more of an issue as their use of technology progresses.

High Schools (9–12)

At this point students are looking toward their future. The vandalism that occurs is usually carried out by a minority, yet it can be more destructive than what occurs in elementary and middle schools. These buildings experience the highest level of use of any of the buildings within the school system. Adult education, extracurricular activities or public functions tend to use the high school building and grounds because of the building's diversity and capacity. It is also a time when some of the best ideas or plans for building use come from the students, who are also the biggest critics of the facility.

Physical aspects of high schools include less traditional classrooms, and more diversity in the labs and grounds. Some of the labs are for chemistry, computers, language, art classrooms, auto shops, wood and metal working. The grounds have athletic fields for competition such as baseball, football and soccer or field hockey. The mechanical systems are consistent with general construction guidelines regarding fixture placement or counter size, and again, the electrical diversity is helpful, as more outlets and capacity are required for new technology.

Physical Plant

Many special considerations are made in the design, repair or renovations to the physical plant of a public school building. Again, the focus is on creating a comfortable, better learning environment.

Locations

Physical plant equipment or mechanical spaces are typically centrally located and are in both common and nonpublic areas. For example, boiler rooms are in basements with an access door in lobby areas. Unit ventilators are in each classroom, yet air handlers are in penthouse or rooftop locations with access from nonpublic areas. This is for the protection of the students and to make a more comfortable learning environment.

Noise or Vibration

In most operations the equipment noise or vibration can be a key factor in design or maintenance because it can adversely affect the students. Protective measures include, for example: insulated mechanical rooms, extra vibration mounts on equipment, wider diffusers on air handlers, oversized ductwork, sheets of lead and rubber under pump mounts, more expansion joints on wall penetrations, etc.

Construction Cost

Typical modern schools are designed for optimum conditions, but due to the fact that committees make decisions, many original plans are modified in the interest of fiscal responsibility. Ideally, all of the previous categories, among others, should be addressed in maintenance planning for educational facilities. However, cost is also a key factor in this world of public bidding, and the low bidder wins. So, for example, the design may show vibration mounting, but cost cutting may eliminate it. These are the kinds of decisions that are made at the bargaining table.

The key to success is fact-based decision-making. When it is time to negotiate costs, ask for samples of the substitutes. Don't be afraid to ask if someone else is using that substitute. Then go visit them and check it out.

Staffing Considerations

Operating Hours and Seasonal Issues

The busiest time of year is, of course, determined by school schedules, which allow little opportunity for major repair or construction projects during the fall, winter and spring.

Staff Understanding

Maintenance and custodial staff need to be diversified because the changing environments of technology will be changing the environments of school facilities. Maintenance and custodial staff are also expected to work in very public areas. An understanding of universal precautions as well as the ability to comply with right-to-know laws must be part of the staff education process. A good sense of humor also will help staff who will be dealing with children.

General Budgeting Approach (Including Capital Planning)

When developing the budget for your operation it is recommended that it be broken down into four categories:

Maintenance Costs: All the costs associated with maintaining the facility and grounds, including labor and contracted services, as well as the supplies needed to complete the tasks.

1. Maintenance Salaries: Salary and overtime for in-house staff and administrative support.
2. Benefits Costs: Cost of health insurance and any other negotiated benefits.
3. Insurance: Workers' Compensation and General Liability.
4. Contracted Services: Outside services that are under contract for part or all of the fiscal year. Examples would be bleacher service contract, boiler service contract or refuse removal.
5. Outside Services: Costs for any service from an outside provider that is called in only when needed and has variable costs.
6. Maintenance Supplies: Expendable supplies used in the general maintenance of equipment or facilities that cannot be capitalized.

Custodial Costs

1. Custodial Salaries: Salary and overtime for in-house staff and administrative support.
2. Benefits Costs: Cost of health insurance and any other negotiated benefits.
3. Insurance: Workers' Compensation and General Liability.
4. Contracted Custodial Services: Outside services that are under contract for part or all of the fiscal year. Examples would include carpet extraction or window washing.
5. Equipment Services: Costs related to the service of custodial equipment.
6. Custodial Supplies: Expendable supplies that are used for cleaning such as chemicals, mop heads, rubber gloves, trash can liners etc.

Utilities: Costs related to the use of services from government-regulated utilities.

1. Electricity: Including demand charge, KWH costs and fuel surcharges.
2. Heat (Steam, Oil or Gas): Depending upon the operation, whatever fuel source is used to heat the facility.
3. Water & Sewer: Potable and process water should be metered separately. Water that is used for irrigation of the grounds should also be metered separately as it may affect the cost of sewerage. Some municipalities only meter the water, and bill for sewer based on water usage.
4. Telephone: Some operations separate telephone costs by department. This is recommended as it reflects true cost generation.
5. Other: Many schools use other utilities, such as cable television, etc.

Capital Improvements and/or Equipment: Many factors govern the development of this budget. Some examples are listed below.

1. Donor issues: Renovation and maintenance priorities can be specified based on a donor's desires. As mentioned in the "Higher

Education" portion of this chapter, some projects are clearly going to be better candidates than others. For example, have you ever heard of the John Smith memorial roof?

2. Function Operations: Functions in school gymnasiums or auditoriums are becoming more prevalent as corporations, businesses and community groups seek space to host large groups or functions and — want to support the school at the same time. Also, community groups are trying to meet their own needs at little or no expense. Taxpayers tend to focus on their individual needs regarding the facility, and capital improvements will be voted on accordingly.

3. Seasonal or Special Operations: Many school sports or activities are seasonally oriented, and operations must plan accordingly. As facility usage is studied, many improvements are developed to enhance seasonal operations. Examples are a renovated football field when the team makes it to the finals, or a larger auditorium for the community drama group's popular spring play.

4. Funding: This is the driving force for all schools, both public and private. If usage or enrollment is low, voter appropriations will reflect it. If enrollment is low, the admission income will affect operations as well as capital budgets, depending on the operation. If the buildings are not maintained well, the comfort of the students may be affected, and since the parents and students are the lifeblood of a healthy operation, the most successful operations strive to satisfy their students and parents.

Conclusion

Educational facilities can be seen as a unique building type in the sense that elementary, middle and high schools have a specific and focused mission, different from other building types. Features such as classrooms and play equipment, and issues such as special safety requirements and smaller-sized fixtures and furnishings for young children, are unique to schools. Maintenance planning must take into consideration the special needs of children and adolescents, and must direct itself to effective support of the educational process. Maintenance of institutions of higher learning share the mission of K–12 — to maintain and support an effective learning environment — but many operate on a completely different scale, addressing a tremendous range of facilities requirements. A large university may resemble a town, incorporating most of the special building types covered in Part 4 of this book, from libraries and museums to apartments and retail facilities, and involving the complexities and responsibilities normally associated with a municipality. In either case, it would be difficult to argue the importance of attaining an effective maintenance program, knowing it is one of the integral contributors to the mission of successful education.

Chapter 23

Health Care Facilities (Hospitals and Clinics)

Phillip DiChiara, CFM, CHE

Health care facilities, while they share many maintenance considerations with other buildings, have a number of unique requirements. In addition, the nature of health care and its governance, management, and operation, have a major influence on how the facility manager and his or her team need to go about the planning and estimating process.

Organization and Funding

The vast majority of hospitals are not-for-profit, community-based and -governed institutions. The Board of Trustees/Directors represents the community and provides strategic direction to the organization, the CEO, and the management team. Accordingly, the facility manager should, ideally, have access to this board in order to effectively plan and implement facility repairs and improvements. While the day-to-day routine of preventive and repair maintenance is not directly influenced by the Board, hospital funding is. In the second half of the 1990s, major changes in the way health care is provided and financed are placing major burdens on hospitals, and it is often the facility planning, maintenance, repair, and capital improvement areas that bear a disproportionate burden of reduced funding.

While costs tend to be better understood in terms of patient care issues, a certain level of sophistication is needed to understand and manage funds for maintenance and repair so as not to ultimately create higher costs to this industry. For example, as capital dollars diminish, the acquisition or first cost of plant infrastructure is typically given greater attention than life cycle, or operating and maintenance, costs. Experienced professionals intuitively understand that this perspective may result in ongoing operational costs that eventually defeat the well-meaning intentions of an uninformed governance group. While the decision for major repairs and improvements is usually made at the governance level, input from the facility manager will aid in making the decision an informed one. The relative proportion of operating funds directed to maintenance and repair is influenced by this governance group, though the Plant and Engineering Department may still have control of most expenditures and the way they are identified and implemented.

An example of emphasis often inappropriately placed is in the area of HVAC infrastructure. Replacement or acquisition cost (the purchase price) becomes more important than the combined cost of purchase plus the operating costs over the expected life of the equipment. Saving $40,000 on a $400,000 acquisition is tempting, but if the ongoing operational costs exceed the savings, it may make more sense to pay more up-front and save the differential over time. Simple enough, but often not factored into the equation unless the facility manager has the ability to influence the decision.

Another simple example that is frequently lost in the larger picture is the use of inappropriately sized copper cabling. While larger diameter cabling has a higher acquisition cost, its use may well make it less expensive to run certain equipment, thereby avoiding energy waste. Choosing smaller diameter cabling is an example of the kind of insensitive attempt at value engineering that may lower the first cost of a project, while adding it back in multiples over ensuing years. Facility managers of health care institutions need to be informed so they can make these judgments carefully, for noncapital recurring costs are a growing concern.

In the case of clinics or other sub-acute health care operations, a different governance arrangement may be in effect. Many hospitals operate such facilities under their license, but physician-owned and -operated clinics are run more like physician practices. A Board of Directors from the community may not exist, and the owners of the facility act of their own accord within the constraints of regulations. Here, too, the facility staff must be present and involved in decisions relating to the plant. However, the nature of their governance (be they for-profit or not) is such that they ordinarily act in their best interest, and their decision making suggests a closer eye to life cycle cost. While it would be unfair to suggest this is not the case in hospitals, consideration for care of indigent patients, and provision of services that are clearly unprofitable but comprise a community's health needs, do typically receive more attention in hospitals, thus impacting facility decisions.

Facility maintenance and repair in general can be given short shrift unless appropriate communication and education efforts are made by the facility manager. In health care, it is much more difficult to sell maintenance and repair over new, revenue-producing, and patient-attracting medical technology. Is there a strategy that ensures that appropriate attention, and therefore funding, is given? While there are no guarantees that every budget provided will allow the cycles of maintenance, improvement, and repair/replacement to be perfectly in balance, it is most assuredly the case that absence of a strategy will result in chaotic facility outcomes.

Maintenance and Repair Strategies

Essential Characteristics

A good underpinning for any health care facility management strategy involves the following:

Data. As noted in the Introduction to Part 2, the facility audit is an essential tool for understanding where the facility stands relative to its condition, capacity, and potential. Effective and complete data collection is imperative for making informed decisions. The potential for problems exists in how we collect and process that data, and, ultimately, present it to our superiors. Health care in particular is sensitive to these issues, as most physicians and health care administrators have strong training and education as scientists. *Analysis,* not intuition, is the foundation of patient care and health care administrative management. Despite the incredible intuitive skills of many facility professionals who are able to identify, diagnose, assess, and resolve problems with high levels of success, the process (which has become intuitive to them through repeated exposure) must be presented logically and quantitatively to others responsible for allocation of funds. Physicians and administrators want and need to see the underlying assumptions and the information contained within the data. Health care administrators will be increasingly reluctant in the future to entrust major dollar responsibilities to any professional who fails to delineate, in a quantitative fashion, the needs of his or her operation. The anecdotal cliché we all hear is the highly competent facility manager who implores his superiors to "see it my way" and repeatedly fails to achieve the desired (and often desirable!) end result. "They just don't understand," is the oft-heard response. True enough. Facility managers are no longer responsible for just knowing the correct direction to take; they must also be able to demonstrate its likelihood of success in meeting objectives. Past accomplishment and strong persuasive skills alone are no longer adequate... SHOW THEM THE DATA.

Credibility. So, we must have data. Data, however, is not knowledge, and knowledge is not wisdom. How can we obtain funding for and properly direct maintenance activities in a period of rapidly diminishing funds? Credibility is built on *strong analysis, a successful track record, ethical conduct, and an understanding of the larger picture.*

Persistence. Since you will not win every battle, even with data and credibility, stick with it. In fact, if you fail to be persistent on an issue, you will most likely be perceived as inconsistent, and leadership will not know which issues or projects are really important to you and the facility's status.

Self-Confidence. Your demeanor must radiate confidence and enthusiasm, as well as respect for the authority your superiors represent. A sense of the appropriate will contribute to the confidence you display.

General Tactics

While a strategy for maintenance and repair must be completed with an eye to the specific circumstances of each facility, many plans utilize similar general tactics to ensure that the support and pace of maintenance meets demand. Some suggestions:

Periodic demonstration, at appropriate forums, that facility maintenance and repair is essential to ongoing service and revenue. This might include annual presentations at department manager meetings, or an "open house" in the department to demonstrate the sophistication and complexity of maintaining a health care facility.

Established criteria for prioritizing work. Any organization, when confronted with a lengthy list of demands and an always limited budget, will view the task as overwhelming and frustrating. Agreeing to certain principles or criteria for classifying need before the list is presented makes the distribution of limited funds more rational. It is desirable for any entity, health care included, to focus on revenue enhancement. The consequences of that focus can easily be the unknowing undermining of the facility's infrastructure. By identifying items by code and safety need, or by presentation of current operations, a division of the available monies can be made more equitably. How else can one weigh the modest $200 cost for painting a physician's office against a $200,000 cost for an elevator upgrade?

Maintenance and repair tasks should be classified into categories such as:
- Code- or safety-related issues
- Preservation of current desirable operation
- Revenue enhancement
- Desirable modifications

A subset of this effort can be the establishment of a prioritization matrix within each category. The values can be negotiated, but agreement on definitions, in advance, can simplify the management task come budget time. Figure 23.1 is a sample. (See Chapter 19, "Deferred Maintenance," for more on prioritization of projects.)

Augmenting the annual budget with the creation of activity-based costing budgets to enhance typical line item supply, expense, and salary budget forms.

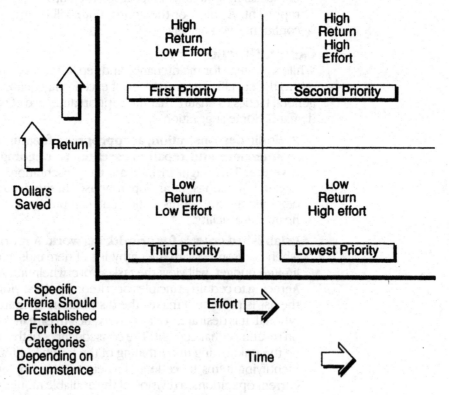

Prioritizing Effort and Return:Deciding What To Do First

High Return Low Effort — **First Priority**

High Return High Effort — **Second Priority**

Low Return Low Effort — **Third Priority**

Low Return High effort — **Lowest Priority**

Return

Dollars Saved

Specific Criteria Should Be Established For these Categories Depending on Circumstance

Effort

Time

Figure 23.1

Maintenance of a master list of all projects proposed, requested, anticipated, and completed, with appropriate costing information. A thorough file and annual summary of all major efforts is fundamental to good plant management.

Creation of a senior management level committee to advise and consult on facility management and repair. Most health care organizations have a committee to address plant issues, though typically emphasis is on new construction. The value of this committee is to remove the politicization of the plant task. Though the reverse can sometimes occur in such a forum, most often the involvement of interested senior management members like the COO, CFO, Physician Chief of Staff, and others, along with two or three board members with private industry experience and, of course, the facility manager, can lead to careful stewardship of the facility function.

Longer budget cycles. Propose budget cycles that are different and longer for maintenance departments (12-month intervals are not compatible with good facility operations). Most institutions view the calendar year's term of 12 months as an appropriate budgeting cycle. Increasingly, this relatively small increment of time blurs the need to think in longer cycles, forcing facility managers into a mode of thinking that places too much emphasis on the short-term fix and not the long-term (and probably more permanent) solution. Confer with the CFO and determine if such a plan can be accommodated for your department.

Weekly or monthly queuing publication (electronically) of the status of all maintenance work orders and projects. Again, maintaining a database that summarizes all current and planned activities can lend vision and order to the task at hand. It's never comfortable when the boss says, "Oh, I didn't realize that was happening," or "When are you going to get to..." The method can be simple and require minimal effort, but it needs to address the key issues, be published faithfully at the given intervals, and be used as a tool for planning. If it is provided but not utilized, the concept isn't wrong, but its execution may be.

Assignment of higher purchasing/signature authority for the facility professional. For example, the typical department head level of signature authority is up to $500, sometimes $1,000. For the facility manager, the need to authorize repairs and other emergent efforts justifies signature authority of at least $5,000.

Contingency funding for unexpected maintenance requirements, and creation of restricted funds. Even with substantial signature authority, the source of funds within the budget cannot always provide for the unexpected. Expect the unexpected. Contingency budgets for a major health care facility in Plant and Engineering operations should run, at a minimum, as follows:

Budget Responsibility of Plant (Salaries & Wages and Supplies & Expenses combined)	Contingency (defined)
under $1,000,000	3% – 7%
$1,000,000 – $2,500,000	$100,000 – $250,000
$2,500,000 – $5,000,000	$250,000 – $500,000
$5,000,000 +	5% – 7%

Creation of mutually agreed upon Key Indicators of Facility Performance (e.g., the churn rate, the rate of infrastructure repair and replacement, etc.). While most health care organizations are meticulous about financial statements and operating budgets, there are statistics and values that are tremendously helpful in understanding why your costs are what they are. Most are obvious; rarely are they committed to a written record and graphically tracked over time. For example, the cost of snow removal is tied to the amount of snow! What is your aggregate cost per inch, and what is your average regional snowfall? If you know these numbers, you can usually cast an impressively accurate budget. When there is unfavorable variance, you can identify its cause, which is usually not mismanagement, but an uncontrollable change in circumstance. Consider the following:

	Key Indicator:
Churn Rate	% of space affected by moving offices or changing functions of the space, usually defined as: $$\frac{employees\ moved}{employees\ total} \times 100$$
Building Efficiency	$$\frac{usable\ space}{gross\ space} \times 100$$
Cost of Providing Fixed Assets	The totals of capital, mortgage cost, plus taxes, insurance, and depreciation. Σ *capital, mortgage, taxes, insurance, depreciation*
Overtime Related to Emergency Repair	Is the trend up or down from previous month/year? Σ OT YTD $\longleftrightarrow$ Σ OT LYTD
Ratio of Preventive Maintenance Time to Repair or Corrective Maintenance	Is corrective maintenance overwhelming preventive maintenance? If so, it may indicate a failing infrastructure. PM/CM $\geq$ 1

Unique Maintenance Requirements of Health Care Facilities

Facilities of every type probably have more in common than not, but the unique attributes are what make each building type the manifestation of the industry it represents. For health care, two key differences are the complexity of overlapping codes and the nature of patient care. Beyond the standard requirements, the demands of accreditation through the Joint Commission on Accreditation of Healthcare Organizations' (JCAHO) routine codes often cite additional standards such as NFPA, despite the fact that JCAHO is not a building code per se. Most hospitals and clinics choose to meet those standards to qualify the institution for Medicare reimbursement. Secondly, patient care affects maintenance and repair because there is no down time in hospitals. Because hospitals are 24-hour operations, any work being done there is bound to affect someone or some service. The increased sensitivity to patients who are ill, aversion to noise and odor, and the life and death nature of many facility infrastructure systems combine to make health care a challenging environment for the engineering and maintenance crews.

JCAHO Standards

Recent revisions to the Accreditation Manual of the JCAHO have placed great emphasis on Continuous Quality Improvement (CQI). Nowhere is this more evident than in the section entitled "Management of the Environment of Care." The former requirements of PTSM (Plant, Technology, and Safety Management) are now included under this broader title, and the impact on the facility maintenance and report effort is significant. By referring to the Life Safety Code, Interim Life Safety Measures, and the Guidelines for Construction and Equipment of Hospitals and Medical Facilities, the Commission has made compliance a bit of a paradox: it is more specific in its intent, but less specific in terms of delineating its own expectations. While the American Society for Healthcare Engineers is working with the Commission to again review certain aspects of their mandate, the potential for a negative citation appears to be increasing. Facility professionals, our work is cut out for us.

The mission of the hospital and the responsibilities of maintenance and engineering come together in JCAHO's Environment of Care Standards. The goal in the management of the environment of care is to "provide a functional and safe environment for patients and other individuals served by or providing services in the organization." [1] To accomplish this, the organization must do the following:

- Plan and design the environment of care in a manner consistent with the mission and vision.
- Regularly orient and educate staff to individual and collective goals to support their environment.
- Develop performance standards and tools to measure and continuously improve performance.
- Collect data to measure, assess, and improve the environment of care.
- Give due consideration to the needs of patients and staff.

The focus is placed on life safety, hazardous materials and waste, utility systems, security, and medical equipment. Programs to maintain each and ensure maintenance of a safe environment are expected to document the protocols established in the organization. Emergency preparedness is the planned response to either internal or external disasters that test the organization's ability to meet its mission under adverse conditions. Health care organizations have special obligations to society in that, during extraordinary events, the fulfillment of the medical mission becomes essential, and appropriate planning must be done so that this obligation can be realized even under circumstances that might close other facility operations.

The new JCAHO standards place tremendous value on constant improvement, and that is achieved through data collection, assessment, and modification of practice to meet community obligations.

Other Hospital-Specific Requirements
While JCAHO requirements create standards for performance, there are many routine, hospital-specific maintenance problems to be faced on a daily basis. Maintenance of special gas systems for oxygen, nitrogen, medical air, and vacuum for surgery assist in creating a preventive maintenance load. Failure of any of these systems is life threatening. Requirements for, or the desirability of, these gases in various areas makes renovation difficult, not to mention expensive.

In recent years, the evolution of antibiotic-resistant bacteria has created the need for larger numbers of negative air pressure rooms. The impact on HVAC systems may be significant. Increased filtering needs are also evident. In some major centers, the advance of surgical practice has become increasingly tied to facility status. The ability to rapidly drop ambient temperatures (from $72°-74°F$ to $60°-64°F$) in operating room suites, particularly for cardiac-related procedures, has gained attention. Often, HVAC issues involve adequate duct cleanliness to minimize iatrogenic disease transfer, and removal of highly toxic Ethylene Oxide (ETO) gases utilized in central sterilization operations.

Plumbing systems have also been affected. Historically, many facilities have utilized mercury-based thermometers and sphygmomanometers, and when they broke, the liquid residue within was simply flushed away. While this practice is no longer accepted, the accumulation of mercury in system piping continues to discharge slowly, long after it is originally deposited. Similarly, those operations with laundries must often create and then maintain neutralization tanks, as some soaps contain mercury residue, and the current EPA standard is below five ppb (parts per billion).

Electrical systems standards, in terms of power distribution, are often straining under the weight of the increasing medical diagnostic technology. The need for "clean" power is a growing financial and maintenance consideration. The sensitivity of computer-based, high technology medical procedures now routinely requires power conditioning equipment. Along with electrical considerations, computers and networks require miles of cabling and telephone connection, all demanding increasingly frequent attention.

Even floor and wall coverings are an important consideration in health care facility management renovation decisions; the spilling of iodine-based products used in patient care, the ejection of fluids from syringes to vacate air, and incontinent patients all influence decision-making on floor coverings. Equipment being moved dents and scars even walls prepared with appropriate coverings, so railings and corner protections are necessary, but are expensive and are often "value-engineered" out of a project. This may be even more true in clinics where capital dollars are more difficult to come by.

ADA Requirements

The Americans with Disabilities Act (ADA) is another critical factor that facility staff must address. While adapting to these updated standards is entirely reasonable, it does place a burden on operations already facing financial strain. Fortunately, the Act provides for "undue hardship." The intent of the Act is to force progress on the removal of architectural barriers when this is "readily achievable." This phrase is evidence of the flexibility built into the Act, though many facility operators would argue with the practical reality of that flexibility. Priority must be placed on making facilities accessible, with emphasis on rest rooms, access to areas where goods and services are available, or the so-called Primary Function Areas. Creation of a path of travel, or a continuous unobstructed pedestrian passage to permit access to the services of the facility, is expected. Congress did recognize that path of travel alterations precipitated by renovations would be considered disproportionate if they exceed 20% of the cost of the alterations. New construction, of course, must be in full compliance.

Continual Changes

In health care, the rate of change is increasing, thus the facility must be modified with increasing frequency. ADA has added a new set of considerations. Organizations that have the capacity to build anew must give greater consideration to the use of more flexible building elements. Interstitial space is a consideration, and increased floor-to-ceiling heights are the norm.

Another example of the kind of changes occuring in the industry that lend new challenge to facility and maintenance managers is the advance of medical technology. Facilities that were built for one use are now being asked to qualify for another. The trend to outpatient surgery is, in part, driving renovation and new construction. Modestly sized and often well-equipped operating rooms are being constructed in sophisticated medical office buildings. The regulatory environment is not keeping pace with the avalanche of technological improvements in health care. Code for hospitals and code for office space do not always reconcile, yet there is a real ability to perform surgical and other types of procedures in locations other than the traditional hospital. The impact on maintenance is a need for specialized expertise. Specialized clinical equipment subcontractors are required for preventive maintenance and corrective maintenance due to high technology equipment. Those same firms are expanding into broader facilities maintenance for sophisticated environments such as free-standing care centers.

The increasing popularity of HMOs is driving the creation of facilities that support their mission of care delivery in the most cost-effective fashion possible. The utilization of the traditional hospital with its high fixed overhead costs (mostly the plant itself) is giving way to community-based and often HMO-operated health centers that create an environment that is not only pleasant for its customers, but cost-effective from a facilities standpoint. Life-cycle maintenance considerations are being factored into the construction planning, and sophisticated maintenance management is expected. Clearly there is a strong opportunity for maintenance professionals to prove their value given the changes the health care industry is undergoing.

As the movement to managed care continues, the structural inefficiencies of the health care delivery system are being scutinized. Excessive beds due to lowered length of stay and need to lower fixed costs and maintain cost efficiency, has created an environment where mergers and acquistions are increasingly frequent. Does this affect maintenance? Certainly—On the positive side, mergers and the ensuing due diligence process often dictate an assessment of the plant. The good news, or bad, is presented for all to see. The opportunity to explain our role in these unique circumstances should not be lost.

In addition, the turbulent nature of health care has created shorter tenures in the senior management staff. Hospital COOs' average tenure is shorter in many circumstances, and the lack of continuity in managers and shorter planning horizons leave the maintenance operation with a heavier responsibility. It is tough to lose a boss who has been a champion for our function, and re-establishing the operating assumptions and trust with a new boss is always a challenge as well as an opportunity. The fact is, facilities have operating lives that extend beyond the service of a single manager or chief executive. Having an operating plan that records pertinent data for handing off to the next regime is increasingly important as transiency increases. It is more than a matter of courtesy. It is a matter of professional behavior and one's reputation in the industry.

Safety Issues

Beyond ADA, the issues of ergonomic design are closely addressed in health care, for it is particularly cogent when a health care employee's status is impaired by working conditions at a hospital. The flow of "clean" vs. "contaminated" materials influences many of the repair and modification strategies envisioned. Further, the requirement for redundancy in certain systems, most notably electrical systems, makes for a crucial and somewhat costly maintenance list. The churn of offices and the relentless modification of the plant to accommodate new technology create increasing burdens on emergency power supply and distribution. Failure to closely monitor this situation is ill-advised. Older facilities are less well prepared to respond to current electrical demands, but even newer facilities may feel falsely confident of their status, and disaster preparedness in hospitals and health care is a baseline expectation. Failure of these systems may prevent an institution from meeting its community obligation.

Ironically, the increasing budget constraints have an impact on maintenance staff responsiveness, both in terms of preventive and repair maintenance issues, and in terms of emergent efforts. The vicious cycle of inadequate capital breeds poor plant infrastructure choice and increased operating costs. Fortunately, facility professionals are responding with increased sophistication as they adapt to the changing environment. Recognizing that we are now working in a time of severe cost cutting, strong arguments can be made for reduction of fixed overhead costs by proper choices in acquisition and maintenance.

Maintenance and Repair Staffing

The basis of staffing is primarily a function of size and sophistication. Additional consideration is given to the philosophy employed by management: some organizations subcontract out a great many tasks, some very few. The concept of contract management has also affected the way departments are staffed and organized, for some executives farm out the entire service, management and all.

In a perfect world, staffing would be decided following an objective assessment of the preventive maintenance required, project work, and expected repair frequency, and formulated into an equation that meets those requirements. That, however, is rarely the case. Either one inherits a staff that only superficially accomplishes the task, or more rarely, one that is just barely adequate in size. In the case of the former, documentation of the impact of truly inadequate staffing can reverse the trend, but that battle is increasingly difficult to win in the cost-constrained '90s. Nonetheless, collection of data and creation of key indices are important ways to verify the adequacy of staffing, define its productivity, and demonstrate its value. Routine skills required include multi-talented generalists with licenses in a trade, as well as specialists in HVAC, electricians, mechanics, carpenters, painters, and even locksmiths. Most moderately sized hospitals expect baccalaureate-qualified and licensed engineers to be included in the overall administration of a department. All of these individuals may be supported by helpers, clerical staff, data managers, and secretaries; again, size is the primary determinant.

Beyond preventive maintenance and repair, project work philosophy influences staffing. Some organizations subcontract all painting and wallpapering, snow removal, and other modest project work such as cabinetry and small renovations. Others feel that retention of a small crew of project people who can handle moderately sized (under $50,000) projects may outweigh the carrying cost of such a crew. In between projects, such a team can be a boon to the primary staff. Without such a crew, sometimes the primary maintenance crew must do these projects at the expense of routine maintenance. This is short-sighted and may indicate a failure on the part of the facility manager to have established a defined program. Regretfully, in some cases it is a matter of financial survival, but the hidden impact of deferred or delayed preventive maintenance is higher replacement costs.

Current IFMA research suggests that overall staffing is as follows, for all skills, whether permanent or temporarily utilized.

Full Time Equivalents (FTE's) of
Maintenance Staffing by Size of Facility*

100,000 S.F.	5.6 FTE
100,001 – 200,000 S.F.	6.0 FTE
200,001 – 500,000 S.F.	21.9 FTE
500,001 – 1,000,000 S.F.	34.7 FTE
1,000,000 S.F.	91.9 FTE

*Source: Benchmarks II, Research Report 13, IFMA 4, 1994. Houston, TX.

General Budgeting Approach for Health Care Facility Maintenance & Repair

Facility maintenance and repair budgeting in health care is becoming an increasingly precise science. In the past, health care fiscal departments simply shipped out the year-to-date figures for the first seven or eight months and expected the facility management staff to extrapolate this number into a new budget by identifying unique requirements that might be coming up, and essentially running that composite average into the future. Some institutions applied various forms of so-called "zero-based" budgeting, which insisted that every expense be re-justified each year. In the case of utilities, the battle wasn't difficult, and did lead to some earnest review of programs that had a positive impact on the budget. Nonetheless, it is now becoming a baseline expectation that facility staffs demonstrate the value added by their efforts.

It is appropriate to comment on the larger context of budgeting so as to demonstrate the information necessary for a maintenance department to successfully sell and receive support for its needs. Beyond the standard Operating Budget, the Basic Financial Statement must be considered, plus the Capital Budget, and numerous management accounting documents that support the estimate for the Operating Budget.

The value of a line-item-based budget is in its ability to track, by expense category, the accumulated costs accrued by the department. It is effective in understanding when and how much expense is accrued, by category, but it does not effectively convey the activities of the department, nor its impact on the operation as a whole. Nonetheless, it is the mainstay of every facility operation. Careful attention must be paid to full utilization of the organization's Chart of Accounts. Identifying a plumbing cost as part of the electrical supply cost is no better than summarizing all expenditures under Miscellaneous, and undermines the major strengths of this budget document. In health care, it is advisable to categorize expenditures as precisely as possible, and even to consider supporting detail documents. For example, the category *Electrical Supplies* is helpful; distinguishing costs from off-site locations may also be desirable in a given operation. Refer to the sample Chart of Accounts in Figure 23.2.

Institutions' numbering systems vary, as do the definitions for each category. The more specific the category, the easier to identify true costs; however, the added detail requires careful monitoring to ensure the assignment of invoices to the appropriate category. Figure 23.2 shows how utility costs can be distinguished, rather than simply lumping all under electricity.

Do we have the cash to support efforts that the facility manager may feel are necessary or desirable? Do improvements require borrowing as a source of funding, or can an internal return on investment be gained by operational savings? Understanding the impact on and connection to the Balance Sheet, Income Statement, and the Sources and Uses Statement is no longer solely the territory of the CFO and the senior management staff. The facility is a large portion of the assets our departments are charged with maintaining.

The Capital Budget

The Capital Budget is the plan for major facility and equipment improvement. Understanding this budgeting process is essential if the objectives of the department are to be achieved. Typically, the process involves development of a program with a preliminary cost estimate. Means' *Facilities Maintenance & Repair Cost Data* and *Facilities Construction Cost Data* are invaluable in assembling this information in an efficient and cost-effective manner. Outside estimates can augment or refine this preliminary estimate so that a valuable budget can be derived. Information is essential to fully informed decision-making, and senior management/boards expect a level of precision that this resource can provide. When estimates are reviewed following completion of the task,

An Excerpt from a Typical Chart of Accounts	
7010100 Maintenance Supplies	Maintenance supplies including hardware, lighting supplies, lumber, etc., used in the general maintenance of the hospital's buildings and equipment. Also includes gas and oil, tires, grease, etc., used to maintain hospital vehicles. Does not include paint or painting supplies.
7010200 Painting Supplies	Cost of paint and painting supplies.
7020100 General Office Supplies	General office supplies such as pens, pencils, stationery, etc., and other items typically used in the operation of a particular department.
7020200 Training Supplies	Any items (including films) used specifically for the training of employees or patients.
7020420 Letterheads, Memo Pads	Cost of outside printing of letterhead stationery and memo forms.
7020430 Other Printing & Forms	Cost of outside printing of forms other than charge tickets, letterhead stationery, and memos.
7021000 PC Software	Cost of proprietary software programs used on personal computers.
Breakdown of Utility Costs	
761100 Electricity, Repair Usage	The portion of the electric bill that is based on the utility's rates as approved by the Dept. of Public Utilities.
7601105 Electricity, Regular Usage, Location A	The portion of the electric bill attributable to Location A.
7601106 Electricity, Within Lease, Location B	Charges for electricity provided by the landlord related to air conditioning and lighting in common areas.
7601200 Electricity—Fuel Adjustment	Additional charge associated with the utility's increased fuel costs.

Figure 23.2

coming in under budget may be worse than coming in over expected costs. What projects were disallowed since dollars were identified, but were not used? Will senior management see this type of "padded" budgeting as a cover for imprecision, or as poor management? What creditable resource was utilized in compiling this budget? In health care, precision is particularly valued, and extends the credibility of the engineering staff.

The supporting documentation that is maintained as management accounting is critical in the health care environment. For example, in the case of an Operating Budget line item such as "Utilities – Oil Consumption," the budget proposed should be backed up with a calculation of the heating load requirement of the facility. There is a need for the facility staff, if not the senior management, to be able to verify the rationale for the request. As the unit cost of oil changes, and the facility expands or contracts in area, the requirement will change. Last year's numbers may have little resemblance to current needs.

The same approach, in concept, should be applied to all maintenance and repair items, as well as the ongoing operating budget. Ideally, the data is maintained in a computer database of a computer-aided facilities management (CAFM) system. Understanding the records of each piece of equipment is necessary to make the strongest case for improvement or replacement. Without this understanding and the ability to convey the documented facts, the request for even reasonable replacements will be questioned.

Again, the decrease in health care revenues is being borne, in large part, by facility operations. As the situation worsens, the apparently incredible demands being placed on maintenance operations can become a sad necessity. The solution is to have the supporting data to aid in making the best case for your budget requests.

The best current thinking in budget management is the utilization of "Activity-Based Costing." The complexity and regulation of health care is, from an operational standpoint, exceeding the capacity of the tools traditionally employed to control an organization's budget. Decreasing revenue and increasing difficulty in obtaining capital funds have made all hospital and clinic operations particularly sensitive to potential failures in plant operations, some of which are life-threatening. As this is not acceptable, yet management is charged with operations closer to the edge in order to survive, good facility operators are now formalizing the process by which they explain and justify their departmental expenditures.

Activity-Based Costing augments the traditional budget forms that list by supply and expense, or salary and wage, categories. Listing expense items does not define the service or program supported. A simple example is the capture of *Maintenance Supplies,* usually a line item identified under the broad titles *Maintenance Supplies – Electrical, Maintenance Supplies – Plumbing, Maintenance Supplies – Mechanical.* As individual line items, they are vulnerable to the routine statement, "cut all expenses by x%." At some point, you're not cutting expense; you're eliminating a service.

Reconciling the total operating cost of maintenance against a cost-based detailing of services can visibly demonstrate the impact of reduction on the operation and viability of the plant. Reduction may still come, but the decision-making process will be a far better informed one as to the impact. Please refer to the two budget overviews in Figures 23.3 and 23.4. Facility managers should consider casting their budget so it can be converted to Activity-Based Costing.

Conclusion

Maintenance and repair in health care facilities reflects the nature of this industry and the service it renders to society. It is increasingly driven by technology and cost. The desire to balance the absence of a true free market motive in pursuit of one's personal health with the limitations of one's resources provides an interesting analogy to the facility operation. We want a plant that is tailored to our desire to achieve excellence, but compromise is necessary. Unfortunately, compromise in this operation has the potential to influence the care given a patient. Health care facility managers must strive to balance these challenging, and opposing, objectives.

Few organizations run a 24-hour operation, 365 days a year, at levels of utility consumption and plant sophistication as high as health care institutions. The complexity of the task is augmented by the knowledge that failure to maintain a viable plant operation is life-threatening to the institution's "customers." The ability to create and maintain an operation that is continually learning how to perform better and more effectively is the challenge of the facility manager. Fortunately, organizations like IFMA and ASHE have provided learning opportunities, certification, and professional colleagueship that allow those in the field to measure themselves and benchmark against other, similar operations.[2] Willingness to adapt to change, and then do it over and over again throughout a career, is the hallmark of the successful health care facility operator.

[1] The Joint Commission on Accreditation of Healthcare Organizations, 1995 Accreditation Manual for Hospitals, Oakbrook Terrace, IL, 1995.

[2] See the "Bibliography/Recommended References" section at the back of the book for information on these and other organizations.

Maintenance Budget Proforma

Staff Salaries & Wages	Last Year	Current Year	Proposed New Year
Mechanic	$ 23,000	$ 24,150	$ 24,875
Lead Mechanic	26,000	27,300	28,119
Groundskeeper	21,000	22,050	22,712
Electrician, Master	52,000	54,600	56,238
Electrician, Staff	34,000	35,700	36,771
Carpenter	36,000	37,800	38,934
Plumber	36,000	37,800	38,934
Locksmith	38,000	39,900	41,097
General Helper	21,000	22,050	22,712
Administration, Secretary	28,000	29,400	30,282
Asst. Mgr., Engr. & Maint.	51,000	53,550	55,157
Director, Engr. & Maint.	75,000	78,750	81,113
Subtotal, Staff Salaries & Wages	**$ 441,000**	**$ 463,050**	**$ 476,942**
Supplies & Expense			
Supplies — Med./Surg. gases	27,600	28,594	29,366
Supplies — Housekeeping	143,500	148,666	152,680
Supplies — Laundry	97,200	100,699	103,418
Maintenance — Building & Carpeting	25,600	26,522	27,238
Maintenance — Plumbing	34,500	35,742	36,707
Maintenance — HVAC	52,100	53,976	55,433
Maintenance — Electrical	76,900	79,668	81,819
Maintenance — Paint	9,400	9,738	10,001
Maintenance — Mechanical	12,600	13,054	13,406
Maintenance — Other Misc.	5,300	5,491	5,639
Office — Forms	1,200	1,243	1,277
Office — Equipment	1,100	1,140	1,170
Office — Other	500	518	532
Repair — Building & Carpeting	18,900	19,580	20,109
Repair — Plumbing	27,300	28,283	29,046
Repair — HVAC	32,400	33,566	34,473
Repair — Electrical	47,050	48,744	50,060
Repair — Paint	6,200	6,423	6,597
Repair — Mechanical	21,900	22,688	23,301
Repair — Other Misc.	8,200	8,495	8,725
Rental — Equipment	6,500	6,734	6,916
Rental — Other	2,000	2,072	2,128
Consulting	25,000	25,900	26,599
Utilities — Electrical	672,500	700,000	711,000
Utilities — Gas & Oil	376,500	400,500	479,000
Purchased Services	77,000	79,772	81,926
Travel	5,000	5,180	5,320
Dues & Registration	2,100	2,176	2,234
Subscriptions	500	518	532
Functions	3,000	3,108	3,192
Postage — Mail	750	777	798
Postage — Freight	1,250	1,295	1,330
Other, Misc.	6,700	6,941	7,129
Subtotal, Supplies & Expense	**$1,828,250**	**$1,907,803**	**$2,019,100**
Grand Total	**$2,269,250**	**$2,370,853**	**$2,496,042**

Figure 23.3

Maintenance Budget Proforma
Activity-Based Projections for New Year

Routine Preventive Maintenance	
Supplies and Expense	$ 42,300
Salary & Wage	176,000
Repair, Routine	
Supplies and Expense	97,400
Salary & Wage	47,200
Repairs, Emergent	
Supplies and Expense	58,600
Salary & Wage	17,500
Plant Operations—Steam Plant	
Supplies and Expense	5,600
Salary & Wage	34,200
Plant Operations—Emergency Generator Plant	
Supplies and Expense	3,700
Salary & Wage	27,800
Isolated Utility Costs	
Supplies and Expense	1,190,000
Salary & Wage	0
Snow Removal	
Supplies and Expense	4,100
Salary & Wage	22,200
Construction Project A	
Supplies and Expense	252,000
Salary & Wage	65,000
Construction Project B	
Supplies and Expense	320,000
Salary & Wage	75,000
Undefined Programs & Services	
Supplies and Expense	45,400
Salary & Wage	12,042
Totals	
Supplies and Expense	$2,019,100
Salary & Wage	$ 476,942
Grand Total	$2,496,042

Note: Overhead and Administrative Expenses are pro-rated.

Figure 23.4

Chapter 24

Historic Buildings

William H. Rowe III, AIA, PE

To those who love buildings, the vistas of Grand Central Station, Market Street in Philadelphia, and New England town commons are outstanding, as are the great houses of the West Coast and the American shingle style. Historic structures are the best of the best. Historic structures, while often expensive to build originally, are often relatively simple to maintain. They offer convincing arguments for quality design and construction since they have often lasted so well with relatively little effort. However, when an upgrade is due, it is often for serious problems and is more expensive than repairs for conventional buildings because of the fine materials and workmanship that are required. Historic structures are overhauled infrequently and for only a fraction of their overall worth. They provide one of the essential ingredients to our culture and sense of place.

Maintenance and Repair Strategies

In the operation and maintenance of historic property, the most significant expenses are the capital improvement costs and annual energy operating costs. In many respects, historic structures operate like most other building types. The basic difficulty in upgrading historic structures is that the high first cost of capital improvements is difficult to finance. In private lending, banks are reluctant to loan for a term of more than ten years, and public funding is very low.

The key to successfully managing a historic property is to break down the projects into financeable bits in the private sector, but to push for an entire full renovation with full funding in the public sector. Thus privately owned buildings are continually upgraded in a series of projects, e.g., first the roof, then mechanical systems, enclosure walls, windows, and building finishes. There are not many opportunities to fund maintenance and repair projects for public buildings, and once funded it is almost impossible to obtain more money if the project was underfunded in the first place. It is generally more feasible to package all of the building improvements to be undertaken all at once. If the project is properly funded, designed, and constructed, no major capital improvements should be necessary for another 50 or 100 years (roofs and mechanical systems excepted).

The National Register of Historic Places

Buildings achieve historic designation in several ways. The most notable is by designation as an Historic Structure by the National Park Service and placement on the National Register of Historic Places. To achieve status on the National Register is a mark of significance that clearly sets a building apart. Buildings are placed on the National Register most often because they are outstanding in design and quality. Many buildings designed by America's best architects, and many of America's most loved buildings, are on the Register. Buildings listed on the Register are protected in that no federal funds may be used to contribute to their demolition. No highways, federal loan programs, or other projects that use federal dollars can threaten the buildings listed. In addition, federal programs have provided grants, loans, and tax credits for the renovation or restoration of historic properties listed on the National Register. Unfortunately, most of these programs are gone, with little chance of being reinstated.

In any work associated with a building on the National Register that involves federal money directly or indirectly, plans must be reviewed and approved by the federal government. This process usually involves acceptance of the design by local and state officials, who then, with the owner and architect, submit the designs to the National Park Service for review, comment, response, and eventually approval. This approval process does add time to the design process and should be anticipated. This process is not an appealing prospect for most private buildings, and is generally sought only for public buildings that need federal funding because renovation work cannot proceed renovation work without the historic work being done.

For property managers, facility managers, and public facility departments that operate public property on the National Register, the current status is as follows. For projects that use federal funding, the design review process is extensive, and there are few additional funds available for the restorative improvements that are generally favored by the National Park Service. As a consequence, improvements on non-federal, public property today are financed by state or local funding only. This avoids federal funds and allows the work to proceed without the restrictions, review, and delays associated with federal approvals. Federally owned property, on the other hand, must follow full design review procedures, and allocate funds for the improvements.

State and Local Designation

Most states, cities, and towns have historic commissions that review and designate significant buildings or districts as historic. Buildings that are listed as historic by a state or local historic commission usually have certain associated benefits and restrictions. The benefits include some forgiveness regarding conformance to current building codes. Except for structural soundness and perhaps fire alarm systems, which are relatively inexpensive and unobtrusive to install, the buildings are allowed to continue "as is" under the general theory that if they have lasted until today (often over 100 years) without a problem, they must be acceptable. When the buildings are significantly renovated or open to the public, however, they must make provisions for the disabled in accordance with the Americans with Disabilities Act (ADA), as well as remove any deficiencies in life safety. Another benefit of historic designation, particularly if the building is in a historic district, is that the ambiance of the area is protected—the value of the property is not brought down by less sympathetic development.

The review process for state and local historic designation is similar to that for the National Register. Improvements to the property must be reviewed by local authorities. In many communities, designation in the historic district means that certain improvements (usually those that are open to public view or that can be seen from a public way) must be approved by the local governing authority—usually the planning board, historic commission, zoning board, or a combination of them. There is very little public money available for private historic building improvements, and there are significant restrictions that follow the award of any public grant. Some states require design review in perpetuity. Again, there are few advantages overall to seeking out such designations.

Common Maintenance and Repair Projects

Historic buildings are special because of their many unique features and materials. Stained glass windows, terracotta tiles, ornamental iron, heavy timber framing, slate shingle roofs, murals, decorative plaster, stonework, and cast iron columns are only a few of the materials encountered in historic buildings. There are artisans and contractors who specialize in the techniques necessary for work in these specialty trades; facility managers are advised to employ them for work on historic buildings. As expensive as they may be, redoing poorly executed work is even more costly.

Exterior Work

Much of the character of historic buildings is derived from the appearance of the exterior. Slate roofs, multi/small-paned windows, brick or clapboard siding are all a part of the exterior fabric that requires special maintenance and repair. Following is a list of major areas requiring exterior work on historic buildings, and the methods that can be used to restore, preserve, or protect them.

Landscaping

Walks, paths, and roads around historic structures are often constructed of brick or gravel. They require some attention to maintain their contour, and repair in the spring after frost heaves.

Street furniture—such as lightpoles, benches, and signage—requires annual inspection to determine the need for touch-up painting, repair, or replacement. Replacements for such items are obtained from companies that specialize in historic reproductions. In some cases, the original companies are still in business.

Because planting is usually a significant feature of an historic setting, it is important to account for lawn and plant care in the budget. The costs are generally the same as for any well-maintained landscape.

Roofing

Many historic buildings have pitched roofs. Such roofs are a significant part of the building's design, incorporating materials such as slate, terracotta, or wood shingles. When repaired, these materials should be replaced in kind. For buildings that have flat, less visible roofs, modern materials are often more suitable than the original tar and gravel.

Slate Roofs: Slate roofs are expensive to repair and replace. They require repair of the flashing, of the slate itself, and of nails and substrate.

Flashing Repair. In older buildings it is common for large sections of the flashing to be pitted, ripped, or have broken seams. Repair of copper flashing often involves removal of the underlying slate or cap stones, removal of the old flashing and the addition of new, as well as removal and replacement of broken slates and those that had to be removed to get at the flashing. New flashing should be as thick as possible—20 gauge if possible—and consideration should be given to using lead-coated copper for longer life. Gutters, leaders, and snow guards are often redone, as well as materials to properly conduct water away from the building. Older buildings often use dry wells, which become clogged and unsuitable. Connection to the storm drain is often the preferred method, which can be incorporated with associated street and sidewalk repairs. Finally, as with many pitched roofs, it is often desirable to provide crickets at chimneys or intersecting valleys to better conduct water off the roof.

Slate Repair. Slates are available in several colors, including red and gray, which are priced differently. Slate roofs are often decorative, with patterns woven into the slate—this level of craftsmanship makes them more expensive. With time, a small number of slates will drop from the roof to the ground. If the entire roof is not replaced, repair of individual slates is possible using babitts (or babies), which are individual copper sheets that are bent to hold individual slates in place. Unfortunately, it is difficult to replace just one slate, since many more become damaged during the repair process when workers walk on or even touch slates that may have concealed cracks.

Nail and Substrate Repair. The nails used in many roofs will corrode long before the slates fail. Moisture condensation under the roof will cause the fasteners to rust and the slates to loosen and fall. Also, it is not uncommon for portions of the substrate to rot, particularly where leaks have developed. Consequently, a loose shingle may be the "tip of the iceberg" in roof repairs that can involve slates, flashing, substrate, crickets, gutters, and leaders with connections to the storm sewer in the street.

Terracotta Roofs: Many of the same principles and problems associated with slate roofs and flashing apply to terracotta roofs. In addition, terracotta—which is manufactured like pottery—is often unique to a building. There are a limited number of terracotta roof tile manufacturers in the United States, and in many cases the particular shapes for a roof must be custom made. This is expensive and takes considerable time, for not only do the tile shapes need to be made, but also the ceramic glaze on the tile must be made to match the existing tiles.

Wood Shingle Roofs: Classic wood shingles often must be used to repair historic buildings. In modern construction, local codes may require fire-retardant shingles. Wood shingles are the least expensive to repair individually. However, because wood is not as durable as other roofing materials, it is more likely that the entire roof rather than individual shingles will begin to cup, warp, and split from excessive sun and heat, requiring full replacement. In wood frame buildings, care should be taken that the attic spaces are properly vented. When buildings are insulated for modern energy efficiency, proper circulation of air from eave and ridge vents is often blocked, which accelerates structural decay.

Enclosure Walls

Brick: Brick is a common material in historic buildings. Older brick was manufactured by hand in wood forms, dried by air, and then baked. The clay and sand composition, size, color, texture, water resistance, and thermal expansion characteristics are different from modern brick. Brick repair is often done at cracks or where bricks have failed, often from water intrusion (see the section on "Mortar" later in this chapter). Matching the mechanical properties of the original brick as well as the aesthetic characteristics is important to ensure that the bricks will be durable and stable over many future weathering cycles.

Historic buildings use more intricate brick details than modern construction. Arches, corbels, bonding, patterns, textures, and colors are all a regular part of historic repair work and require special care and detailing. Matching brickwork with newly manufactured products requires effort and patience.

Stone: Stone masonry is almost a forgotten art — repair of stonework is usually done by masons. Cleaning and patching are the most common maintenance activities. When cleaning stone, it is important to do test patches. Certain stones, such as marble and limestone, will be aggressively destroyed by cleaning agents. Patching stonework often requires additional work to locate the original quarry to determine whether matching stone can be obtained. If this is not possible, the building itself may be scavenged in less prominent areas for stone to use for required patches.

Mortar: The most critical aspect of masonry (stone, brick, or cement units) is the proper choice of mortar. Older mortars are "soft." They contain larger quantities of lime and are structurally weaker than modern mortars. It is very important to use these lime-rich mortars on historic buildings, even though they may seem to be weaker and less structurally suitable. The older bricks are also weaker than modern counterparts. Properly soft mortar will "bed" the older brickwork and move (infinitessimally) with it. In addition, the lime mortar has a "self healing" property that will permit resealing of microcracks over time. Cement-rich mortars have no give and will cause older bricks to crack. Cracked bricks are subject to accelerated deterioration from water intrusion and freeze-thaw cycles.

Many masonry failures result from water intrusion, which washes out the mortar behind the masonry wall. In repairing a failed wall, loose bricks — even entire sections of wall — can be removed from mortar that has the consistency of loose sand. Such work is often difficult to anticipate without extensive testing or investigation, but should be repaired as soon as it is discovered.

Wood: Wood rots. At eaves, near the ground, at leaks, and where infested by termites, carpenter ants, or powder post beetles, wood needs more constant attention than masonry. Annual inspection for potential rot, observation of insects in the spring, and visual observation of peeling paint or rotted wood are effective in limiting repair and maintenance costs. Wood will last a long time with minor incremental maintenance. Removing earth and leaves to 6" below wood sills, cleaning gutters in the late fall, and properly controlling pests are low-cost or no-cost items that will lower overall maintenance expenses.

Windows

Windows in historic buildings are commonly single-pane, divided-lite units with operable sashes on sash cords with counterweights. Over time, the wood deteriorates, caulking and broken glass require replacement, and trim requires repainting. For ordinary repairs such as painting, caulking, and glass replacement, crews that are generally familiar with such work are satisfactory. Special consideration, however, must be given to lead paint and asbestos caulking. Lead paint on windows is to be expected, and special precautions in the removal of the paint may be necessary. Such precautions will add to the painting cost, depending on the regulations of the local authority. Some older windows contain asbestos in caulking materials, especially the caulking between the windows and masonry walls. Such windows may need to have the asbestos caulking abated, which will also add to the repair costs.

When replacing windows in older buildings, it is often energy efficient to install windows that are double glazed. Many manufacturers produce windows that match the mullions and muntins that are often featured in historic buildings. These double-glazed windows also avoid "frosting" in winter, so that condensation does not collect and rot out the interior sash. Unfortunately, some historic commissions do not readily accept such windows — if grant monies or public approvals are required, some convincing by the building owner or facility manager may be necessary. One argument in favor of double-glazed windows in historic buildings is that because storm windows are not necessary, the new windows will better show the building's original character.

Windows in historic structures are generally more expensive to repair and maintain than those in modern buildings. Older buildings tend to have more custom windows, a greater variety of window types, and millwork details, and trim that must be fabricated to match existing conditions.

Waterproofing

In historic structures, waterproofing (other than roofing) often is concentrated in the basement, where ground water, surface water, or water from improperly drained rain leaders can enter the building — often at holes in masonry or stone walls. Simple solutions such as regrading the outside may better conduct surface and rain water away from the building. When water cannot be stopped from entering the building by regrading, it may be necessary to provide trench drains at entries, dig around the entire perimeter to the depth of the footings and install a perimeter drain, foundation waterproofing, or other similar protection.

Painting

Painting of historic structures is similar to other buildings, with the added concerns related to lead paint and asbestos caulking noted in the discussion of windows. Because of the fine detail of many older buildings, the preparation for painting is more costly in terms of properly scraping at crevices and mouldings. It is also expected that the painted surface will be scraped to obtain samples of the original paint to test for color and composition.

Interior Work

Historic properties have above-average operations costs because of the special character of many features and the unavailability of parts, particularly for finish materials such as mouldings, hardware, windows, and various woods. Mechanical costs are also higher because of the difficulty in running MEP (mechanical, electrical, plumbing) equipment through walls, unknown conditions, and the work in design and construction necessary to conceal these systems and maintain the historic integrity of the building.

Demolition costs such as the following add to the overall cost of interior construction:

- Asbestos removal, particularly at boilers, piping, plaster, and concealed spaces.
- Paint stripping and lead paint removal.
- Unknown conditions, e.g., rot, structural damage.
- Phasing of demolition in occupied buildings, which may need to be scheduled during off-peak hours.

Code Upgrades for Life Safety

Much repair and maintenance work on historic buildings consists of upgrades required for life safety as required by code. Fire alarm upgrades include the installation of smoke detectors in storage rooms, basement areas, corridors, and at the tops of elevator and stair shafts. Fire alarm systems are comparatively inexpensive safety devices that are often permitted in historic structures in lieu of other code-conforming devices such as sprinkler systems, widened exit stairs, or non-combustible finish materials.

HVAC Systems

HVAC systems in older buildings normally exist without current code-conforming ventilation or air conditioning. Incorporation of such systems into historic buildings is more expensive than conventional construction because suspended ceilings (which are generally inappropriate anyway) or chase spaces are not available to conceal the systems. Perimeter cast iron baseboards, fan coils, and historic-looking grilles and diffusers are generally more expensive to install and detail.

Boilers and other systems are often upgraded to smaller pieces of equipment and control systems that offer improved energy efficiency.

Some older buildings are characterized by open plans that include large staircases and expanses of space that are not subdivided to contain fire. In any repair or renovation work, the building official may ask that portions of the building be partitioned from others to better contain the spread of fire. The introduction of fire or smoke walls can be expensive. A useful solution is the installation of automatically closing smoke doors that, when open, are generally transparent to the building user, but when the building goes into alarm, will close to inhibit smoke between sections of the building.

Fire protection systems such as sprinklers are commonly required in historic buildings because of their effectiveness and because most historic buildings contain significant quantities of flammable materials. Sprinkler systems, which are dimensionally small, can often be effectively concealed. The cost for such systems is more than conventional construction, but overall they are not considered expensive considering their overall effectiveness in preserving and protecting such property. Compartmentalization is also used in historic buildings to prevent the spread of fire.

Handcrafted Materials

There are many materials that were used on historic buildings that are not readily available today. Cast iron facades, tin ceilings, leaded and stained glass, brass railings, marble balustrades, brick carved mouldings, and printed wallcoverings are all examples of items not commonly manufactured. It is encouraging that there is a small but effective industry of building artisans available that are providing these materials for historic work.

Accessibility Upgrades

Historic structures are often public buildings that are expected to be accessible to all. Consequently, reasonable upgrades are expected to be in place as a matter of course—the discussion of ADA requirements in Chapter 6 is particularly applicable for historic structures.

In many cases, the costs of accessibility upgrades are prohibitive. Cutting marble trim or modifying significant architectural detailing to meet accessibility standards can be at odds with maintaining the building's character. In such circumstances, building owners and managers can solicit advice from both local and state historic agencies and agencies that serve the disabled, who are aware of such conflicts. They can often provide alternative solutions, variances, or methods to resolve otherwise costly modifications that might destroy aspects of the building's historic fabric.

The following is a list of the items that often need to be upgraded per ADA:

- **Doors, thresholds, and hardware.** Doors need to be sufficiently wide (36″) to accommodate a wheelchair. At least one door to all public spaces and to any space used by an employee in a wheelchair may need to be modified. Doors to accessible spaces require door handles that can be operated by someone with a closed fist — i.e., lever handles. Lever handles were common historically, but not for all types of buildings. Changing hardware, locksets, and thresholds in historic structures is more expensive than average due to the aesthetic requirements of matching existing trim and hardware. As an alternative, automatic opening devices provide access while allowing historic hardware to remain in place.

- **Toilets.** Toilets for the disabled are a standard in building construction today; facilities for the handicapped are expected to be provided in historic buildings as well. Often the biggest problem is finding spaces large enough to meet the spatial requirements for handicapped toilets. Demolition and reconstruction costs are high. Existing toilet rooms, typically underfixtured by modern standards, will often lose fixtures in order to provide handicapped stalls, thus necessitating discussions with the building official and possibly the historic commission and representatives of the disabled community.

- **Elevators.** Elevators are an essential key to providing access to the disabled. In historic buildings, the placement of elevators requires skill to minimize structural damage and other disruption to the floor plan. Once installed, however, the annual repair and maintenance costs of elevators are likely to be consistent with conventional construction.

- **Signage.** Appropriate signage must be provided for the disabled. Fortunately, once provided, signage needs little maintenance or repair other than changing names of spaces, which generally is a minimal cost.

Upgrades to Improve Building Performance

HVAC and energy improvements have been discussed. Historic buildings are especially prone to the need for energy improvements. Lighting fixtures in large spaces with high ceilings and large windows often must be modified to reduce annual energy costs.

Lighting upgrades are accomplished with increased use of compact fluorescent fixtures to replace incandescent fixtures where appropriate, as well as occupancy sensors that control the need for lighting when spaces are occupied.

Low voltage systems for data and communication are particularly useful in historic buildings. Systems commonly provided include:

- Telephone
- Security
- CATV
- Computer — LAN (local area networks)
- Sound systems (e.g., intercoms)
- Wireless systems

The use of wireless systems can provide considerable cost savings. Such systems are currently being proposed for fire alarms, telephones, data systems, and intercoms. The benefits in terms of both cost and appropriateness in historic structures is obvious.

Conclusion

Maintaining historic structures provides both challenge and reward. The owner, facility manager, and designer must have an appreciation, knowledge and understanding of both the historic and technical considerations of historical buildings. Many of today's functions are new to historic buildings and these facilities must be upgraded to comply with current needs. Child day care, computer rooms, sprinkler systems, and small cafeterias are a few of the many features that were not originally part of, but must now be incorporated into, historic buildings. The planning and technical considerations in the maintenance and repair of historic structures require a coordinated effort to achieve solutions that are cost effective, yet maintain the character and detailing that represent the essential identity of these structures.

Chapter 25

Hotels and Convention Centers

William H. Rowe III, AIA, PE

Hotels and convention centers are among the most complex buildings to operate and maintain because they can include aspects of many different building types. Hotels contain sleeping and eating facilities at a minimum. Hotels and convention centers may also include some or all of the following features:

- Banquet facilities
- Conference rooms for lectures and workshops
- Entertainment (often a full stage)
- Exhibit halls
- Large dining areas
- Office space for patrons and staff
- Rooms with audio, visual, and teleconferencing capabilities
- Shops
- Swimming pools
- Tennis courts
- Workout rooms

Hotels and convention centers also contain a full range of support spaces and activities, such as:

- Atriums
- Commercial kitchens
- Concierge services
- Elevators
- Landscaped grounds
- Laundry rooms
- Loading docks
- Lobbies
- Mechanical and electrical equipment
- Public toilets
- Repair shops
- Staff lounges

Finally, these facilities often seek affiliations with specialty restaurants, local attractions, teleconferencing services, skywalks, heliports, riverboats, and seasonal or historical events. All in all, the running of a hotel or convention center requires both skill and grace. Much of what happens is routine, but in the background is a continuous state of crisis management on which the success of the hotel or convention center depends.

Operating Costs

After hospitals, hotels and convention centers are the most costly facilities to operate and maintain. A wide variety of systems must be kept operative, and backup systems with quick responses must be as complete and timely as possible. This means 24-hour, 7-day service year-round on all building systems.

Construction costs for these facilities are high, whether measured on a dollars per square foot or a dollars per room basis. In addition, hotels and convention centers incur costs for land, pre-opening costs, furniture, annual capital costs, and annual maintenance and repair.

Once the facility is built, the annual costs include those detailed on the proforma (see Chapter 1). The expenses also include staff salaries, license fees, franchise costs, and supplies, as well as similar costs that are unique to the hospitality industry. Energy costs can include electricity, fuel, steam, water, and gas. Electricity accounts for approximately 70% of the energy budget in hotels and convention centers, which is a good reason to implement an effective energy management system.

Factors Affecting Operating Costs

The annual costs of operating a hotel or convention center can be affected by the following:

Programs

The range of program offerings adds to the annual budget. The number of dining rooms or kitchens, the range of outdoor sports (e.g., golf course maintenance), the extent and expanse of public spaces, and the staff required to meet guests' needs all influence the costs related to the facility.

Facility Design

The architectural layout of the facility affects the overall economy of operation. The smoothness of traffic flow, separation of building service functions from customer service functions, and proper positioning of spaces for functional adjacency all affect the cost of operation. High-rise buildings incur increased costs for elevators and mechanical systems. Low-rise buildings are likely to have higher land and landscaping costs, depending on the location.

Building Systems

Proper design and selection of building systems must provide the necessary structural, mechanical, and electrical functions without burdening the facility with higher than average first costs or annual costs. The goal should be easily diagnosed, easily repaired systems that do not require a large on-site staff. Integrated state-of-the-art building control systems work with the check-in desk to activate a room's HVAC system when a guest checks in, and return it to the unoccupied mode when a guest checks out. Such systems can also personalize computer key locks, initiate phone and cable TV billing, utilize the TV for in-house communications and teleconferencing, and initialize electronic mail to the LAN connection in each room.

Construction Quality

Properly built, quality building components ensure that annual expenditures are dedicated to maintenance and repair rather than to the correction of construction defects.

Finish Material Selection

Quality finish materials ensure durability and reduce annual maintenance and repair costs. Higher than average annual costs are incurred in buildings that require exterior painting, fail to provide suspended ceilings to easily route electrical and mechanical systems, or have improperly insulated roofs, walls, windows, and piping. On the other hand, even at the cost of higher maintenance, expensive materials may be preferred—such as wood instead of vinyl flooring—when a richer look is desired. A careful review of finish material choices should include a forecast of related annual maintenance and other costs.

Equipment Selection

Hotels and convention centers require a great deal of equipment for kitchens, laundries, swimming pools, electrical generation, sewage disposal, water treatment, and telecommunications systems. Such equipment should be selected for durability, ease of repair, and overall efficiency. The equipment should also be located and designed to be accessible to the disabled.

Maintenance and Repair Strategies

The facility manager of a hotel or convention center encounters a full range of immediate and long range problems and opportunities. The goals include an operation in which the staff works as a team with little friction, costs are controlled when possible, and guests' needs are accommodated.

The overwhelming majority of hotels and convention centers are privately owned and operated. The few that are publicly owned are almost always contracted to private vendors. Governments and public bodies do, however, influence hotels and convention centers. Facilities that cater to the public are subject to a number of regulations, particularly those concerning public accommodation such as specified in the Americans with Disabilities Act (ADA). Further, they are often taxed to cover the costs of providing goods and services to the public that might not otherwise be necessary, such as public transportation, terminals, and related amenities. Because such facilities are associated with increased traffic and heightened activity at shops, theaters, and restaurants, they will often participate in the formulation of public policy concerning development of such areas.

Funding for operating expenses of hotels and convention centers comes from the income generated by the services they offer. For hotels and convention centers in high-rent, well-located areas with high occupancy rates, the revenues are strong. As neighborhoods or transportation patterns change, the effect on hotels can be dramatic.

A hotel or convention center is often comprised of a collection of franchises. The franchise agreements obligate the parent hotel and the franchisee to provide certain levels of ongoing repair and maintenance. Both parties must agree on the monies to be allocated for emergency conditions or capital reserve for future projects to avoid disputes when the funds are appropriated.

Common Repairs and Maintenance of Building Elements

The repair and maintenance program for a hotel or convention facility involves teams that are assigned to different tasks. The maintenance and repair functions are structured to suit the expected demand. The major aspects are as follows:

- **Housekeeping and groundskeeping:** Cleaning, janitorial, minor custodial work; lawn and garden care (including snowplowing and maintenance of walkways, driveways, and parking areas); trash removal; and carpet cleaning are all scheduled routine functions of the housekeeping and groundskeeping staff.

- **Building maintenance:** Minor carpentry, repair of floors and doors, patching and paint touch-up of walls are done in guest rooms and customer areas to maintain the appearance of a well-run facility.

Preventive and Scheduled Maintenance

The facility manager will have a prescribed schedule for the maintenance of each piece of equipment, room, or activity assigned to a team. Maintenance activities may be scheduled daily, weekly, monthly, semiannually, or annually. They can involve working on equipment, turning mattresses, changing filters, and checking door closers, locksets, and exit lights. Scheduled maintenance activities cover all the items in the facility, room by room, and week by week. Some activities may be done by staff; others may be done by outside vendors who lubricate, drain, or balance certain pieces of equipment depending on the season or performance requirements.

Exterior features that require maintenance and repair include the building's structure, roof, enclosure walls, windows, waterproofing, and landscaping.

Structure

Building structures tend to be relatively trouble-free unless they have design deficiencies or have been subjected to unusual forces or inappropriate renovations. An annual inspection is useful, as is a review of the structure after a major calamity such as a fire, flood, hurricane, or earthquake.

Under normal circumstances, cracks on building walls or water leakage should warrant a more detailed inspection to ensure that the problems are not significant. Serious problems are less costly if they are addressed immediately. Misaligned doors, cracking structural members, accelerated rusting of members, widening of cracks, or noticeable vibrations are signals to call in a structural engineer for an evaluation.

Roofs

Roofs are typically not repaired until they leak. Many roofs last a long time without a problem; some leak into mechanical or service areas where they are (inappropriately) more likely to be tolerated. Over time, all roofs deteriorate. The causes of roof failure include brittleness resulting from low temperatures, flow of materials such as asphalt under high temperatures, wear and tear from walking and servicing, holes in membranes, tears at the joints, or damage from excessive vibration.

Most of the time, roof maintenance is a simple process of inspections to look for cracks, bulges, blisters, leaks, stains, or other signs of distress at regular intervals and after major events such as storms. Periodic maintenance includes repair of simple defects, removal of debris, clearing of drains, correction of equipment problems, and patching of weak spots in membrane or flashing.

Enclosure Walls

With time, exterior walls may show signs of wear. Caulking may become brittle, bricks may show signs of efflorescence, paint may peel, and surfaces may be marred by graffiti or staining. Older buildings may need insulation in walls or thermal pane glass. Structural deficiencies may result from settlement cracks or other problems that should be corrected.

Periodic maintenance for exterior walls usually involves cleaning, preparation, and painting or refinishing. Recaulking of joints is also usually provided by the painter. Proper cleaning and preparation is the largest part of the job and should be done carefully and correctly to ensure value in the work. Masonry walls often need repointing; care should be taken that the work is more than "skin deep." Loose mortar should be raked out, and root causes of extreme deterioration, such as leaking interior drains, should be addressed.

Windows

Windows in hotels and conference centers are not subject to excessive abuse because they are often kept closed. Over time, they may develop objectionable leaks, become inoperable, or have excessive glazing or caulking problems that favor replacement with newer double-glazed thermal windows.

Annual inspections of windows should be conducted to ensure that windows open without difficulty, counterweights and balances work properly, weatherstripping is secure, hardware is operative, and glass and glazing are in an acceptable condition.

Waterproofing

Water in buildings can be a serious problem. The basic solution is to either divert the water before it gets in, or pump it out afterwards. Regrading to divert ground water and installing perimeter drains are usually effective measures. Often, water penetrates buildings because of blocked dry wells or clogged drains. Whenever the foundation wall is dug up, full waterproofing prior to backfilling is advised.

Annual inspection to look for evidence of water intrusion, leaks, blocked drains, mildew, or foundation cracks is recommended.

Landscaping

In most facilities, proper landscaping provides exceptional value. Landscaping is labor intensive and not specifically income producing, but has a dramatically positive effect on patrons and should be maintained at a high level. Plants and grounds continually change and need to be refreshed, mowed, pruned, fertilized, or replaced.

The frequency of mowing depends on rainfall. Pruning must be done properly and carefully for each plant. Lawns are also irrigated in conjunction with need. In some localities they can be watered with "grey water" from the sewage treatment system. Irrigation systems commonly include spray heads imbedded in the lawn, which are usually activated in the evening to minimize evaporation.

Mechanical and Electrical Systems

In hotels, mechanical and electrical systems must be extremely reliable, as well as safe, capable of maintaining comfortable conditions, and as energy efficient as possible. Predictive, preventive, and general maintenence, as well as repairs, must be done efficiently and expediently to maintain a high level of service to the hotel's customers.

Water

Water is essential for hotel operation. It is so important that many facilities consider providing reserve tanks or extra wells as a back-up, and always provide a significant reserve in domestic hot water systems. Cold showers are very unpleasant for both the guest and the hotel manager.

A hotel's actual water consumption varies from 100 to 250 gallons per day per room. Large deluxe hotels use more than smaller economy hotels.

As the costs of water rise, hotels are initiating some controls to reduce consumption. When possible, hotels install their own wells to avoid higher water rate charges. Costs for heating domestic water can be reduced with preheat systems that utilize waste water or waste heat to preheat incoming water through a properly designed system of heat exchangers.

Common repairs associated with water systems include repair of leaks, unclogging drains, and preventive maintenance on pumps and heating coils. Other considerations for water systems are discussed below.

Meters: Not all water that enters the facility will leave by the sewer system—some is used by cooling towers, swimming pools, or the irrigation system. These are usually metered separately so that the incoming water quantities are adjusted to account for the water that is not discharged into the sewer system, thereby reducing sewer charges. When permitted by municipalities, separate meters are effective in saving considerable costs. Also, recreational water in pools can be saved by use of a pool cover, which will result in considerable savings when the pool water is also heated.

Grease Traps: Grease traps must be provided in kitchens and other grease-producing sources. Grease is not accepted by municipal systems and must be removed prior to discharge into the sewer system.

Water Temperature: Water temperatures must not exceed norms for scalding in guest rooms, yet must be sufficiently boosted for proper dishwashing and laundry purposes.

Tags and Identification: Proper identification of water systems aids greatly in improved maintenance. Labeling of valves with tags and charts, identifying domestic cold and hot water systems, and clearly marking shut-off valves will help to facilitate maintenance procedures.

Irrigation: Water for irrigation consumes large quantities. These quantities can be better controlled by paying attention to actual weather conditions, rather than relying on a timer.

Storm Water: Storm water systems conduct large quantities of water away from buildings during rain and storms. For facilities with basements with below-grade spaces, on hills, or in areas where water can naturally drain toward the building, proper storm drainage is important to avoid flooding in the building itself. Cleaning drains and removing leaves, checking that backflow preventers are operating, and checking that the pipes are in fact conducting water freely are useful measures.

Electricity

Electricity is the principal energy source for hotels and convention centers, accounting for as much as 90% of the total energy bill in some facilities. Constant, reliable power is essential to proper operation. Kitchen functions, air conditioning, and hot water are crucial to hotel operation—secondary sources of power, such as from a separate source, a power grid, or an emergency generator, are always considered. Electrical power is often purchased at 480 volts or higher, and always as three-phase, which is required to run motors. Step-down transformers are used to provide power for lights, equipment, and receptacles. Different rate structures are usually available for a particular location. These rate structures should be analyzed to verify that efforts are being made to attain all possible savings.

Electricity is used at higher voltages to run motors for HVAC equipment, elevators, and pumps. Electricity is also used to provide lighting as well as power to low-voltage systems such as fire alarms, telecommunications, and security systems.

Emergency power, in the form of a backup emergency generator or battery packs, is required for emergency lighting, fire alarms, fire pumps, and life safety devices such as smoke evacuation. Emergency power is often expanded to include freezers, elevators, hot water heaters, and HVAC controls.

Repairs and maintenance on electrical equipment, except for changing ordinary light bulbs, must be done by following safety procedures. Proper lock-out and tag-out of equipment, shut-down of active circuits, and safety checks before and after the work is done are necessary to prevent injury. Preventive maintenance on motors and controllers, as well as inspections for power surges, short circuits, and cross wiring, are routine activities for the house electrician.

Heating, Ventilating, and Air Conditioning

Building comfort is achieved through properly designed and maintained HVAC systems. Since HVAC systems consume one of the largest portions of the facility maintenance budget, keeping the systems in good repair is key to controlling costs.

The large pieces of equipment that are primarily responsible for heating and cooling are termed *generation equipment*. Boilers, furnaces, cooling towers, chillers, and primary circulating pumps are examples of generation equipment that require the most attention. Proper annual tune-ups for the fuel-burning equipment ensure clean combustion, higher efficiency, and lower costs. Heat transfer surface cleaning, nozzle adjustment, proper water treatment to avoid scale, and similar maintenance activities are discussed in Chapter 11, along with similar procedures for the distribution systems and terminal equipment.

In the hospitality industry, quick response is important. If a problem cannot be solved quickly, alternate rooms may be used to relocate guests or activities until the problem is solved. Staff trained to perform steam trap repair, valve repair, control resetting, and fan belt replacement can make a difference when it comes to quick action required to ensure proper building performance.

Maintenance of HVAC equipment lengthens its life, improves its performance, and lowers costs. Money-saving measures include cleaning filters, coils, and lamps; tuning up boilers; and checking fan belts. See Chapter 11 for detailed maintenance procedures for HVAC equipment. On a daily basis, good maintenance management will see to the following:

- Leaks are fixed.
- Water is not wasted.
- Doors to freezers remain closed when not in use.
- Outside doors function properly.
- Lights are turned off in unoccupied rooms.
- Temperatures are set back in unused rooms.
- Ovens are not left on all day.
- Dishwashers and washing machines are run with full loads.
- Items in need of attention are reported by staff.

Energy Management: Energy use is under the control of the facility manager. When possible, energy costs can be minimized with a single contract with the power company. When the facility has subtenants, they should be sub-metered and backcharged, or billed directly to encourage good energy management. By carefully reviewing the local rate structure, it may be possible to shift demand charges and time-of-day rates. Computer programs can be used to adjust demand so that when elevators are running during peak times, water coolers, secondary fans, and other devices are off to keep the instantaneous demand charges low. In large facilities, considerable savings are realized by off-peak running of ice-making machines to generate ice for air-conditioning applications.

Cost Saving Techniques: To save money, facilities with emergency generators can apply for cogeneration or hookups to power companies to generate power into the grid during brown-outs. The following additional measures can also save energy costs:

- Interruptable gas purchased at lower rates.
- Larger fuel tanks, which may pay for themselves in lower prices per gallon because of the large deliveries possible.
- Fixed windows rather than operable units.
- Adequate — not excessive — bathroom exhaust (under 75 cfm per guest room).
- Occupancy sensors.
- Automatic controls to central check-in/check-out.
- Lockouts that limit the range of temperature guests can control.
- Water-saving shower heads, water closets, and faucets.
- Heat recovery systems that scavenge heat from dryers, waste water from dishwashers and washing machines, or clean process water for cooling towers.
- Water-cooled refrigeration equipment that can preheat domestic water systems.
- Atriums (in winter the warm air at the top is returned to the system, and in summer the cool air at the bottom is returned).

Interior Work

Interior features that require maintenance and repair include safety and security systems, laundries, lighting, telephone systems, waste management, and food service.

Safety and Security

Guests and staff expect a reasonable level of security at hotels and conference centers. This includes life safety concerns related to accidents and fires, as well as personal and property protection from criminal activity.

Worker Safety: OSHAct provides a wide range of protections that are discussed in Chapter 6. Prevention of injury to workers requires constant monitoring, safety officers to review accident reports, programs to encourage worker safety awareness, and management programs and procedures that promote a safe environment.

Guest Safety: There are many areas related to guest safety to consider. In baths, safety is promoted by proper provision of nonslip surfaces, grab bars, monitored hot water temperature (120°F in most jurisdictions), ground fault interrupting (GFI) electrical receptacles, and temperature-compensating shower valves.

Building Safety: Building safety is established by codes and is discussed more extensively in Chapter 6. Hotels and convention centers — as public buildings and as spaces where people sleep — are subject to the most stringent rules and have some unique characteristics. Sprinkler systems are generally required along with a complete fire alarm system, both of which are tied to an emergency generator. Careful maintenance of these facilities includes avoiding trash accumulation that blocks exits; regular cleaning of dryer ductwork; replacing filters; and actively cleaning and managing the kitchen equipment, particularly range hoods and grease traps.

Security: Security is a key element in hotel and convention center selection. If there is a perception or reputation of danger, then the facility's survival is in danger. To ensure a sense of well being, the entry should clearly restrict access to guests and staff only. Supervision of entries, good lighting, and access to staff are essential. Guest room doors should close and lock automatically and have peepholes, bolt locks, and programmable lock codes. Locks must be changed immediately when keys are suspected missing or off premises, and a regular system of key control should be in effect. Indoor and outdoor spaces should be organized to eliminate areas that can be threatening or can provide a place for criminals to conceal themselves.

The level of reasonable care that needs to be provided is being determined in the courts. Any patron who suffers an injury on hotel or convention premises may have a basis for a lawsuit. Proper records of management directives, training sessions, and responses to accidents and incident reports will help show a positive record of care and diligence that can mitigate an otherwise disastrous claim.

Laundries

The need for laundry facilities in the hospitality industry is undisputed. Guest and dining linens, as well as staff uniforms, make up a significant laundry load each day. Facilities that have the space to provide for an adequate on-site laundry will experience lower costs, better control, less loss, and longer wear. Alternatively, outside vendors can provide laundry services.

The transport of laundry to the washrooms is done by staff with carts or, when possible, down laundry chutes. Chutes are an advantage because they save time. Automated overhead transport systems in the washroom allow staff to sort the laundry by type into bags that are automatically fed into the washer. In some cases, computers can be used to track the completion and distribution of laundry.

Washing machines for larger hotels can handle up to 700 pounds of wash and include cool-down cycles that avoid wrinkling fabrics. Tunnel washers, which are actually a series of interconnected machines, wash predetermined loads of wash in batches. In top-transfer tunnel washers, the load is automatically transferred from one batch to the next, allowing the water to be reused on the successive batch. In bottom-transfer machines, the load and the water are transferred to the next machine until the process is complete. In either system, the load transfer is automatic, fewer workers are needed, and water can be recycled for maximum effect.

Cost Savings: The facility manager can find cost savings in more efficient laundry operations. The quantities of water, the need to heat the water to proper temperature, and the energy used to dry laundry all represent possible reductions in quantity of water, heat reclamation, and reduction in staff. Computers are often used to cycle load times, adjust temperatures in the cycle, control the length of time the washers and dryers run, and operate the energy reclamation systems.

The first step in improving the overall efficiency in the laundry is to review the inventory. If it is too small, there are fewer machines to absorb the rush jobs. Rush jobs affect overall production, and can result in overtime and wear on fabrics. A review of controls may result in improved inventory or less damage. A careful inventory may also show that some equipment (e.g., extra carts) that take up space are in fact surplus. If space limitations prohibit laundry expansion, it may be possible to expand the number of shifts to handle the load.

When upgrading equipment, be sure to allow for breakdowns. The benefit of using two smaller machines is that if one breaks down, the other will still be running. If space or costs prohibit this, then some arrangement with an outside vendor can be useful.

Lighting

There have been dramatic changes in lighting fixtures in the recent past. Fluorescent lighting has always been inexpensive to run, because it provides much more light per watt than incandescent lamps. Only recently, however, has the quality of the fluorescent light and other types, such as halogen, made low-cost lighting an attractive alternative to incandescent. Other lighting types, such as metal halide, mercury vapor, high-pressure sodium, and low-pressure sodium, are used in areas such as open storage rooms and parking areas, where a large area must be lit but lighting color and variability are less important.

Relamping and cleaning comprise a significant portion of the lighting budget. Cleaning lamps semiannually is typical and will result in more light to the spaces. Depending on the facility and the accessibility of the lamps, bulbs are replaced individually when they burn out, or replaced wholesale in groups at predetermined times. For example, a group replacement of a few high ceiling ballroom lights, which requires a crew and a moveable stage, is more efficient than bringing in the portable stage when each bulb burns out. On the other hand, replacement of individual bulbs in guest rooms is a simple matter and can be done on an as-needed basis.

To reduce the overall lighting budget, facility managers can use more efficient lamps, use fewer lamps, and/or install a control program to limit the hours of operation of each lamp. Occupancy sensors, twist timers, and photocell sensors for exterior lights are effective in controlling lighting costs.

Telephones

Telephones and telecommunications are expanding the types and levels of available service. To a large extent, telephones are a net income stream, since customers are generally billed for most calls at a rate higher than the direct cost to the hotel. This is possible because hotels are usually able to negotiate discounted rates. Because of the rapid advances being made, it is possible to spend more than necessary on a phone system and still have one that falls short in some critical design features.

Types of Calling: There are many types of phone service available to guests. They include:

- Local calls—often provided with the room at no charge.
- Direct dial long distance calls—the "1+..." calls billed to the guest by the hotel.
- Credit card calls—the "0+..." calls billed to the guest at rates negotiated by the hotel, which may receive a fee for each call placed.
- Operator-assisted calls—using the hotel operator or "0+...", which can include third party billing, person to person, billed to room, international, 800, and 900 numbers.

The hotel pays to make all of these types of calls available; the mechanisms by which the expense is redeemed should be verified. Often, telephone service is a substantial cost that is inadequately reimbursed. It is also important to avoid inordinately high charges. The actual telephone bill is often paid by the customers later, when they receive their credit card bill at home. If the charges are more than they feel are reasonable, they will complain and potentially shift business.

Types of Phone Systems: For hotels and convention centers with more than 40 lines, a PBX (phone branch exchange) is usually installed. A PBX allows for the hotel to have a limited number of outbound lines but a much larger number of in-house lines. When coupled with a call accounting system (CAS), guest calls can be monitored for billing and the charges automatically placed on the bill at the completion of the call. CAS systems need to be updated to be current with existing phone rates.

PBX systems should have a feature that verifies when a call begins and ends. A common way to set up the billing is to have the faceplate on the phone direct the customer to dial "8 + 1", "9 + 1," or "0 + 1," which will trigger the CAS system and initiate the PBX system to route the call.

Conferences or other large groups may request that special trunk lines be set up. It is important that these lines be disconnected after the convention leaves to avoid charges. Also, PBX systems can place calls on the wrong outgoing line, resulting in higher charges or even charges that bypass the CAS system. Current technology is still perfecting calling systems that properly track customer calls and accurately bill them.

Modern guest room phones have many features, such as built-in data lines for connection to computers. Some rooms have two phones, which guests find useful, and features such as conference calling, speed dial, hold, voice mail, fax, speakers, and message waiting. Additional phone features may include controls for television, cable TV, up links and down links to satellite systems, tie-ins to Internet, lighting, radio, and HVAC systems.

Regulations: Current requirements under ADA require that hotels, as public accommodations, provide public phones, guest room phones, and telecommunications devices for the deaf. The front desk should have a TTD to handle incoming and guest calls. Telephones are also required to be compatible with hearing aids.

Guests must be allowed to select an operator other than that provided by the hotel.

Waste Management

When people go to a hotel or convention center, they generate an enormous amount of trash. The disposal of trash is more and more difficult each day. There are some areas where trash to energy conversion is more feasible, but waste management is a serious problem for society in general, and hotels and convention centers in particular. For the most part, outside vendors are engaged to haul trash away to disposal sites.

The need to reduce, or at least slow, the accelerating cost of waste management has caused a reevaluation of the way we make and dispose of trash. Currently the following methods are used to manage the costs of trash removal:

Conservation: By decreasing the amount of products bought or used that need to be disposed of, the actual source of waste is reduced. Use of fewer disposable items, more bulk buying with less packaging, and stricter inventory controls reduce both the cost of the purchase and the disposal of the goods. Suppliers who stock shelves or dispensers directly can avoid containers entirely. Glass, china, and linen do need to be washed, but in the long run are less expensive than disposable styrofoam, plastic, and paper.

Food portions should be monitored to verify that they are consumed. If menu items are regularly not eaten, their portions should be promptly reduced or removed from the fare. Some hotels are affiliated with charitable organizations who pick up unused portions that can be used by various shelters.

Reuse: Certain items can be reused. Containers, boxes, and trays might be used for more than one application, even by someone outside the hotel. Even Henry Ford specified the dimensions of the wooden crates in which batteries were delivered so that they could be placed directly into the automobiles as seats for his cars.

Recycling: Recycling involves sorting materials and selling them if possible to secondary manufacturers, who use them as raw materials for their production. Paper, cardboard, metals, plastics, and yard waste are the major materials currently being recycled. For each material, a recycling vendor can be found and the procedures worked out. Since up to 60% of the trash generated can be either reused or recycled effectively, there is significant cost savings available to the facility manager in this area.

Food Service

Food is a central feature in hotels and convention centers. Buying, processing, serving, cleaning, and disposing of food, along with maintenance of food service equipment, represent a major operating cost. Kitchen equipment includes machines that mix, cut, slice, cook, fry, reconstitute, radiate, microwave, infrared, broil, rotisserie, charbroil, braise, griddle, deep-fat fry, steam, and warm food. Food is stored in refrigerators, freezers, and on holding tables. Ice makers, dishwashers, compactors, and garbage disposals are also commonly found in commercial kitchens.

Maintenance of equipment for overall smooth operations is essential. The food service and building maintenance staffs need to work together to keep equipment in top working order. Equipment that begins to fail will only cost more to fix later, with a longer time of inefficient operation. Coordination between the two teams is also important to ensure that new purchases have the proper utility connections and service spaces around them for proper operation. Cleaning standards and maintenance manuals need to be kept up to date so that when emergency repairs are necessary, the materials to fix them are readily available.

Food service staff must also be properly trained in the correct operation of kitchen equipment. This training, together with close supervision of equipment operation, is key to preventing excessive repair costs.

Money-saving maintenance and monitoring of food service equipment includes:

- Reading meters for air velocity.
- Detecting leaks in microwaves.
- Checking water temperature in hot water systems.
- Checking steam pressure.
- Using thermometers in freezers, refrigerators, and ovens.
- Cleaning grease from refrigerator and freezer air-cooled condensers.
- Using split systems to place condensers outside.
- Avoiding water-cooled condensers that use city water.
- Using well water in properly designed water-cooled systems.
- Constant checking for leaks, clogged drains, poor seals on equipment or doors, and odors from drains, particularly in food storage areas.
- Following an effective cleaning program for kitchen equipment and spaces.

Conclusion

Hotels and convention centers are among the most intensive facilities to operate. They can involve hundreds of staff members working in a great variety of activities. They provide a place to host, serve, and display the best of what we have to offer—this in itself demands great ability in and understanding of human relations, strong technical skills, and crisis management. There are many opportunities to save costs and improve the overall operations.

Chapter 26

Museums
and Libraries

James Armstrong

The primary mission of museums and libraries is preservation and education. Both collect artifacts to preserve the past, and provide inspiration for the future. Both strive to educate visitors by helping them interpret art, science, history and literature through books and artifacts.

Maintenance and Repair Strategies

Funding for museums and libraries is similar in that they have steady resources in the form of tax dollars, endowments, or private funding. Museums generally differ in that, to a greater extent, they depend on income from visitors in the form of donations, admissions, or museum store or restaurant sales. Entertainment and functions have become a key income source for both of these facility types, which are also used for public and private meeting places.

Most museums and libraries are considered either public or not-for-profit entities. Trustees or the general public have a large say in their operation and expenses. Careful documentation of contracts for material purchases and services is vital. In many cases, bidding is required for all purchases, or at least three quotes must be obtained. The processing for payment can be negotiated during the bidding process, but usually not afterwards. Most operations run on a net 30-day payment process, with some going as far as net 45 days.

Administrators of museums and libraries often have a background focused on education rather than operations. The most successful administrators tend to spend quite a bit of their effort on development for grants or gifts to their respective organizations. Facilities maintenance and repair are usually not part of the organization's mission of education, and may therefore be considered as overhead or, in a sense, a necessary evil. From the maintenance manager's point of view, the key to success in maintaining this type of facility is providing comfortable surroundings that create a more effective learning environment.

Key Building Elements

The following items tend to be the focus for museums and libraries in terms of frequent, costly, or crucial maintenance and repair.

Physical Plant

Many special considerations are made for museums and libraries in the design, repair, or renovations to the physical plant. Once again, the goal is to create a comfortable environment that enhances learning and viewing.

Locations

Physical plant equipment or mechanical spaces in museums and libraries are usually hidden behind exhibits, under floors, or above ceilings. Features such as hidden door handles or remote-operated access are common, and have their own maintenance considerations. Boiler rooms are typically housed in basements with an access door in nonpublic areas. Air handlers are usually in penthouse or rooftop locations with access from nonpublic areas.

Noise or Vibration

Equipment noise or vibration can be a key factor in design and maintenance because it can adversely affect visitors or exhibits. Some of the ways of controlling noise and vibration include: insulated mechanical rooms, extra vibration mounts on equipment, wider diffusers on air handlers, oversized ductwork, sheets of lead and rubber under pump mounts, and more expansion joints on wall penetrations.

Electrical and Mechanical Diversity

Most museums and many libraries need to regularly change the physical layout of both exhibits and learning areas. Furthermore, the media in which the educational process is carried out changes as technology changes, and in accordance with the type of display. Temporary exhibits may require plumbing or electricity to meet specific requirements.

Fire Suppression

The preferred method of fire suppression is a dry system with sprinklers. Remember, one of the primary missions of a library or museum is to preserve, so efforts must be made to reduce any false release of water yet ensure that the system will suppress any fire as quickly as possible, and with a minimum of damage.

Operating Hours

Many maintenance projects must be performed after operating hours to help maintain museum illusions and reduce any distractions to library users. Group functions that take place in the facility often occur after hours. Preparation for and cleanup following these events must also be scheduled.

Environmental Conditions

A variety of environmental conditions can contribute to the deterioration of collections, whether books, art, or antiquities. Humidity, sunlight, and extreme temperatures are all factors, depending on the items.

Air Quality — Many facilities use clean room conditions including HEPA, ionization, and activated charcoal filtration to protect their collections. Many items in the collections contain potentially hazardous materials, such as arsenic or paradichlorbenzene (mothballs) used in the preservation of animal specimens. Care must be taken in performing maintenance activities, and staff educated on the handling of these materials.

Sunlight — Ultraviolet light contributes to the deterioration of paintings, furniture finishes, fabric, and printed items. Protective window treatments or drapes are often used, and these have special maintenance requirements.

Humidity—Excessive moisture can deteriorate porous items by breaking down bonding agents and promoting bacteria growth. Excessively dry conditions can remove inherent moisture in some artifacts, causing flaking, cracking and splintering. Maintenance programs should be designed to address these factors.

Extreme Temperatures—These conditions can cause melting of some materials or promote drying. Temperature change can cause potentially destructive expansion and contraction.

Staffing Considerations

Operating Hours and Seasonal Issues

The busiest time of year for visitors generally coincides with public school scheduled vacations and holidays. School groups tend to visit more toward the end of the semester or school year. Weather can affect visitation to both museums and libraries in that moderately adverse weather tends to bring people inside. Most of these facilities are climate controlled because of the nature of the collection.

Staff Understanding

Maintenance and custodial staff need to be diversified in that the changing exhibits, technology, and special events continually alter the environments of the facilities. When exhibits or displays are set up, maintenance will be called on not only to make the design work from conception to operation, but to maintain them once in place, and then assist with dismantling and related tasks. Maintenance and custodial staff are also expected to work in very public areas, and to be fully apprised of security requirements to protect valuable collections. Some of the collections contain hazardous or unknown substances, necessitating an understanding of universal precautions and compliance with Right-to-Know laws.

All of these special facility features require appropriate training programs. Staff education is clearly a key factor in maintenance management for museums and libraries.

General Budgeting Approach (Including Capital Planning)

When developing the budget for your operation it is recommended that it be broken down into four categories:

Maintenance Costs: All the costs associated with maintaining the facility and grounds, including labor and contracted services, as well as the supplies needed to complete the tasks.

1. Maintenance Salaries—Salary and overtime for in-house staff and administrative support.
2. Benefits Costs—Cost of health insurance and any other negotiated benefits.
3. Insurance—Workers' Compensation and general liability.
4. Contracted Services—Outside services that are under contract for part or all of the fiscal year. Examples are computer service contracts, boiler service contracts, or refuse removal.
5. Outside Services—Cost for any service from an outside provider that is called in only when needed and involves variable costs.
6. Maintenance Supplies—Expendable supplies used in the general maintenance of equipment or facilities that cannot be capitalized.

Custodial Costs:

1. Custodial Salaries—Salary and overtime for in-house staff and administrative support.

2. Benefits Costs — Cost of health insurance and any other negotiated benefits.
3. Insurance — Workers' Compensation and general liability.
4. Contracted Custodial Services — Outside services that are under contract for part or all of the fiscal year. Examples include carpet extraction or window washing.
5. Equipment Services — Costs related to the service of custodial equipment.
6. Custodial Supplies — Expendable supplies that are used for cleaning, such as chemicals, mop heads, rubber gloves, and trash can liners.

Utility Costs:
1. Electricity — Including demand charge, KWH costs, and fuel surcharges.
2. Heat (steam, oil or gas) — Depending on your operations, whatever fuel source is used to heat the facility.
3. Water and Sewer — Potable and process water should be metered separately. Water that is used for irrigation of the grounds should also be metered separately, as it may affect the cost of sewerage. Some municipalities meter the water only and bill for the sewerage based on water usage.
4. Telephone — Some operations separate telephone costs by department. This is recommended, as it reflects true cost generation.
5. Other — Some operations use other commodities, such as chill water.

Capital Improvements and/or Equipment: Many factors govern the development of this budget. Trustee and/or member/visitor programs have a major influence. Exhibits or parts of the facility that bring in more visitors or users naturally tend to be awarded more capital improvements. Some examples are listed below:
1. Developmental Issues — Renovation and maintenance priorities can be directed according to issues expressed by donors' desires. Some building components are more popular than others. Donors' contributions are usually associated with highly visible displays or building elements, such as fountains, sculptures or, on a larger scale, building additions, rather than essential but less glamorous items like a new roof.
2. Function Operations — Special functions in the exhibit halls of library spaces are more frequent as municipal and state governments, corporations, and businesses seek attractive space for large groups, with the public relations benefit of supporting a public institution. Many museums and libraries supplement their own fundraising efforts through functions for members.
3. Temporary Exhibits — In the drive to attract visitors, more museums are using temporary or rented exhibits. These exhibits help increase visitation to other exhibits and inform the general public of the total operations. Some of these are categorized as "blockbusters." These can break a budget if not planned well, just like in show business.
4. Seasonal Operations — Many libraries or specialty museums are seasonally oriented and must plan accordingly. As visitation patterns are studied, many improvements are developed to

enhance seasonal operations. Some examples are a new walkway by the river, an atrium on the side of the building with the view, or courtyard dining facilities.

5. Admissions and/or Funding—This is the driving force for museums in particular. If visitation is low, voter budget appropriations will reflect accordingly. Low admissions income will affect daily operations as well as capital budgets. If the building is not maintained well, it can affect the comfort of the visitors, and visitors are the lifeblood of a healthy operation. Successful facilities of this type place a high priority on satisfying their customers.

Conclusion

Like educational facilities, museums and libraries share the mission to educate. (In fact, nearly every school and university has a library, and many have museums as part of their facility.) Public museums and libraries have the challenge of maintaining a high level of interest and use from their customers—members of the community—in order to ensure continued funding from public appropriations and private donations. This goal is accomplished in a variety of ways, from timely, stimulating exhibits, to an attractive atmosphere that supports the learning process. Maintenance and repair are key components in achieving a high level of success. These operations must be unobtrusive, and attentive to the needs of both valuable display items and the people who visit them.

Chapter 27

Manufacturing Facilities

James E. Armstrong

Maintenance and Repair Strategies

When working in a manufacturing environment, the main focus is the bottom line. Maintenance is a key component in delivering product and, therefore, profit. There have been many innovations in management over the last few years regarding manufacturing. To understand repair strategies in manufacturing, let's explore some current management philosophies.

TQM (Total Quality Management)

Total quality management has evolved from many of the original concepts regarding employee involvement. In theory, TQM is re-engineering that focuses on the different processes in assembly or manufacturing. The front line employee, through involvement teams, looks at and learns the total process that he or she is a part of in producing the end product. Each person on the team develops a better understanding of others jobs; together they work toward the elimination of repetitive steps or unnecessary procedures. The process of TQM constantly evolves as the team members develop a sense of ownership in the machines they operate and the process of which they are a part. Also, as products change, the processes evolve with the new or improved product.

JIT (Just In Time)

Just In Time is a process in which the warehousing of products is minimized by focusing on manufacturing the correct quantity of product to fill the orders without a surplus. In the cash-strapped eighties this process became essential for small companies to survive. Competition became strong in many markets, so profit margins were reduced. JIT gave many small and large operations the cash they desperately needed.

Flow Management

As part of the re-engineering process, one of the key issues is flow management. This is what happens from the time the raw materials come in the door to when they go out the door as a product. The questions raised involve such issues as how many times the material is handled, and what can be done to minimize the handling. Many approaches are taken, such as establishing different small processing areas that allow for the most efficient use of space, machines, and labor. Optimum flow management constantly evolves as product lines change. Some facilities use the same machines for many different products.

Manufacturing Diversity

As stated previously, many product lines change as quality is upgraded or as sales require, so the facility must have the ability to react to changes. In facilities where diversity is a key attribute, most energy sources, such as electricity, compressed air, steam, water, etc., come from above, as opposed to below the floor. This allows for easy movement of machines and/or production lines.

Automated Manufacturing

With the advent of computers, automation seems to be changing operations on a daily basis. As more powerful technology is developed, certain processes can be programmed to be more accurate than if they were performed by a human. They also require a higher level of training and knowledge for the maintenance staff. Processors are being used to control robots in repetitive motion processes. Some can manage and store information on material flow, quality assurance assessment, cost management, equipment reliability, and life cycle costing of equipment. In many operations, this is all being done in real time. This provides for automatic re-ordering of parts or raw materials as inventory is consumed. When the parts are received, the automated system schedules equipment use to ensure optimum through-put. Machines are scheduled for preventive maintenance during off or off-peak production hours. Some machines that use computer diagnostics on-line can predict failure or wear, thus allowing for the planning of maintenance.

Employee Ergonomics

One of the fastest growing concerns in the Workers' Compensation field is repetitive motion-inflicted injuries, such as carpel tunnel syndrome. As part of the manufacturing engineers' concerns when developing machines for manufacturing, the human factor must be considered. Special tooling and/or safety features are being developed to reduce the repetitive motion and increase the safety of machines for employees. Improvements of this type may fall within the responsibility of the maintenance department.

Dollars per Square Foot—Parts per Labor-Hour

Fact-based decision-making is one of the keys to success in manufacturing today. As part of the decision process, a quantifiable analysis allows for true fact-finding. In today's market, dollars/S.F. and parts/labor-hour are commonly used formulas to show improvement. Your facility may have other similar quantifiable equations. Use of fact-finding allows for unbiased processing and also helps with life cycle costing of equipment and end products.

Contracts for services and purchases are subject to close scrutiny in fact-based decision-making, and must be carefully documented. In many organizations bidding or at least three quotes is required for all purchases. The terms of payment can be negotiated during the bidding process, but usually not afterwards. Most operations run on net 30-day payment terms; some extend as far as net 45 days.

Manufacturing Facility Categories

- *Manufacturing Mills:* This category includes the mills that were built at the turn of the century and have been used for many types of production over the years. Many of these facilities were built adjacent to rivers and used hydropower to turn hundreds of shafts and pulleys throughout the plant. In later years, large electric motors or steam engines replaced the water wheels.

- *Food Processing Plants:* This category includes food product and/or animal food processing plants. These facilities typically need access to a significant supply of water, as well as large refrigerated operations for food preservation.
- *Assembly Plants:* In many operations, parts are made in many specialty plants and assembled closer to the end user.
- *Light Manufacturing/Assembly:* These facilities are similar to an assembly plant, yet they tend to be more of a warehousing and distribution center for an end product.
- *Product Processing Operation:* These are specialized treatment plants used by many different companies. Operations might include galvanizing, painting, or plating. Most of these operations have many environmental issues and outsource to specialists to distribute the liability.

Manufacturing Operations History

There have been two main manufacturing eras in this country: the immigrant boom from the turn of the century until the depression, and during World War II, for the war effort through its aftermath, to support the "Baby Boom" needs.

Because many mills were built in the teens to the twenties, most of these facilities have long ago had to begin dealing with deferred maintenance. This, of course, is also true of operations built after World War II. After twenty years of operation, many major maintenance issues occur, such as roofs, heating oil tanks, brickwork, and boilers. During the fifties and sixties, energy was relatively cheap, so the construction and renovations that took place during this period were not energy-conscious. Uninsulated roofs, oversized boilers, aluminum or wood single-pane windows, asbestos insulation, and inefficient lighting have presented a challenge to cost-conscious facility managers over the past thirty years. The Americans with Disabilities Act (ADA) has been an issue because many buildings sprawl over a large area and have more than one story, and must now meet accessibility requirements. Plumbing construction of the fifties and sixties used galvanized pipe for drains, and these eventually rust and fail. The preferred piping insulation was asbestos, a good product for the cost at that time. Now many plants are dealing with or have dealt with asbestos abatement issues.

For the plants that were built during the teens and twenties, maintenance of historical integrity has become a costly issue. These buildings tend to have character, yet are often suffering from a lack of proper and protective maintenance. The repairs that are needed are costly, and many operations are looking at cheaper alternatives. These include abandoning the building and replacing it with new construction, or looking in other geographic locations.

Other issues on older plant sites are ground contamination from food and manufacturing processing wastes, fuel oil tanks, and asbestos and lead paint removal. The process of transferring property and receiving a clean 21E is becoming very complicated. [A 21E is a Federal (EPA) environmental form required for transfer of property. Sellers typically hire a consultant (an environmental specialist) to inspect the property and produce a report, which must be approved by the EPA.] Companies are finding it cheaper to hang on to the property.

Maintenance and the Physical Plant

There are maintenance and repair issues in manufacturing or processing operations that differ from those of other facilities. Many special considerations are made in the design, repair or renovations to the physical plant of each individual manufacturing or processing facility. Unique equipment is required to produce each product, and the environmental and maintenance requirements will also vary from plant to plant. A key factor is supporting a diversified environment that is adaptable to change and allows for an efficient work flow.

Locations

Physical plant equipment or mechanical spaces can be found almost anywhere in a processing facility. For example, a boiler that is used to produce process steam tends to be located near the process equipment so energy waste can be minimized. Equipment for other energy sources tends to be overhead, with multiple tapping points to allow for a flexible operation.

Noise or Vibration

In most operations, equipment noise or vibration can be a key factor in design or maintenance because it can adversely affect the quality of product. Some solutions to these problems include: insulated mechanical rooms, extra vibration mounts on equipment, wider diffusers on air handlers, oversized ductwork, sheets of lead and rubber under pump mounts, and more expansion joints on wall penetrations.

Manufacturing or Process Equipment

Manufacturing or process equipment is the lifeblood of the maintenance department. The equipment must be maintained with the minimum downtime. Depending on the operation, the machinery can be as simple as drill presses and pumps, or as complicated as robots with automated material flow equipment. The more complicated the equipment, the more diversified the operators and maintenance staff must become.

Preventive Maintenance

Preventive maintenance is a particularly crucial factor in successful management of manufacturing facilities. Many manufacturing processes are continuous, with some plants operating 24 hours a day. Product production, delivery, and quality can all be affected by a problem on the line.

Predicting and taking steps to avoid equipment failures is key to preventive maintenance. Complete documentation is, naturally, an important component, as is an effective reporting system. Regularly scheduled maintenance and routine inspections, such as vibration analysis, are part of the program.

A Computerized Maintenance Management System (CMMS) can be extremely useful for managing an effective preventive maintenance program — for reporting, recording, and evaluating equipment condition, and for planning and scheduling preventive maintenance tasks. Other technology is available to monitor and analyze other functions, such as ultrasonic scanners that can detect valve leaks in tanks or blockages in turbine meters.

Staffing Considerations

Operating Hours

Operating hours of the maintenance operation within manufacturing and/or processing plants vary depending on the operation. Some manufacturing operations have the ability to perform maintenance during off-hours so as to minimize production's downtime. Others operate 24 hours a day, and address major maintenance by shutting down the plant at different times of the year for maintenance or production line upgrades. Some 24-hour operations perform maintenance between processes or when the machine fails.

Staff Understanding

Maintenance and custodial staff need to be diversified, since changing technology is constantly changing the facilities' environments and the knowledge required to manage them. To maintain qualified, versatile staff, facility managers must be committed to ongoing training in systems analysis and management. Maintenance and custodial staff have different work demands than in other types of facilities because of the high productivity required. An understanding of universal precautions, as well as an understanding of the ability to comply with Right-to-Know laws, must be part of the staff education process.

A good sense of quality product and pride in their work are necessary qualities for maintenance staff. They work hand-in-hand with the engineering staff to design and assemble new production lines or production line upgrades, and many mechanics can make machines work when the engineers can't. Maintenance is a central and vital function in the operation of a profitable plant.

General Budgeting Approach (Including Capital Planning)

When developing a budget for a manufacturing operation, it is recommended that it be broken down into four categories:

Maintenance—All costs associated with maintaining the facility and grounds, including labor and contracted services and the supplies needed to complete the tasks.

1. Maintenance Salaries: Salary and overtime for in-house staff and administrative support.
2. Benefits Costs: Health insurance and any other negotiated benefits.
3. Insurance: Workers' Compensation and general liability.
4. Contracted Services: Outside services that are under contract for part or all of the fiscal year. Examples include bleacher service contract, boiler service contract, and refuse removal.
5. Outside Services: Outside services used only when needed (variable costs).
6. Maintenance Supplies: Expendable supplies used in the general maintenance of equipment or facilities; cannot be capitalized.

Custodial

1. Custodial Salaries: Salary and overtime for in-house staff and administrative support.
2. Benefits Costs: Health insurance and any other negotiated benefits.
3. Insurance: Workers' Compensation and general liability.
4. Contracted Custodial Services: Outside services that are under contract for part or all of the fiscal year. Examples include carpet extraction and window washing.
5. Equipment Services: Costs related to the service of custodial equipment.
6. Custodial Supplies: Expendable supplies that are used for cleaning, such as chemicals, mop heads, rubber gloves, and trash can liners.

Utilities — Costs related to the use of services from government-regulated utilities.

1. Electricity: Including demand charge, KWH costs, and fuel surcharges.
2. Heat (steam, oil, or gas): Will depend on the operation. If steam is used in manufacturing, the fuel source is often separately metered, for tax reasons.
3. Water and Sewer: Potable and process water should be metered separately. Water that is used for irrigation of the grounds should also be metered separately, as it may affect the cost of sewerage. Some municipalities meter only the water and bill for the sewer based on water usage.
4. Telephone: Some operations separate telephone costs by department to have an accurate picture of cost generation.
5. Other: Some operations use other commodities, such as chill water.

Capital Improvements and/or Equipment — Many factors govern the development of this budget. The predominant influence is the need for production improvements. Some examples are listed below:

1. Production Machines: Well-run engineering and maintenance operations maintain a standard for life cycle costing. This is a vital part of the maintenance and repair management within the manufacturing or production operations.
2. Energy Sources: Source for production steam or air. Do they adequately meet the needs of production, and are they reliable?
3. Major Building Repairs: These can be identified and prioritized using Chapter 19, "Deferred Maintenance."
4. Facility Upgrades or Expansion: These are based on the facility's needs as determined with feedback from involvement teams, including marketing and production. The options can be reviewed using the project matrix such as that shown in Chapter 23, "Health Care Facilities."

Chapter 28

Office Buildings

William H. Rowe III, AIA, PE

Office buildings are money-making machines. More than any other type of building, they are built to rent space for profit. Traditionally, they are also the easiest to build. Office buildings have the simplest of layouts, consisting for the most part of two exit stairs; a men's and women's toilet; incidental electrical and mechanical closets; and an open, uncluttered floor space for subdivision by tenants. When compared to the intricacies of residential buildings, or the complexity and cost of manufacturing plants, it is not difficult to see why office buildings are the building type most favored by developers.

Because they are so desirable to build, there can sometimes be an over-abundance of office buildings available—thus they are also the most competitive to rent. Traditionally, office buildings were in cities. After World War II, suburban office parks were established, and when many families began to own automobiles, office buildings outside cities in the suburban rim became a significant portion of the office market. As the result of advances in telecommunications and computer systems, workers and employers are able to communicate regardless of location. With the growth of the home office industry, the character, size, and usefulness of the office building is under strong re-evaluation.

Office buildings are often multi-use facilities. The ground floor is typically rented for retail activities, and the upper floors are rented for offices. To provide spaces that will draw tenants, office buildings are expanding the types of services they provide, resulting in increased work for facility managers. To make their buildings attractive to prime tenants, office building managers are seeking complementary tenants: day care centers, cafeterias, health clubs, restaurants, and support services such as messenger services, valet parking, and taxi stands. In addition, a full range of telecommunication services including satellite dishes, cable, fiber optic, and other building systems that can support computer equipment and flexible office environments are part of the package offered to many tenants.

Public vs. Private Ownership

Most office buildings are privately owned and operated. The owners rent to both private and public users at market rates. Rental rates depend on supply and demand. Funds for upkeep, maintenance, and repair are dependent on a healthy cash flow. When office buildings are situated in a larger office market, added building features are commonly provided to entice higher-rent tenants. Public office buildings depend on yearly appropriations to fund annual budgets for maintenance and repair. Many public office buildings have the special maintenance and repair issues of older, historic structures (see Chapter 24). Public buildings such as city halls and libraries often establish the image of the office buildings around them.

Much of the maintenance and repair work in office buildings, including public buildings, is predictable and performed by staff. Cleaning, repair of minor leaks, touch-up painting, and small jobs in electrical and mechanical maintenance tend to be done by full-time staff. This works best in companies or organizations that employ a crew of workers for a number of buildings. Outside vendors are typically used for larger than ordinary repairs and for capital improvements. When the need for certain vendors is known in advance—such as for cleaning filters, starting up the cooling system, or cleaning windows—funds can be appropriated annually and bids can be requested for the work.

Maintenance and Repair Strategies

Much repair and maintenance work in office buildings is routine. The routine is interrupted primarily by changes in tenants and by capital improvements. Once built, most building systems remain unaltered and are maintained until they are replaced. If particular systems do need upgrading, such as increased power to a particular floor for computers, it is often in response to tenant requirements.

Exterior Work

Exterior features that require maintenance and repair include landscaping, roofing, enclosure walls, windows, and walkways.

Landscaping

Landscaping is a low- or no-cost item for many urban buildings, but can be a significant expense for suburban office parks where a major selling point is often the plantings one encounters on the approach.

Maintenance of indoor plants, particularly in lobbies, can be done by staff or as a vendor service to firms that will guarantee the plants and replace them as necessary.

Roofing

Roof repairs are common in the vicinity of rooftop equipment that must be serviced. Cooling towers and rooftop package units that require seasonal adjustment can be a source of wear on the roofs. Roof pavers are helpful in minimizing roof leaks. Other than such repairs, roof replacement is programmed into the capital improvement budget for the office building.

Enclosure Walls

Exterior walls of older office buildings can experience leaks from caulking and sealants that have become brittle. Curtain walls, stone veneers, and pre-cast concrete buildings can require replacement of failed sealants. Another requirement is frequent cleaning of enclosure walls to keep the building looking fresh. Painting of wood facades and repointing of masonry walls are also expected.

Windows

In some larger buildings, window cleaning is an ongoing maintenance operation. In multi-use buildings, the retail (ground floor) level is often glazed with single-pane windows to facilitate overnight repair in the case of vandalism.

Other Exterior Features

Exterior ramps, walks, paths, and signage require minimal maintenance involving snow and debris removal and annual inspection of areas needing touch-up work.

Interior Work

Office buildings are subject to significant costs to remodel interior space for compliance with government-mandated programs. Many recent laws require that office buildings (especially high-rises) be upgraded for fire safety or to provide access to the disabled.

Code Upgrades for Life Safety

Fire alarm systems are routinely upgraded in office buildings. Code requirements for fire safety are updated regularly and are often the most likely upgrades to be required, even in existing buildings. Any modification to this alarm system often triggers upgrades such as automatic fire detection in duct systems, storage rooms, corridors, and elevators. The older an existing fire alarm system, the more likely it is that it will cost more to provide enough detection points or a sophisticated enough logic system for the mandated requirements.

Elevators

Elevator maintenance is usually handled on a service contract basis. Minimum sizes for elevators are now larger because of the Americans with Disabilities Act (ADA), and the costs of upgrading elevator systems in older buildings can be extensive. Most elevator codes require annual inspections to check for safety and identify areas needing repair.

HVAC Systems

HVAC systems in office buildings are easiest to maintain when each tenant has individual controls. The greatest determinant in tenant satisfaction is often the quality of the environmental control system. Poor HVAC performance is a serious issue, and much of the maintenance and repair budget is spent ensuring the proper operation of HVAC systems. Direct digital control (DDC) systems, which permit more rapid diagnosis and response, as well as direct remote control, are useful in maintaining overall building comfort.

Fire Protection Systems

Fire protection systems are typically upgraded as part of tenant improvements. Some jurisdictions permit existing systems to remain, and delay any mandated retroactive upgrades until the tenant leases expire or until existing tenants substantially upgrade their space. However, some jurisdictions have mandated that fire protection systems be retrofitted into existing buildings (See Chapter 6, "Codes and Regulations").

Upgrades for the Disabled

The ADA and most local codes require that buildings be made accessible to the disabled. The required modifications to existing buildings can be phased with other work—except for certain low-cost items that must be addressed right away. (See Chapter 6 for additional information regarding provisions for accessibility.) Generally, the first step is to upgrade access from the outside into the building. It is expected that public walks and lobbies be accessible to those with physical impairments. Doors, thresholds, and hardware to tenant spaces are modified to allow wheelchair passage, minimize threshold height, and provide lever handles. Again, when permitted, this work is deferred to coincide with tenant improvements.

Public toilet revisions are among the most costly accessibility upgrades, particularly when existing walls and plumbing must be relocated to accommodate the larger stalls.

Elevators are required to provide access to floors used by the public. While most office buildings already have elevators, two- and three-story buildings that heretofore had only stair access may now be required to provide elevators — a significant expense requiring some careful analysis.

Upgrades to Improve Building Performance

Because HVAC systems are critical to tenant satisfaction and represent a significant annual cost in fuel, building managers tend to devote considerable effort to improving the energy efficiency of their buildings. The introduction of advanced remote control systems is effective for managers of several buildings; and modifications to more energy-efficient equipment such as boilers, chillers, and economizers, are ongoing capital improvements calculated to save money on the long-range bottom line.

In office buildings as well as other building types, conversion to energy-efficient lighting has an immediate payback. Many utility companies offer rebates to offset the cost to improve lighting efficiency. Offices, which have large areas of artificial lighting, are prime candidates for such improvements.

Low voltage systems such as telephones, security, CATV, computer – LAN, sound systems (e.g., intercoms), and wireless systems are increasingly being introduced into office environments. In some instances, the repair and maintenance of such systems is entirely the responsibility of the tenant. In any case, this type of maintenance is usually performed by vendors specializing in such systems.

Special Requirements of High-Rise Buildings

High-rise buildings have particular life safety requirements, including smokeproof enclosures, fire protection, smoke detection, elevator recall, fire safety plans, fire alarm command center, smoke evacuation, and emergency power systems.

Ventilation

Both ASHRAE and BOCA have revised standards for ventilation, principally in response to concerns about indoor air quality (IAQ) and sick building syndrome. Issues concerning IAQ and sick building syndrome surfaced after the energy crisis in the 1970s, when lower ventilation standards were implemented to save energy. The effect of the new ventilation standards on energy consumption is considerable, since the values have generally doubled. Ventilation air minimums are largely based on providing a certain amount of fresh outdoor air per person.

A summary of the health and safety standards is shown in Figure 28.1.

Health and Life Safety Standards					
	ASHRAE	BOCA	NFPA 90A	CABO-M	Remarks
	62-1989	Mechanical Code 1993	1985	EC 1992	
Natural Ventilation (Outdoor Air)		An area equal to 4% of the floor areas must be operable[1].			Toilets and kitchens are required to have mech. vent. in some codes.
Mechanical Ventilation (Outdoor Air)	10-35 CFM per occupant. Refer to Figure 2.5[2].	10-35 CFM per occupant. Refer to Figure 2.4[1, 2].		References ASHRAE 62	Required if natural ventilation is not satisfied.
Natural Light		An area equal to 8% of the floor area must be glazed[1].			
Artificial Light		[1]		General Light 20 FC[3]	
Smoke Detectors in Air Handling Units >2000 CFM		Supply Duct	Supply Duct		
Smoke Detectors in Air Handling Units >15,000 CFM		Supply and Return Ducts. Also Supply and Return Ducts when >50% of system air is exhausted.	Supply and Return Ducts		
Sprinklers		Both the 1990 BOCA Fire Prevention Code and NFPA 96 require sprinklers in commercial kitchen hoods and ducts. The intent is to prevent grease fires.			
(1) Natural Light, Artificial Light and Natural Ventilation data is from the 1993 BOCA Building Code. The equivalent in artificial light is an acceptable substitute for natural light. The 1993 BOCA Mechanical Code is referenced for mechanical ventilation. (2) Higher values are required for special occupancies. (3) IES Lighting Handbook is referenced for specific lighting.					
Additional reference standards: CABO-MEC (Council of American Building Officials-Model Energy Code) 1992 (BOCA reference CABO-93) ASHRAE 62-Ventilation for Acceptable Indoor Air Quality 1989 NFPA 96-Vapor Removal from Cooking Equipment 1991					

Figure 28.1

Chapter 29

Retail Facilities

Stephen C. Plotner

Customers are generally attracted to retail facilities by choice rather than by necessity. Unlike being tied down to a particular office building, apartment building, hospital, or school because of where they live and work, shoppers have the freedom to choose the retail establishments from which to buy goods and services. True, they all must shop at supermarkets, department stores, clothing stores, restaurants, book stores, gas stations, and others to obtain the variety of goods they need and want, but their choice of establishment will be driven in large part by such things as location, adequate parking, accessibility, and the appearance of a well-maintained facility. The retail mission, then, is to attract as many customers as possible, have them part with their money in exchange for a good value, and then have them return as repeat customers. If a retail facility is maintained well, it will affect the perception of its customers — and customers are the lifeblood of any retailer.

Maintenance and Repair Strategies

In the retail environment there are market forces at work that influence maintenance and repair strategies. The marketplace is extremely competitive, which forces retailers to operate on slimmer and slimmer profit margins. Lower revenues will affect daily operations as well as capital budgets; to keep more revenues flowing in, retailers are staying open to the public for longer hours, which places an extra burden on facility infrastructures. At the corporate level, maintenance and repairs are not usually part of the retail mission, and may therefore be considered as overhead, or even as a necessary evil. Also, in an effort to improve their bottom line each month, local retail operations managers are tempted to indefinitely defer or not disclose the need for maintenance and repairs. This strategy could lead to premature equipment and component failures or a full or partial shut-down of the facility, causing customer inconvenience, while a major unexpected repair or replacement is performed.

The rising costs of new construction and equipment point to the increasing necessity of ensuring that existing buildings, systems, and equipment are maintained and repaired on a regular basis, and that repairs are made to last over the long term. An effective maintenance and repair program is the only answer to extending the life of existing capital assets. While there are no guarantees that every retailer's budget will allow the cycles of maintenance, improvements, and repairs/replacements to be perfectly in balance, the absence of any strategy will surely lead to a more rapid deterioration of the facility infrastructure, greater exposure of retail customers and employees to the risk of injury, lower employee morale, increased customer inconvenience, and a decline in revenues.

Particularly Relevant Building Elements

The following sections outline the retail facility elements that most commonly require maintenance and repair.

Elements of Importance to Customers

The following items tend to be of importance to retail customers and are, therefore, the center of their focus. In addition to general maintenance (housekeeping), there are sometimes some particular concerns.

Exterior Signage (Retail Name)

First impressions are lasting impressions. The exterior lighted sign is the nameplate and identity of a retail facility — it is the first element that a current or new customer will see. Lighted signs need to be lit from dusk to just past closing time, with the use of a local timer or as part of an energy management system. Burned-out fluorescent lamps, neon elements, ballasts, and transformers need to be replaced. Replacing blinking or flashing lamps and neon elements as soon as they become apparent will reduce an undue burden on their associated ballasts and transformers and possibly prevent a costly replacement. Loose faces and trim, if not repaired immediately, can be the source of severe personal injury and property damage as parts become detached during even moderate winds.

Parking Area

Retail customers want parking that is accessible, convenient, clean, well-marked, and safe. Bituminous asphalt surfaces must be regularly maintained and seal-coated, and any cracks or holes in asphalt or concrete lots must be repaired to prevent damage and deterioration of the surface from penetrating into the base course or subgrade. Concrete, granite, and asphalt curbs must be repaired or replaced to control and contain surface water runoff; damage to curbs caused by winter snow plowing and snow storage will become apparent after the spring thaw. Catch basins, storm drains, and detention ponds must be cleaned of debris, sand, and mud, which will enable storm water runoff to flow as designed. Litter collection and power sweeping should be performed on a regular basis. Parking space lines, traffic lines, and symbols must be repainted every few years, and any obstacles such as curbs, light pole bases, and protective guard posts should be painted safety yellow and maintained.

Snow Plowing, Storage, Removal, and De-icing

In winter climates, retail customers expect normal parking spaces to be available year-round. Accumulating snow must be plowed and stored, with thought given to the storage of future snowfalls and the path of snow-melt sheet water flow. When plowing, care must be taken to minimize snowplow damage of curbs, landscaping, and other obstacles. Spreading sand or a sand/salt mixture on icy paved surfaces will help prevent skidding accidents between automobiles and property damage to fixed objects.

Exterior Lighting

The amount of parking lot and building illumination at night can either invite customers into a retail parking lot or cause them to drive on. Customers need to be able to see to locate an available parking space at night, and they want to feel confident that they are parking and leaving their automobile in a safe location. Again, lights need to be lit from dusk to just past closing time with the use of circuit timers or as part of an energy management system. Burned-out lamps and ballasts need to be replaced, and blinking or flashing lamps need to be replaced as soon as possible to minimize the burden on their associated ballasts and possibly prevent costlier ballast replacements. Photoelectric cells that control separate night light circuits must be maintained and replaced as necessary.

Paths, Walks, and Ramps

Paths, walks, and ramps must be structurally sound and well-maintained so as to eliminate the exposure of customers to the risk of tripping and falling. They must also be well-delineated, lighted, and separated as much as possible from the flow of vehicular traffic. In winter climates, they must be kept free of snow and ice.

Directional Signage

Vehicular traffic signs and pedestrian signs must be in place and maintained at all times, and they must be replaced when damaged or missing so as to reduce the risk of vehicular or pedestrian accidents. Directional and informational signs for the main entrance, auxiliary entrances, or other areas must be in place, well-maintained, and in compliance with standards specified in the Americans with Disabilities Act (ADA).

Landscaping

Customers usually don't care whether a retail facility has well-maintained landscaped areas or none at all, but they will remember, in a negative way, poorly maintained landscaped areas. Lawns and fields should be regularly mowed, trimmed, edged, fertilized, and raked of any leaves or other debris. Trees and shrubs should be pruned and trimmed, and planters and flower beds must be weeded, fertilized, and replanted as needed. In some areas, natural rainfall must be supplemented with automatic or manual irrigation to keep foliage, vegetation, and flowers at their peak.

Exterior Walls

Again, one of the first impressions of customers who are about to enter a retail facility will be the condition and appearance of the front walk, walls, windows, and doors. Exterior walls should be structurally sound and in good repair. Physical barriers such as wheelstops, guard rails, fences, and guard posts should be in place and maintained at parking spaces immediately adjacent to the building to keep vehicles from hitting building walls or other important components, systems, or equipment. Because customers want to park as close to the building and entrances as possible, these barriers may have to be installed at other places as well, even where there are no designated parking spaces. Walls should be clean, with all graffiti, mold, mildew, and fungus removed as necessary, and outdated temporary promotional signs, banners, and graphics should be removed along with fasteners and tie-downs.

Front Doors

Public entrance and exit doors of some retail facilities experience extremely high frequency cycle counts, are subject to higher than normal wear and tear, and must be maintained regularly. Automatic sliding and swinging door controls must be checked regularly to make sure that they provide adequate delay before closing, as provided in the codes, so as to minimize the risk of personal injury. Doors must be clearly marked *exit* or *entrance*, especially automatic out-swinging exit doors where a customer could be injured when approaching the wrong door. Vestibules and the area in front of doors, particularly automatic sliding doors with an emergency breakout feature, must be kept clear of merchandise and displays that will impede emergency egress.

Front Windows

Windows of retail facilities are heavily used as a space for promotional signs, banners, and graphics. Some local codes or bylaws have provisions that prohibit or restrict the amount of glass coverage of this type of signage. Outdated materials should be removed along with their fasteners and tape. Protection must be provided, both exterior and interior, to prevent window frames and glazing from being impacted by people, shopping carts, equipment, or other objects.

Flooring and Floor Finishes

Tripping hazards such as uneven floor joints, electrical outlets, covers for junction boxes and access holes, broken or missing floor tiles, torn or worn carpeting, missing or damaged transition strips or moldings, damaged thresholds, and structural floor damage must be repaired or replaced immediately. Floor surfaces must be even, smooth, and as slip-resistant as possible. Door mats, floor mats, and carpet runners, when used seasonally or temporarily during inclement weather, must be cleaned and dried regularly and may have to be exchanged during open hours as they become heavily soiled or saturated. In winter climates, any sand or sand/salt mixture that is spread outside on walks and drives will be tracked indoors, causing carpeting and door mats to become soiled sooner, which will require more frequent vacuuming and deep cleaning. The sand also has an abrasive effect on hard floor finishes, which will require more frequent stripping and refinishing.

Walls and Wall Finishes

Interior wall surfaces must be sound, in good repair, and free of sharp edges and protrusions that could cause damage to clothing or personal injury to customers and employees. Outside corners and vulnerable walls should be protected by corner and wall guards to minimize damage from impact by shopping carts and other objects. Walls should be kept clean and periodically painted, wallpapered, or otherwise refinished to keep them looking new and fresh to customers.

Interior Signage and Graphics

The signs and graphics inside a retail facility should direct the flow of customer traffic—not impede it. Egress aisles must be kept clear, and signs hung from low ceilings must be placed high enough to avoid contact with merchandise being carried by customers.

Furnishings

Shelving, racks, gondolas, and other fixtures used for storing and displaying merchandise must be appropriate for their intended use. They must be strong, stable, in good repair, and free of any sharp edges or protrusions that could cause damage to clothing or personal injury to customers and employees. Merchandise must be stored in these fixtures in a safe manner that will prevent merchandise from shifting and toppling, causing loss of product or even personal injury to customers. Temporary and portable merchandisers and display units must not be placed in aisles or other places where they would impede the path of egress during an emergency.

Rest Rooms

Most retail facilities aren't known for their rest rooms, but an unpleasant experience will be remembered. Even though the employees and managers of a retail facility may not have occasion to enter or use the rest rooms designated for public use, the same attention to detail must be given to them as to the rest rooms for employees. The ventilation system must be functional to keep unpleasant odors to a minimum. Plumbing fixtures, toilet partitions, countertops, and accessories must be intact, functional, and clean. The floors must be cleaned, several times a day if needed, and paper products and soap must be kept in full supply.

Lighting

Interior lighting is most important in a retail facility. Customers need a well-lighted environment where they can see the merchandise that they are considering purchasing. Lamps usually are carefully chosen for a retail facility to achieve the optimum color rendition of the merchandise being sold. When spot-replacing burned-out lamps, it is important to install lamps of like type, wattage, and color rendition (temperature). When group-relamping a section or the entire facility, corporate administrators and merchandisers should be consulted prior to making a change in the original lamp type.

Elements of Importance to Local Operations Management

In addition to giving their attention to the items listed above, which are important to their customers, local operations managers need to focus their efforts on the following elements.

Cash Registers

Point-of-sale cash registers and all peripheral equipment such as terminals, keyboards, printers, scan guns, card readers, and cash drawers must be kept in working condition, and all cable connections must be tight and inaccessible to children. There should be a manual or electronic backup system in place for times when the cash registers and computer system fail to operate properly.

Computer System

Most retail facilities have an in-house mainframe computer, which may be connected to an uninterruptible power supply (UPS) system that provides emergency battery power backup during an electrical power failure and smooths out any power spikes or valleys. It is imperative that the UPS be serviced and maintained on a regular basis, usually by an outside contractor under a maintenance agreement, so that the system is available and functional when needed. Otherwise, the retail facility risks the loss of crucial data during an unexpected power failure. Most small computer systems can operate in an office-like environment, but older and larger systems must be located in a sealed room where temperature and humidity can be controlled within very narrow setpoints. The integrity of the door seals must be maintained in such computer rooms and the heating and air-conditioning equipment, usually redundant split systems, must be maintained on a regular basis.

Satellite Link-up

If a retail facility is part of a larger organization, its computer system will usually be connected to the corporate mainframe computer via satellite link-up. It will also have a backup connection via telephone land lines and a modem. The satellite dish is usually located on the roof of the building on a free-standing frame, weighted down to prevent it from sliding or overturning, or it may be pole-mounted on the side of the building or somewhere else on the site if the roof is unsuitable. The satellite dish is usually maintenance-free, but may require occasional realignment by the installer if it has shifted or its outdoor sending/receiving unit has become disabled.

Telephone System

A retail facility's telephone system is a vital communication link between its customers and its managers and employees. Nothing causes more frustration among customers than to hear repeated busy signals, to be put on hold interminably, or to be bounced around the system unnecessarily. Most of these problems can be overcome by adequate orientation and training of employees to the telephone system and telephone etiquette, but some must be overcome in other ways. If the facility has several outside telephone lines bunched together into a hunting group under its main telephone number, a simple diagnostic test should be performed from the main operator's console every day to check each outside line. Any trouble can be reported to the local telephone company for repair.

For retail companies that depend heavily on telephone use, it might be prudent to have a service and maintenance contract with an outside third-party vendor. This firm can respond quickly to in-house service problems and can perform any moves, adds, or changes to in-house equipment that may be requested by store management.

Anti-theft Alarm System

Most large- and medium-sized retail facilities have some type of system to constantly monitor the public exit and entrance doors for the possible theft of merchandise. Most of these systems utilize a pair of sensing devices at each door opening to detect the passing of special magnetized strips or spots that have been permanently fastened to product packaging. If the magnetized strips have not been desensitized by a cashier at the time of payment, they will set off an audible and visual alarm to alert employees. Other areas of a retail facility may have open displays of small out-of-the-box high-ticket items that will require alarming individually or in groups. These alarm systems must be tested periodically to make sure that they are functional and that sensitivity levels are appropriate. Employees must be trained to recognize the sound of the audible alarms and must be able to both respond in an appropriate manner and reset the systems.

Intrusion Alarm System

Every retail facility should be equipped with an intrusion alarm system that will automatically signal an off-hours break-in to alert a contracted security alarm company, who will then notify a store manager from a call list as well as the local authorities. Some larger retail facilities may employ their own trained, 24-hour security force. These alarm systems typically incorporate door contacts, motion detectors, concealed floor mats, electric eyes, and other triggering devices that will, when tripped, send a signal to a central processor located in the facility, which in turn will notify the central office of the security alarm company via a dedicated telephone land line or radio signal. The security company under contract will furnish a written report that includes opening and closing times as well as other events. These systems are basically maintenance-free, but there are occasions when the alarm system cannot be activated at closing time and the reason needs to be investigated. Any damaged or non-functional equipment will need to be repaired or replaced by the security company.

Trash Compaction and Removal

Retail facilities must deal with disposing packaging and shipping materials in addition to normal facility trash. Depending on the types and quantities of waste in the waste stream, recycling arrangements can be voluntarily worked out with the contract trash hauler if not already mandated by law or code. For instance, corrugated cardboard can be separated out of the waste stream, compacted, and then picked up when the container is full. The equipment used would be a self-contained hydraulic compacting unit that is bolted down to a concrete pad on grade, and a detachable closed container to hold the waste. Under a typical arrangement, the facility owner has the concrete pad installed beside the building, and the electric power brought to a service disconnect adjacent to the pad. The contract trash hauler would furnish and install the compactor and container, including the final electrical hookup for no charge. The trash hauler would maintain his or her own equipment, pick up the waste as needed, invoice for the normal tipping fee and, depending on market conditions for that type of waste, issue a credit for the recycled tonnage instead of an invoice for disposal of the waste into a landfill. Over time, recycling can provide substantial savings.

Loading Dock and Equipment

The loading dock of a retail facility is the gateway through which all its merchandise must pass and, therefore, should be functionally operational at all times. Dock bumpers must be in place to prevent damage to the concrete dock or building walls. If equipped with dock levelers, they should be serviced at regular intervals. If there is no mechanical means of locking a truck to the loading dock, then wheel chocks must be provided. Any lines that can be painted on the ground to help a driver back up to the dock will result in less damage to the building, dock, and dock shelters.

Water Supply Testing

Some retail facilities that are located in rural areas, or in areas where there is no available municipal water supply, must supply their own potable water, usually from a well. As such, the federal Environmental Protection Agency (EPA) would classify the facility as a public water supplier and the facility would then be subject to the same water quality standards as municipal water suppliers. Each state has promulgated its own regulations in accordance with the EPA's Safe Drinking Water Act, which mandates that the water supply be sampled at regular intervals and tested for a long list of contaminants by a certified licensed laboratory. Each year, the EPA has added to the growing list of possible contaminants, requiring that more and more expensive tests be performed. If contaminants were found during the testing procedures, the facility would be liable to put its water supply through any number of mandated treatment processes under the direction of the state's environmental department.

Septic System

If a retail facility is not connected to a municipal sewer, it must have its own septic system, usually consisting of a septic tank, a leaching field and, if required due to site grades, a pump chamber containing ejection or lift pumps. Contrary to the old saying, "Out of sight, out of mind," the septic tank must be pumped out at regular intervals to remove accumulated solids, sludge, and scum before their level reaches the invert of the septic tank outflow. Otherwise, these materials will be forced out into the leaching field, causing premature failure, or they will back up into the building. The pumps, if any, and their controls should be checked out at the same time.

Energy Conservation Measures

A retail facility can realize substantial cost savings from energy conservation measures such as a lighting retrofit program, retrofitting or replacing HVAC equipment with variable air volume blowers and dampers, and installing or updating an energy management system. In the area of lighting, for example, existing fluorescent fixtures can be retrofitted with electronic ballasts, T-8 lamps, and specular reflectors. Existing incandescent fixtures can be replaced with appropriate compact fluorescent fixtures. With these measures, up to 50% savings can be achieved on the portion of the electric utility bill allocated to lighting. In some areas, the electric utility company will offset the cost of qualifying energy conservation measures with rebates ranging from 25% to 100%. If the cost is to be expensed, it may be prudent to have the measures installed during the first quarter of the fiscal year. Some installation contractors, however, will allow the net cost to be spread out over a period of several months so that a positive cash flow can be realized immediately.

Elements of Life Safety Importance

The following items need the attention and efforts of both the local operations management and the corporate structure because the very life and safety of large numbers of people are at issue.

ADA Compliance

The Americans with Disabilities Act (ADA) addresses a long-standing need to overcome inaccessible architecture. Barrier removal is not complicated, and removing a barrier for a particular group of people removes it for everyone. Modifications to elements such as parking, walkways and curb cuts, ramps, stairs, lifts, elevators, doorways and entrances, public telephones, drinking fountains, alarm switches and controls, signage, rest rooms, aisles and pathways, service and information counters, and checkout and cashier counters will make a retail facility safer and easier to use for those with disabilities and for all people, and they will help make the facility more compliant with ADA.

Fire Alarm System

Every retail facility should be equipped with some type of fire alarm system to warn employees and customers of an emergency situation. Depending on codes or regulations and on the size of the facility, the system can be as simple as one pull station with a horn and strobe or as complicated as a multi-zone system including pull stations, room heat and smoke detectors, duct smoke detectors, sprinkler flow and tamper switches, horns and strobes, annunciator panels, and a master control panel with battery backup, which supervises all device loops, shuts down any HVAC or other air moving equipment, closes fire doors, and notifies the local authorities via telephone line or radio signal. If not already mandated by code or regulation, periodic inspections and testing of the fire alarm system in accordance with NFPA standards should be performed, preferably with the local fire department. Any deficiencies found must be corrected immediately.

Fire Suppression System

A fire suppression system in a retail facility may be forgotten by managers and employees simply because it is an automatic system. Its purpose is to be on standby 24 hours per day to knock down and put out a small fire before it can grow and involve the whole facility. The system includes a water source, a wet or dry piping network to distribute the water, and various controls and alarm devices. It may be as simple as a standpipe system with hose cabinets or as complicated as an automatic multi-zone sprinkler system complete with valves, gauges, flow switches, tamper switches, and sprinkler heads. The water source may be a municipal water supply or an on-site or off-site source such as a pond, lake, river, or reservoir. The water pressure and volume can be derived from or augmented by a fire pump driven by an electric motor or diesel engine.

If not already mandated by code or regulation, all the elements of a fire suppression system must be visually checked periodically and must be tested at regular intervals for proper functioning so that it can be relied on during an emergency. Retail managers and certain key employees must be well-versed in its operation and must be able to react quickly to non-fire-related accidents that could cause water flow and extensive water damage.

In any fire suppression system that is supplied by the same water source that supplies the facility's potable water and contains a fire department connection, there exists the potential for a "cross-connection." The federal Environmental Protection Agency (EPA) has mandated that a "cross-connection" be removed by the use of some type of backflow prevention device, which will keep water from escaping from a pressurized fire suppression system and flowing back into the potable water supply. These backflow prevention devices must also be inspected and tested periodically by a certified inspector.

Fire Extinguishers

Fire extinguishers are used by employees as a first-line defense against a fire. Therefore, they must be located at regular intervals on the sales floor in prominent, easily accessible places. Even though they don't fit in with the carefully planned decor of a retail facility, fire extinguishers and their stark red-and-white identifying signs must be highly visible, not hidden or buried among the merchandise. Fire extinguishers must be serviced and recharged at regular intervals and must be completely overhauled or replaced every few years. Employees should be trained in their proper use.

Emergency Lighting System

Smaller retail facilities may utilize several individual battery packs located throughout the facility to provide emergency power to both attached and remote light heads and exit signs during an electric power failure. Larger facilities may utilize one large, centralized battery and charging system to supply emergency power to a vast array of remote light heads and exit signs. These exit signs and lights are provided to clearly show and illuminate the various paths of emergency egress during emergency situations when normal power has failed and the facility lights are off. The emergency lighting system must be completely checked at regular intervals to ensure that it will operate as intended, and any deficiencies such as damaged cases or housings, low batteries, burned-out lamps, or malfunctioning chargers must be corrected immediately.

Emergency Egress

An emergency egress is just that—the way out during an emergency. Retail facilities in particular must ensure that aisles, which will be used as egresses or pathways in an emergency situation, are not clogged with merchandise displays that will impede the flow of traffic. This potentially hazardous situation usually occurs in retail facilities during holidays, seasonal promotions, and heavily advertised sales. Vestibules must be kept clear, and their exit doors clearly marked and highly visible in both lighted and dark conditions. The area outside any vestibule exit door that has an emergency breakout feature must be kept clear of any obstacles that will prevent the door from operating properly. Other emergency exit doors must be equipped with appropriate panic hardware, must be able to open freely from the inside, and must be easily located by appropriate signs. The pathways leading to these doors must be kept free and clear of any merchandise, trash, or other obstacles. Usually these emergency exit doors are equipped with a local audible alarm to discourage non-emergency egress. Retail facility managers and employees should be trained to assist the public out of the facility and to a designated safe location during an emergency.

Spill Cleanup

Any retail facility that sells merchandise containing hazardous materials must have a policy and plan to deal with hazardous waste. Hazardous materials are not hazardous waste as long as they are fully contained and can be sold and used for their intended purpose. These materials become hazardous waste as soon as they are spilled or when their containers can no longer contain them. Employees must be fully trained and equipped to clean up spills and temporarily store hazardous waste. Material Safety Data Sheets (MSDS) for all retail merchandise and facility maintenance products containing hazardous materials should be available in one central location to provide additional information. Pre-packaged spill packs, personal protective equipment (PPE), and other equipment must be readily available. Once they are used they must be sealed in bags and properly labeled. Various types of cleaned up hazardous waste must then be separated into categories or compatibility groups—such as flammable/combustible liquids, corrosives, reactive chemicals, and toxic chemicals—then temporarily stored in separate containers.

Under no circumstance should hazardous waste be merely washed away, flushed or thrown out with normal trash. This is not only illegal but could be toxic and have harmful effects on the health of people and on the environment.

Hazardous Waste Storage and Pickup

Each retail facility that could potentially generate hazardous waste must have a segregated area designated as the hazardous waste storage area. This area must be kept dry, well ventilated, and secure, and could be located in an area under a roof overhang at the side of a building or in a separate shed or building manufactured for this purpose. The area must include a container, usually a barrel or drum with a locking cover or lid, for each type of category of hazardous waste at the facility, secondary containment for each container, proper signs and labels, and a cabinet containing a logbook. Each time a sealed, labeled spill pack is placed in a hazardous waste container, it must be properly recorded in the logbook. When any container reaches full capacity, arrangements must be made to have the hazardous waste picked up and transported under a hazardous waste manifest to a licensed facility by a licensed hazardous waste hauler. Each pickup must also be recorded in the log book.

Every facility that generates any type of hazardous waste has cradle-to-grave responsibility for that waste and must properly clean up, store, transport, and dispose of that waste, and must keep accurate records. Ignoring a hazardous waste problem will not make it go away, and an improper approach to hazardous waste management can lead to significant future liability.

Hazardous Materials Permits

Any retail facility that stores or sells merchandise containing hazardous materials may be required to obtain a permit from the local authorities to store such materials on the premises. Included in the permit application may be the written policy and plan for hazardous materials and waste management, Material Safety Data Sheets (MSDS) for all products and merchandise containing hazardous materials, and a floor plan depicting the areas and locations where these products are stored or shelved. The retail facility may also be subjected to an annual inspection and policy review by the local authorities to renew the permit.

Inflammable Materials Permits

Any retail facility that stores or sells merchandise containing inflammable materials may be required to obtain a permit from the local authorities to store such materials on the premises and may be restricted as to the quantity of such materials that can be stored. Included in the permit application may be the written policy and plan for storing and handling these products, Material Safety Data Sheets (MSDS) for these products, and a floor plan depicting the areas and locations where these products are stored or shelved. The retail facility may also be subjected to an annual inspection and policy review by the local authorities to renew the permit.

Inspections of Weighing and Measuring Devices

Any retail facility that utilizes a device to weigh or measure merchandise it sells must have that device inspected and sealed by a certified and licensed inspector, usually on an annual basis. The device may also have to be registered and permitted by the local authorities to be used in the sale of merchandise.

Ergonomic Cashier Stations

Any activity that requires the repetition of certain movements or a sustained posture could be the potential source of neurological and/or physical pain, suffering, or damage. Most of us recognize this and are accustomed to efforts to provide ergonomic furniture and equipment in such places as offices, factories, and warehouses. Cashiers in retail facilities, who must bend and twist while lifting or moving merchandise and operate scanning devices and keyboards of computers or registers, must be provided with ergonomic cashier stations that minimize repetitive and otherwise potentially harmful movements and activities. Attention must be paid to the heights and locations of counters and equipment as well as the distance of reaches and the weights of merchandise.

Staffing Considerations

Maintenance work includes activities that preserve or restore the function of a facility and can be subdivided into categories of ascending order of complexity such as housekeeping, general maintenance, preventive maintenance, repairs, replacements, improvements, modifications, and additions. Elements to consider when planning for maintenance staff are the category of maintenance activity, the size and sophistication of the facility, availability of resources, hours of operation, as well as the philosophy of maintenance espoused by the corporation's board, executives, administration, and/or management.

Most retail facilities are of a small to medium size and, therefore, will probably outsource or contract out most, if not all, of their maintenance and repair activities. Even when a retail organization is comprised of several stores spread out over a wide geographical area, it would be difficult to justify keeping cleaners, licensed and unlicensed generalists, trained and certified specialists, their managers, and administrative support staff on the payroll. Not only would they have to be kept fully occupied, utilized, and trained, they would also have to be furnished with the equipment and vehicles necessary to perform their in-house maintenance and repair work, and that equipment would also have to be maintained. All of this activity would divert the energy of the retail organization away from its core business of selling goods and services to customers. Nevertheless, local retail managers need to be informed of and involved with all the elements and considerations of maintenance and repair work that were mentioned earlier in this chapter so that they can inspect, keep control of, and pro-actively manage the subcontracted activities. They will also be better informed as to costs and can actively participate in the budgeting process each year.

Larger and more sophisticated retail facilities—whether single stores, retail chains, or a mall—tend to have a centralized facilities staff that will budget, schedule, and manage maintenance and repair activities. Even though a majority of the activities and projects may continue to be subcontracted out, this centralized staff will provide the support and administrative oversight that is needed to keep projects and activities within budget and in conformity to expressed company goals and specifications. They must invite the participation of and work closely with their local retail managers to develop facility audits, budgets, objectives, and goals.

Another important consideration when scheduling and staffing maintenance and repair activities for a retail facility is the potential level of customer inconvenience. Some activities can be performed with very little inconvenience or even without the customers knowing about it. Other activities may have to be scheduled during off hours when the store is closed or during off-peak hours when customer shopping is at its lowest. If a decision has been made to perform a project or activity at night, additional crew members or staff may have to be brought in to concentrate on the project to ensure that it is completed before the doors open to the public the next day.

General Budgeting Approach

All maintenance activities require the expenditure of funds for such elements as labor, materials, supplies, spare parts, equipment, and/or overhead support. A first step in controlling the cost of maintenance, repairs, and improvements is to establish realistic cost estimates for each individual activity that incorporates these necessary elements. Different techniques can be employed to develop accurate cost estimates, such as the facility audit, historical records, comparison of past budgets to actual expenditures, and consulting cost guides such as *Means Facilities Maintenance & Repair Cost Data.*

The overall cost of any maintenance, repair, and improvement program falls into one of two types of budgets. The first, the Facility Maintenance Budget, consists of operating expenses such as housekeeping, general maintenance, preventive maintenance, minor repairs, and minor replacements. The second type, the Facility Capital Budget, consists of major repairs, major replacements, improvements, modifications, and additions. Considerations such as the point or dollar amount at which a minor repair or replacement becomes major, and which types of projects or activities will be expensed or capitalized, should be specified within the corporation's capitalization policy guidelines.

Retail operations managers at the local store level need to know that the corporation's board, officers, administration, and/or management are as committed to the concept of an effective maintenance and repair program as they are to revenue-producing capital improvements, and that they are committed to funding the program through adequate expense and capital budgets. Also, local retail managers will feel that they have an ownership stake in any maintenance program involving their facility if they can participate in an annual facility audit. Any maintenance program must include criteria whereby needs can be classified and prioritized, which will allow rational decisions to be made concerning the distribution of limited funds.

Facility Maintenance Budget

Projects and activities for retail facilities in the areas of housekeeping, general maintenance, preventive maintenance, minor repairs, and minor replacements will be performed, depending on the organization's maintenance philosophy and strategy, by in-house staff, outside contractors, or both. If performed by in-house staff, full consideration must be given to all costs associated with labor such as wages and salaries, premium and overtime pay, FICA taxes, Medicare taxes, unemployment insurance, Workers' Compensation insurance, medical and dental insurance, retirement contributions, excess travel reimbursement, and other miscellaneous benefits. In addition, allowances must be made for employee absences as well as for employee morale and productivity. All these labor factors will be included in an outside contractor's price for performing these same activities and projects.

As mentioned previously, all operating expenses associated with maintenance of a retail facility will affect the bottom line of the facility's profit and loss sheet, and if a project or activity has not been included in the budget, it may be ignored or indefinitely deferred by the local store manager. Poor maintenance or slow repairs can lead to the shortened life of a facility component or system or even to the partial or full loss of the use of a retail facility. This in turn causes hidden costs and loss of profit which, although hard to measure or predict, are very real indeed. One tool that can enable retail organizations to include all possible activities and projects within a maintenance budget is a *chart of accounts*. If all incurred operating costs associated with maintenance and repairs are assigned to an appropriate account, this can be the basis of historical data from which budgets can be built and tracked. A suggested list of accounts or projects and activities is shown in Figure 29.1.

Facility Capital Budget

As stated previously, the capital budget is the plan for major repairs, major replacements, improvements, modifications, and additions. An annual facility audit is one method used to assess the current status of elements and components of grounds, building substructures, exterior enclosures, interior spaces, mechanical and electrical systems, and other equipment. Developing a cost estimate for each project is the next step and can be achieved utilizing *Means Facilities Maintenance & Repair Cost Data*, *Means Facilities Construction Cost Data*, or *Means Assemblies Cost Data*, which can be augmented by outside estimates. Each project should then be classified into categories of descending importance such as:

1. Implementation of safety- or code-related measures
2. Protection of existing fixed assets
3. Preservation of current beneficial operations
4. Special initiatives for the enhancement of revenues
5. Desirable modifications for comfort and convenience

Within each category, needs should be prioritized with collaboration between local operations managers and regional or corporate managers or executives. This will ease the pain of corporate budget cuts made during the budget review and approval process. Allowances should also be made for budget reserves with an eye toward future implementation of deferred capital improvements. A suggested checklist for an annual retail facility audit is shown in Figure 29.2.

Conclusion

Maintenance and repair of retail facilities must be given due consideration even in times of economic hardship and intense competition among retailers. An effective and efficient maintenance program is the only way to allow for continuing the use and extending the life of existing capital investments.

Sample Chart of Accounts

Permits, fees, and code compliance
- Hazardous materials permit
- Inflammable materials permit
- Weights and measures inspection
- Water supply testing
- Oil/gas tank tightness testing

Utilities
- Electricity
- Natural gas
- Heating oil
- Water
- Sewer/septic system
- Steam

Site maintenance
- Parking lot
  - Sealcoating
  - Line painting
  - Snow plowing/sanding
  - Power sweeping
  - Storm drain cleaning
- Litter control
  - Landscaping
- Misc. signs and painting
- Exterior lights and lighted signs
- Fences, gates, guardrails, and posts

Building maintenance and repairs
- Exterior enclosure
  - Walls
  - Roof
  - Windows
  - Doors and hardware
  - Loading dock
- Interior
  - Floors
  - Walls
  - Ceilings

- Mechanical
  - Elevators/escalators
  - Plumbing and fixtures
  - Fire suppression system
  - HVAC
  - Refrigeration
- Electrical
  - Power and distribution
  - Lighting
  - Emergency lighting
  - Fire alarm system

Equipment maintenance and repairs
- Furnishings
  - Furniture
  - Store fixtures
  - Cashier stations
  - Information and service counters
- Operating systems
  - Cash registers
  - Computers
  - Satellite
  - Telephones and P/A Systems
  - Anti-theft alarms
  - Intrusion alarms
- Other equipment
  - Fire extinguishers
  - Shopping and other customer carts
  - Receiving and warehouse equipment
  - Conveyors

Janitorial services
- Floor care
- Rest rooms
- Windows and other surfaces

Rubbish removal
- Trash hauling
- Recycling
- Hazardous waste

Figure 29.1

Suggested Checklist for an Annual Retail Facility Audit

Utilities
- Relocations
- Septic system upgrade

Site work improvements
- Asphalt repairs
- Concrete repairs
- Storm drainage upgrade
- Landscaping upgrade
- Exterior sign upgrade
- Fences and gates upgrade

Building improvements/replacements
- Exterior enclosure upgrades
  - Walls
  - Roof
  - Windows
  - Doors and hardware
  - Loading dock
- Interior upgrades
  - Floors
  - Walls
  - Ceilings
- Mechanical upgrades
  - Elevators/escalators
  - Plumbing and fixtures
  - Fire suppression system
  - HVAC
  - Refrigeration
- Electrical upgrades
  - Power and distribution
  - Lighting
  - Emergency lighting
  - Fire alarm system

Equipment additions/replacements
- Furnishings
  - Furniture
  - Store fixtures
  - Cashier stations
  - Information and service counters
- Operating systems
  - Cash registers
  - Computers
  - Satellite
  - Telephones and P/A Systems
  - Anti-theft alarms
  - Intrusion alarms
- Other equipment
  - Fire extinguishers
  - Shopping and other customer carts
  - Receiving and warehouse equipment
  - Conveyors
  - Trash compaction, recycling, hazardous waste
  - Janitorial

Figure 29.2

Part 5

Putting It All Together

Part 5

Introduction

This part of the book offers guidance for the budgeting process from the perspective of an experienced facilities professional. It describes how all of the tools and elements in the planning and evaluation processes covered in Parts 1–4 come together to create an annual or multi-year plan. The goal is to produce a defensible plan that not only sets forth an efficient solution to a facility's maintenance and repair needs, but can also be supported by an appropriate and complete collection of detail sheets. A good plan should also show evidence of analysis for potential improvement, using tools such as benchmarking surveys, and the resulting steps that should be taken to create new efficiencies, through vehicles such as outsourcing or implementation of a computerized maintenance management system.

The author discusses establishing a historical base for budgeting, the chronology of budget planning, and the essential data that will be required to prepare the budget, from salary and wage, to supply and expense, to plant infrastructure. Further, since certain assumptions must be made in order to establish a budget, it is important to clarify them and make sure they are appropriate. Contingency planning is another key item covered in this section, which also includes a sample budget format, with advice on how to determine the effect of anticipated changes on future costs.

In essence, the author of Part 5 connects the entire collection of information in Parts 1 through 4 to the real world of facilities management, where responsibilities are diverse and where documentation and communication are so essential to success.

Chapter 30

The Maintenance Budget

Phillip J. DiChiara, CFM, CHE

A budget is a formal statement of expectations expressed in monetary terms. The maintenance budget is but one portion of an organization's annual operating plan. A manager's success in meeting the plant's requirements is in large part dependent on how accurately he/she can project need and cost, and then effectively and efficiently implement the resulting plan.

Although the components that ultimately comprise the budget are distinct planning activities within themselves, they do form a continuum. Capital planning, benchmarking, or planning a reserve fund are, hopefully, happening continuously and are encapsulated into the organization's larger master budget at the time of its creation. While it is likely that a reserve fund will be created immediately upon acquisition of a new facility in anticipation of future needs, it is also very possible that an ongoing operation does not have such a plan. This is true for other budget components as well. It is up to a prudent manager to assist in their completion before the beginning of the next budget cycle.

The fact is that one can enter the budget process from any number of points. The ideal chronological process may look somewhat like Figure 30.1. There will always be exceptions to the rule, but good maintenance planning depends on the smooth transition of information from and to the larger budget process. It influences and is influenced by company outcomes. Budget integrity is earned when there is high predictability of cost and minimal unexpected variations from the plan. The maintenance operation, like all ongoing activities, can make use of virtually all of the funds it is provided, but the basic economic fact is that there are always limits to the source of those funds. Underfunding a maintenance operation has implications for cost in the future. Overfunding, though we hate to face up to it, can be equally deleterious. Certainly it is to the owner/organization who might use the dollars in an alternate, more prudent, and potentially profitable way. Hopefully, the budget process is, at its most fundamental level, a manifestation of effective communication and decision-making.

Good budgets, while time-consuming and hard work, indeed, flow from good information and flow easily into effective day-to-day activities. Consistency in their application and realism in their outlook makes the process easier each year, though the information may be increasingly

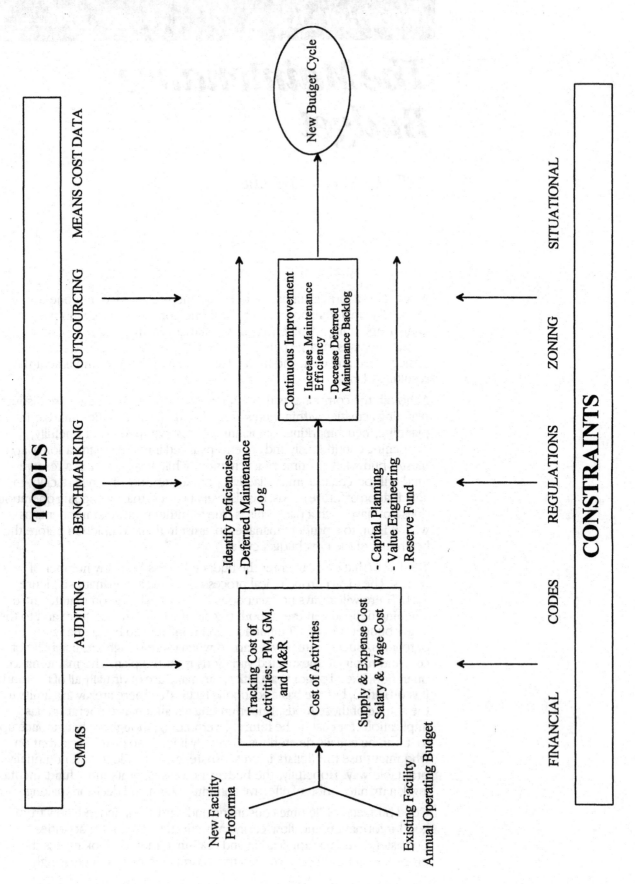

Figure 30.1

complex. Facility maintenance managers can take pride in the completion of a sound budget, for budgets are not an overlay of another responsibility, but rather the foundation upon which effective operations are built.

Proforma

The Proforma is the ideal place to begin in one's assessment of a new or existing building. Understanding initial costs and planned revenue can set the stage for planning how to effectively manage the plant and its systems. Much of the basis for the annual operating plan is contained herein. However, decisions about the application of maintenance resources will still rest with the maintenance and facility manager. Levels of performance expectations need to be set at this stage. At the same time, it is likely that some generalized information must be applied to create the budget. The specific circumstances of the first year of operation are without any historical or experiential base to work from other than generalized data from the industry. Data such as that available in *Means Facilities Maintenance & Repair Cost Data*, as well as one's experience and intuitive sense, can be essential in creating a first year budget. The concepts of benchmarking and Continuous Quality Improvement are well applied; subsequent years' budgets will benefit from their application.

Preventive, Corrective and Deferred Maintenance as Represented by the Annual Budget

The building infrastructure and the combination of performance expectations and O&M requirements will set forth the numerous tasks that the maintenance manager must translate into the job routines of his/her staff. The relationship of preventive to corrective maintenance is usually close to optimal in a new building, other than random failures which we can only loosely predict. While it is a truism that failure to perform PM will result in unnecessary and disproportionate corrective and repair maintenance at a later date, the maintenance manager must get agreement from the facility manager or senior manager as to what can be expected as a building ages. The staffing budget must reflect this agreement.

Managers who inherit existing facilities with a recorded history of the maintenance effort have the opportunity to audit and benchmark the performance of this building against others, and re-establish the amount of effort required to maintain the facility. The case for protecting one's capital investment, and thus rationalizing the cost of the maintenance operation, is sometimes overlooked in the crush of day-to-day activity. Entrance onto the new scene permits, as well as necessitates, a re-establishment of the operating budget requirements.

The new maintenance manager will look to the data available and create an operating plan that will respond to questions such as these:

How much staff do I need to meet the objectives we've agreed to?
- In what specialties?
- At what cost?

What supporting supplies and how much expense will these efforts require?
- What kind of work teams?
- Functional by specialty?
- Cross-functional?
- PM vs. CM

What hours do we need to operate?
- 24 hours a day?
- Monday–Friday ?

What kind of information will management need to be kept adequately informed?

What goals are established, or should be created?

What do I anticipate as future needs?

What processes for communication, up and down, exist or are needed?

The answers to these questions and many more feed information into the budget. The two major categories of every budget — *Supply and Expense* and *Salary and Wage* — get further broken down into categories such as *Utilities* as a part of Supply & Expense. Utilities will get broken down further by type, such as electrical, steam, and gas. These may be further broken down by location. In effect, the budget is a list of all expected costs in a category, and this is projected out over time (usually monthly).

The source of this information is drawn from other documents and historical expense. It is projected by arithmetic calculation, by expected circumstance, or (in as few cases as possible) one's best guess!

Setting Up Systems

Whether starting from scratch as a newcomer to your firm, or as a manager with long tenure, understanding the combined performance objectives under which you are to operate is critical. These will evolve over time, as company finances change, as your plant ages, and as externally mandated regulation evolves. Understanding where your operation has been, through analysis of past budgets and other documents, may provide some sense of those objectives. Discussions and directives from management and those whose work you serve and support provide direction to those objectives you should seek in the future. Following are some of the methods you could employ to collect critical information for the budgeting process:

- Compare actual versus budgeted costs for the past several years to discover trends or patterns. Are there areas of the budget that are consistently being overrun? What are the most notable differences in trends? Which maintenance activities showed increases? Decreases? Why?
- Compare material costs for previous years. Can you identify the high-cost and high-volume items? Can your vendors carry inventory for you? Can you combine shipments of small orders from the same vendor?
- Compare costs by the month and/or quarter. Are there cyclical trends based on seasonality? What are the five highest-cost months? Why? Are the reasons for the higher costs controllable by better planning and scheduling of maintenance work?
- Discussions with the Accounting Department of your organization can reveal their insights about improvements and cost trends. How do the costs of materials, labor, overhead, and contracted services compare with recent years? What are the ten highest cost maintenance activities or services? How can we reduce the cost?
- Discussions with the Operations Department of your organization will reveal their plans for the upcoming year. What expansions, modifications or overhauls are planned? What are the ten most troublesome and costly operations? What equipment has proven to be repair-prone? How can costs be reduced?
- Discussions with the Sales and/or Merchandizing Departments of your organization will reveal their forecasts for revenues by

department or product. Which departments will be busier than last year? Less busy? Can new equipment support greater production? Will more shifts be required? Will more extensive maintenance and repairs be required?

Collecting the following, at a minimum, is essential to creating a formal plan, whether it is titled a *budget* or *annual operating plan.*

I. Salary and Wage Information

  A. Number of employees by wage, title, date of hire.

  B. Staffing schedules.

  C. Benefit requirements, vacation, holiday, and non-production time, including average use of sick time.

II. Supply and Expense Information

  A. Current categories of expense by type (electrical supply, plumbing supply, etc.).

  B. Listing of expenses by task or project (repair of #1 generator, replacement of motor, etc.).

III. Status of Current Plant Infrastructure

  A. Listing of current equipment, its condition, and record of preventive maintenance and repair.

  B. Current and past plans for replacement or repair.

  C. Anticipated changes in staffing, company economics, and product.

  D. All data related to the plant maintenance and repair function.

Each of these individual documents is not unlike a subassembly in a manufacturing plant. In the automotive industry, the big auto-makers subcontract out many components of their automobiles. Braking systems may be made by one manufacturer, transmissions by another, and even engines by still another. These components, however, come together in an efficient way to make the final product—one that meets the needs previously established. In other words, each of these data assemblies can be combined and recombined to fit your design of how the maintenance and repair function can be effectively administered in your operations. It also implies that such a plan is not created in a single sitting, at one point in time. Each of these data assemblies is best done over time by collecting and updating your facility database. To the extent that your annual operating plan and budget is your best current thinking at a single point in time, and one upon which you regularly stake your management reputation, do not underestimate the importance of creating this equation and continuously feeding it data.

Assumptions

While there are several methods used to assemble an annual operating budget (basing it on past experiential trends, zero-based, etc.), most firms pre-establish these parameters for department managers. It is then up to the manager to complete a simple form. It is the supporting data that the manager uses to make his decisions that is critical, yet is not often requested by those reviewing the budget form. As the assumptions upon which you make your plans vary, so can the contents. Testing the veracity of your assumptions against those of senior management is essential. Filling in the blanks on a form is not difficult, but defending them with supporting documentation is what separates the amateurs from the real pros.

As discussed in Chapter 23, despite the incredible intuitive skills of the many facility professionals who are able to identify, diagnose, assess, and resolve problems with high levels of success, the budgeting process (which has become intuitive to them through repeated exposure) must be presented logically and quantitatively to others responsible for the allocation of funds. They will sometimes want and need to see the underlying assumptions and the information contained within the supporting data.

An example of an assumption is the anticipated versus acceptable downtime permitted on a given piece of equipment. If a tool that is essential to the product your company counts on to stay in business is out of action, what should your response be? Planned downtime for maintenance may be one day per month, but unexpected failures greater than three days may jeopardize customer commitments. What are your department's financial obligations and constraints in assuring that this equipment is functioning? Do you have the staff to meet this need and untold other competing needs that may be equally critical? If the answer is to always have adequate staffing for every possible situation, will the cost of operating your department (in what would be a risk-averse mode) be an acceptable cost to the organization as a whole? Confirming the assumptions upon which you cast your budget is essential, and an early step in the process.

Figure 30.2 shows a Proposed Facility Operating Budget. While most organizations have an established budget format which can go into greater or lesser detail than that shown here, this sample budget will serve to illustrate some of the topics discussed in this book. Earlier in this current year the Facility Manager undertook a benchmarking study with a consulting firm which resulted in several recommendations that are contained in next year's proposed budget:

- Building C is in the process of being sold along with the property on which it sits. This will reduce the amount of required maintenance activities, which will allow for the elimination of some staff positions.
- The Assistant Facility Manager will retire at the end of the current year. With the proposed reductions in staff, it was decided that this position will no longer be needed and that the Facility Manager can take up the slack.
- The Facility Manager will need to purchase a CMMS software package to enable him to more efficiently schedule maintenance activities, track maintenance costs, and keep accurate files on the condition and service history of systems and equipment. This will now be essential with the elimination of the Assistant FM position and the other staff reductions.
- A janitor's position can be eliminated because Building C will be sold and it will no longer be economically feasible to keep a janitor on full-time status for Buildings A and B. A cost analysis has shown that money can be saved by outsourcing the custodial services to a cleaning contractor while still supplying the consumable paper goods.
- More than half of the grounds maintenance was concentrated around Building C, which is being sold. A groundskeeper's position can be eliminated because the remaining work around Buildings A and B won't keep him busy on a full-time basis. Furthermore, a cost analysis has shown that money can be saved by outsourcing the grounds maintenance to a landscaping contractor, and the winter snowplowing and sanding activities to another contractor on an as-needed basis.

PROPOSED FACILITY OPERATIONS BUDGET

Description	Last Year	This Year	Proposed New Year	Comments
Administrative				
Salaries, wages & benefits	$154,000	$159,390	$110,340	Asst. manager retires, do not hire
Office supplies	1,700	1,760	1,820	
Office equipment	1,100	1,140	1,180	
Telephone	1,200	1,240	1,280	
Computer services	500	520	4,540	Purchase CMMS software
Postage	750	780	810	
Training	2,500	2,590	2,680	
Dues & registrations	2,600	2,690	2,780	
Travel & per diem	5,000	5,180	3,780	
Consulting services	0	20,000	0	Benchmarking study
Subtotals	$169,350	$195,290	$129,210	33.8% decrease
General Maintenance				
Custodial Services				
Wages & benefits	$27,000	$27,950	$2,000	Bldg. C sold, eliminate position
Materials & supplies	15,000	15,530	8,000	
Equipment	5,000	5,180	200	
Purchased services	0	0	15,000	Contract custodial services
Grounds Maintenance				
Wages & benefits	27,000	27,950	2,000	Eliminate groundskeeper position
Materials & supplies	3,000	3,110	500	
Equipment	5,000	5,180	200	
Purchased services	0	0	22,500	Contract maint., snow plowing
Subtotals	$82,000	$84,900	$50,400	40.6% decrease
Preventive Maintenance				
Wages & benefits	$184,000	$190,440	$132,830	(2) positions eliminated
Materials & supplies	20,000	20,700	12,000	
Equipment & tools	7,000	7,250	5,000	
Safety clothing & equip.	3,000	3,110	2,000	
Purchased services	0	0	32,000	Contract HVAC preventive maint.
Subtotals	$214,000	$221,500	$183,830	17.0% decrease

Figure 30.2

Description	Last Year	This Year	Proposed New Year	Comments
Minor Repairs (In-House)				
Wages & benefits	$49,000	$50,720	$26,250	Eliminate mechanic position
Materials & supplies	12,000	12,420	7,710	New package units = fewer repairs
Equipment & tools	5,000	5,180	5,360	
Safety clothing & equip.	1,000	1,040	1,080	
Purchased services	0	0	20,000	Contract all HVAC repairs
Subtotals	$67,000	$69,360	$60,400	12.9% decrease
Major Repairs (Contract)				
Construction Project A		$160,000		Replace package units #1, 2, 3, 4
Construction Project B			$7,500	Remove underground oil tank
Construction Project C			20,000	Replace oil boiler with gas
Subtotals	$0	$160,000	$27,500	
Reserves for Deferred Maintenance & Capital Projects				
Roof Replacements				
Building A			$7,630	Replace in 10 years
Building B			4,470	Replace in 14 years
HVAC Replacements				
Chiller #1			5,170	Replace in 6 years
Package Units #5,7,9,11			4,670	Replace in 12 years
Package Units #6,8,10,12			2,390	Replace in 16 years
Resurface Parking Lot			4,490	Replace in 8 years
Subtotals	$0	$0	$28,820	
Utilities				
Electricity	$320,180	$339,400	$251,840	Building C sold
Natural gas	156,250	165,630	125,340	Building C sold, add gas boiler
Heating oil	50,500	53,530	0	Remove oil-fired boiler
Water	5,000	5,150	3,820	Building C sold
Sewer	4,000	4,120	3,060	Building C sold
Subtotals	$535,930	$567,830	$384,060	32.3% decrease
TOTALS	$1,068,280	$1,298,880	$864,220	33.5% decrease

Figure 30.2 (cont.)

- With the sale of Building C, and by outsourcing the remaining HVAC preventive maintenance and minor repairs for Buildings A and B, two more staff positions can be eliminated.
- Replacing four old HVAC package units in the Fall of the current year will result in fewer required repairs in subsequent years.
- The benchmarking analysis has shown that an old, inefficient oil-fired boiler can be replaced by an energy efficient gas-fired boiler. Gas piping of sufficient size is nearby in the same boiler room and the conversion can go quite smoothly. This will allow for the removal of an aging underground oil storage tank which has been subject to an increasing amount of state regulations.
- Up to this point, setting aside reserve funds for the future replacement of obsolete systems or equipment has not been part of the budgeting process. With the proposed reduction of five staff positions, the Facility Manager can set aside operating funds into a reserve fund and still present a budget that is lower than that for the current year.
- With the sale of Building C, the overall utility costs will be lower for the facility plant.

All the data that has been collected by the Facility Manager, and upon which he has based his assumptions and figures, may be requested by his superiors who are responsible for approving budgets and allocating operating funds. Copies of this data should be bound in one volume in anticipation of a formal presentation to the budget committee and, in this case, should contain the following as a minimum:

The report of the benchmarking study by the consulting firm which analyzes:
- Material, labor, equipment, and overhead costs for all maintenance activities.
- The impact of selling Building C.
- The impact of outsourcing on in-house staff positions.

Proposals by outside contractors for:
- General maintenance custodial services (with an option to supply consumable paper goods).
- Groundskeeping maintenance program.
- Unit costs for winter snowplowing and sanding activities.
- HVAC preventive maintenance program and unit costs for repairs.
- Removal of the underground heating oil storage tank.
- Replacement of the oil-fired boiler with a gas-fired boiler.
- Future roof replacements for Buildings A and B (in today's dollars).
- Future HVAC replacements (in today's dollars).
- Future parking lot resurfacing (in today's dollars).

Proposals by software companies for CMMS packages.

An analysis of utility usage and costs per building that will show the impact of selling Building C and replacing the oil-fired boiler with a more efficient gas-fired boiler.

Effective and complete data collection is essential for making informed decisions. A facility manager will gain credibility and earn the respect of superiors and department staff by presenting ideas and proposing budgets based on accurate historical data that he or she has methodically and meticulously gathered, filed, and presented when requested.

Routine, Risk and Randomness

Most facility maintenance and repair operations have a plan that operates according to a routine. Staff have their assignments for preventive maintenance (PM) and corrective or repair maintenance (CM or M & R). This work may be performed by a separate crew, or it may be outsourced. The fact is that eventually a corrective repair or even catastrophic failure of one element may occur. Depending on the circumstances of a plant's age, capacity, and intensity of use, the annual operating plan assumes that you can maintain certain minimum levels of output. A highly efficient PM system may be the antidote to excessive unplanned repair, but it is not a guarantee. Sometimes, and despite good systems and focused efforts, failures in plant infrastructure occur. The extent to which one anticipates and prepares for the unexpected defines the level of risk he/she assumes on behalf of the firm. Random failures—the unexpected—do occur. The equation of your department—or its cost to operate—can be lowered if one accepts greater risk. The key is to operate an optimally responsive department and negotiate with senior management the acceptable level of risk, or downtime, associated with it.

Firms that are economically healthy and risk-averse may be more amenable to creating facility maintenance and repair operations with more depth and funding. On the other hand, firms that are equally strong from a financial standpoint may also be willing to assume additional risk and provide somewhat fewer resources for M&R operations.

Ironically, firms that are less economically stable are just as likely to go in either direction. Moreover, the potential for these firms to be tightly or even under-funded generally increases. The consequences add a great deal of pressure to facility and maintenance managers to direct effort away from items that might seem unimportant on the surface. It therefore becomes even more important that the effort at planning become more sophisticated, which means more data and better analysis. While this takes time, it is the only way out of the problem. Abandoning planning simply because there is not enough time is a recipe for disaster—both for the operation and for one's career.

Capital Planning

The relationship of the Proforma to the operating budget is critical in the transition of a new facility from construction/acquisition to ongoing operation. The budget cycle is typically set on 12-month increments, but facilities have needs that show up over much longer time periods. Increasingly, this relatively small increment of 12 months blurs the need to think in longer cycles, forcing facility managers into a mode of thinking that places too much emphasis on the short-term fix and not the long-term (and probably more permanent) solution. A good example is the roof of a structure, which may be one of the lower maintenance consumers from your budget, but does require total replacement at some point in the future. (See Chapter 8 for specific roofing issues.) Capital planning acknowledges the need for a longer planning cycle, and formulates, according to a prioritization, what major costs need to be considered in the future. While it would be simplistic to suggest that maintenance items of a certain financial magnitude automatically become part of the capital planning process, it is true that certain costs are viewed as major and non-cyclical and must be parceled out according to the financial ability of the organization, outside of the routine budget cycle. Criteria for deciding which maintenance efforts need to be bumped into the planning process are established by the facility manager and senior management, but the maintenance manager must be involved with and act as an advocate for modifications, improvements, or replacements that are essential to the smooth ongoing operation of the plant in conformance with codes, safety requirements, and prudent business practice. While all building elements deserve attention in terms of their long-term capital needs, mechanical and electrical systems are of particular concern in capital planning. This is because they are so fundamental to a functioning plant and are subject to the changing demands of the business, new technology, and simply wearing out. (See Chapters 11 and 12 for specific details.)

Outsourcing

In developing the budget, it may become apparent through a benchmarking study that some activities are best provided using an outside contract management firm. Outsourcing non-core activities may be the most effective way of optimizing scarce budget resources. As detailed in Chapter 3, however, outsourcing is not a panacea, and the tasks that seem to be candidates need to be carefully evaluated to ensure that this is the case. Entire maintenance functions may be outsourced, but it is more frequently the case that certain specialties are the best candidates for consideration. The janitorial function is one good example, with a further subset example being external window washing. While the process of identifying and selecting a source for each demands careful and planned consideration, it adds rationality to the budget development process. It is an issue of efficiency and effectiveness for a given dollar, as much as it is to achieve the least possible cost.

Benchmarking

The budgeting process is universal, but that universality lends itself as the tool for comparison to other, similar operations. As a result, if the boss is presented with costs that are dramatically different from his or her own past experience, they will be subject to scrutiny. ("Why is it that we need so many electricians?") The benchmarking process provides a tool that can be used to compare, on equal footing, the standards and methods observed by other organizations, and aid in the improvement and understanding of why our costs are where they are. It should, therefore, not be feared as an unfair comparison, but rather used as a tool for continuous improvement. A key by-product of this effort is creating an understanding at all levels of management of the uniqueness of every operation. Benchmarking should be done on an ongoing basis to continuously refine and improve the outcomes that the budget represents in monetary terms.

CMMS

Computerized Maintenance Management Systems (CMMS) are a wonderful resource in the creation of a budget. The ability to assemble and sort detailed facts and formulate them into genuinely meaningful information for inclusion in the budget is a boon to managers. While there are mostly only advantages to having this type of system, it is worth noting that it is best utilized as a source of information for the budget, and ideally should be responsive to the budget, as opposed to modifying rationalized budgeting processes.

Codes & Regulations

There is clearly a budget impact from meeting codes and regulations, and the cost of compliance must be factored into the budget plan. Most of the impact is of a capital nature, so it is best part of longer-term planning.

Safety-related issues, however, are annual operating cost considerations, and ideally are identified as such in the narrative budget explanation that typically accompanies a budget submission to senior management.

Value Engineering

If budgeting is a listing of costs, then value engineering is the decision of which costs to list. At its ideal, value engineering is good budgeting. It implies that all costs are considered when maintaining or upgrading a facility. In Chapter 18, there are several examples that demonstrate the kind of thinking that goes into value engineering.

Value engineering distinguishes the routine from the extraordinary. When problems are simple and straightforward, the decisions can be pre-programmed. If the plumbing in the rest room leaks, we fix it. There are a few questions to ask, and generally supervisors or sometimes even staff can make the call to take the appropriate action. These problems are relatively discreet and do not impinge on other areas or systems. This approach works fine as long as all maintenance and repair problems are this simple. Unfortunately, as management systems and building systems become more complex, this type of thinking will not suffice. True systems thinking assumes and understands that problems are complex and related to other aspects of the operation. These kinds of issues cannot be pre-programmed. They often require choices that are riskier and less than perfect. The kinds of analyses, planning, and detail required to make these decisions must often have review at the highest levels in an organization. This is where the data maintained by you comes to play a most important role. Value engineering frames the issues to ensure that we consider alternate costs such as replacement and improvement, not merely maintenance of obsolete systems.

Deferred Maintenance

Given that the facility maintenance budget is but one part of a larger whole, it is inevitable that there will be limits and constraints placed on our expenditures. We may agree to defer activities and place them in queue for attention in another budget cycle, keeping in mind the fact that deferred maintenance is a particular concern within the budgeting process. Failure to capture and identify it at a later date is a major error to be avoided at all costs. Managing it within the context of ongoing maintenance considerations is essential. It is inevitable that some activities will be deferred. It is the continuity and consistency established within the budgeting process that ensures that deferred maintenance items will resurface to be reconsidered, and ultimately addressed. Sometimes items are deferred because the value engineering process did not prioritize them highly against current needs and existing resources. Obviously an excessive amount of deferred maintenance implies inadequate funding on both a budget cycle basis and potentially within reserve funds.

Reserve Funds

Annual operating funds can be supplemented by the creation of a reserve fund account. In the past, building owners tended to view both routine and extraordinary long cycle costs as part of an annual budget. This is no longer the case. Setting aside monies each year in expectation of a system's obsolescence or even catastrophic failure permits even the unexpected to be handled in a relatively routine fashion. Reserve funds are frequently required by lenders. For the maintenance and facility manager, such a fund acknowledges that many costs have long cycles, and that replacement must be calculated and planned for right from the outset.

Contingency funds are not to be confused with reserves. The former are usually moderate sums of money set aside within the budget for unplanned and unexpected activities, whereas the latter, by their nature, expect conditions requiring their utilization.

Contingency Planning

While contingencies are stated in financial terms, the need for equivalent operational planning, such as disaster planning, downtime exercises, and other means to anticipate how one will react in unforeseen situations, are an important way to ensure that your facility department is prepared. Weather emergencies are the subject of most common contingency planning. The exercise itself has a budget cost which should be included in routine budgeting. Real emergencies, however, have a tendency to use dollars out of normal operating proportion. For example, a hurricane or massive snowstorm should be viewed as a variance from your budget plan, whereas normal snow removal should have a cost assigned in the plan. If your geographic area expects an average annual snowfall of 42″, then variances of 10–15% are to be expected. In the year that we have a "One-Hundred-Year" storm, and receive twice the normal or expected amount of snow, we may call on contingency dollars set aside for the truly unpredictable. Contingency planning should address all types of natural and man-made disasters relevant to your region.

Continuous Improvement

The nature of the facility maintenance and repair task is that it is never complete, never as precise as we would like, and always provides an opportunity for improvement. Despite the knowledge, resources, and expertise reflected in this text, it is unlikely that a perfect system in its totality can be implemented at once. It is important to work from a vision to a plan. Incremental change, hopefully utilizing some of the tools identified, is the best way to meet the goals you set for your operation. While our superiors are often criticized for not knowing the link between cause and effect for problems confronting them, so too are facility managers guilty of being distracted by day-to-day problems and failing to take a long-term perspective.

The basic nature of the facility maintenance and repair function is one of service, and that will not change. Poor systems, lack of data from which to draw informed conclusions, and a weak intellectual framework from which to operate this service become tolerated because of an excessive focus on the crisis of the day, as opposed to a forward-looking plan. Balancing the two is difficult. Failure to address the need for such a plan will only increase an organization's inability to address routine, daily demands.

Credibility and Negotiation

Organizations learn, as do individuals. Facility managers can educate organizations as to how best to achieve a balance between plant needs and output. Resources such as R.S. Means' cost data manuals and reference books demonstrate that there are patterns and examples that can be applied to understand costs. That information can be utilized in presenting the broader needs of maintaining an ongoing operation, thereby enhancing the facility or maintenance manager's credibility. That credibility is key to negotiating a balance so that everyone understands the risks, tradeoffs, and objectives of an integrated facility maintenance and repair program. The guidance presented here in this text can be applied to reality. It's simply a matter of discipline and consistent effort over time.

Putting It All Together

On any given morning, the facility manager walks into work and sits down to a day of routine rituals and, for the most part, expected problems. While the occasional unexpected or even urgent issue adds drama, a reasonably well-managed maintenance and repair operation should flow on with a life of its own. The effect is subtle, and can lull us into a dangerous satisfaction with the status quo. On a positive note, most facility professionals are proud of their responsibilities. They fight the tedium with an enthusiasm for new ideas and approaches, a commitment to continuous quality improvement, and a willingness to break out of patterns that insidiously, almost continuously, offer reassurance that everything is going reasonably well.

Those that need and acquire a book such as this one are rarely seeking affirmation that they are on the right track. They are the ones who want to "do it a little better" — a little more precisely — and with a kind of professionalism of which they can be proud. What makes their work interesting is that while it has an element of the routine and repetitive, it also includes widely diverse tasks and responsibilities.

Planning for and operating a facility maintenance and repair effort is a balancing act. There is the plant itself, with its unique needs in order to function. There is the safety of those it employs and supports. There is the economic reality of the firm or business that draws some portion of its income from the use of the plant. There is the community of which it is a part. It is true, if not often apparent, that each of these arenas is involved every time a plant component is repaired. It is up to the manager to balance this frequently contradictory and sometimes even irreconcilable equation, which is not written solely in economic terms.

It gets more challenging, in the sense that the facility manager, to a large extent, writes the equation for his or her own responsibilities. Yes, it is true that general guidelines are set by most organizations (e.g., budget forms, operating policies, safety and profitability targets), but achieving the balance mentioned above is the essence of the job. There are incentives to do it well: salary, recognition, professional fulfillment, and continued employment. Failure to achieve balance jeopardizes the goals and ideals we value as facility and maintenance managers, and undermines our personal potential.

Most managers, therefore, work hard at creating the most sophisticated "equation" they can without creating a level of complexity that is unmanageable within their circumstances. What are the building blocks of this equation? They are budgets and any data that can provide additional knowledge. They are codes and regulations. They are the organization's values and policies. And they are discussions between people both in and outside the company. Communication and the meaningful information it provides are assimilated by good managers, and the outcome is efficient, effective operating systems, created purposefully and with an understanding of both their capabilities and limitations.

Appendix

Facilities Associations

The American Society for Healthcare Engineering
1 North Franklin, Chicago, IL 60606
Phone: (312) 422-3800 Fax: (312) 422-4571

The American Society for Healthcare Engineering represents approximately 6,000 facilities managers, engineers, clinical engineering/medical equipment managers, and professionals in the fields of health care design and construction and safety. With the many government regulations and other codes and standards affecting health care facilities, ASHE's efforts have focused on providing its members with information and education through its publications and educational programming.

The American Society for Healthcare Engineering is dedicated to helping health care facilities managers meet the challenge of health care systems in transition, as a resource for information exchange, education and professional development.

The Association for Facilities Engineering
(formerly the American Institute of Plant Engineers)
8180 Corporate Park Drive, Suite 305, Cincinnati, Ohio 45242
Phone: (513) 489-2473, Fax: (513) 247-7422, E-mail: aipe@ix.netcom.com.

Since 1954, the Association for Facilities Engineering, formerly the American Institute of Plant Engineers (AIPE), has provided its members with the resources they need to meet the challenges of facilities engineering. AFE is the largest and oldest professional association devoted exclusively to facilities engineering professionals in manufacturing, health services, education, government and other institutions. Membership benefits include:

- *AIPE Facilities* Magazine — A bi-monthly technical journal that provides information on facilities engineering and management issues.
- Professional development — Education and information on facilities issues, and professional development offered at four national and sixteen regional conferences and trade shows.
- Networking — Members meeting regularly with their peers at local programs sponsored by AFE's 136 chapters throughout the country.
- Professional recognition — AFE's Certified Plant Engineer (CPE), Certified Facilities Environmental Professional (CFEP) and

Facilities Management Excellence (FAME) awards, created to demonstrate professional competence and recognize outstanding on-the-job accomplishments.

- Reference publications — Publications available through AFE's Facilities Management Library and Information Services Catalog.
- Career net — Job listings published in each issue of *AFE Newsline*, the member newsletter.
- Member resource directory — A list of members, cross-referenced to help locate fellow professionals with specific expertise.
- Research — The AIPE Foundation conducts research on topics such as the "Definition of a Plant/Facilities Engineer," and compensation of the facilities engineer.

Each year AFE conducts four national conferences and seminars in conjunction with major facilities engineering trade shows. Educational programs offered cover a broad spectrum of facilities issues. Technical programs, combined with management and self improvement tracks, provide the facilities engineering professional opportunities to customize continuing education based on individual needs.

APPA: The Association of Higher Education Facilities Officers

1643 Prince Street, Alexandria, VA 22314-2818
Phone: (703) 684-1446, Fax: (703) 549-2772, Internet: info@appa.org, World Wide Web: http://www.appa.org

APPA provides facilities professionals with information and a network of professionals. The following list includes services APPA provides their members.

- *Facilities Manager* — A quarterly magazine that provides case study solutions to management and technical problems, feature articles, regulatory issues, association news, and book and software reviews. It's the only publication devoted to facilities managers working in higher education.
- *Inside APPA* — A newsletter published eight times a year, with news on the profession, the association, regulatory and other issues.
- Information Services — Unlimited access to the information services department. These resources include the International Experience Exchange; Comparative Costs & Staffing Report; Database for Stored Refrigerants; National Census for District Heating, Cooling, and Cogeneration; Regulatory Compliance Survey; and The Decaying American Campus Report.
- Recognition — APPA awards programs, including an institutional award for excellence in facilities management, as well as individual honors such as the Meritorious Service Award.
- Professional Development and Education — A variety of educational programs, available at special member rates. The Educational Conference & Annual Meeting features presentations on leadership, best practices in the profession, and a trade show.

The Institute for Facilities Management is a three-track training program designed to teach basic facilities operations. Special programs supplement the training by focusing on specific areas such as capital project planning, construction, and energy and utilities management.

Seminars focus on specific topic areas, including contract administration, regulatory compliance, and building commissioning.

- Publications — A collection of resources on facilities management in higher education.
- Government Relations — Access to legislative and regulatory information, as well as a contact person for key departments and others who influence education, regulatory legislation, energy policy, and funding issues.
- Membership Directory — Includes Subscribing Members to help to locate products and services.
- Facilities Management Evaluation Program — A service to help members measure and assess the effectiveness of their facilities operation through self-evaluation and a peer-to-peer review program.

Building Owners and Managers Association (BOMA) International

1201 New York Avenue, NW, Suite 300, Washington, DC 20005
Phone: (202) 408-2662, Fax: (202) 371-0181

BOMA has a network of 98 North American local associations representing more than 6 billion square feet of office space, and 101 BOMA associations worldwide. BOMA International's 15,000 members are building owners, managers, developers, corporate facility managers, leasing professionals, service providers, suppliers and a host of related industry professionals, from both public and private sectors.

This association is a source of information on alternative uses of conventional office space and the emerging technologies that are transforming the way in which business is transacted, ideas exchanged, and work managed. The so-called "virtual offices" utilizing these new technologies are being explored at BOMA. Keeping its membership abreast of the impact of these changes, and of other issues affecting the industry is BOMA's primary mission.

BOMA has testified before all levels of government on property management issues and has become a leading source of information about all aspects of commercial real estate.

BOMA also publishes ten issues of *Skylines*, a magazine for members. Topics include products, legislation, and other building management issues.

The International Facility Management Association

1 E. Greenway Plaza, Suite 1100, Houston, TX 77046-0194
Phone: (713) 623-4362 or (800) 359-4362, Fax: (713) 623-6124

The International Facility Management Association (IFMA) identifies trends, conducts research, provides educational programs and assists facility managers worldwide in developing strategies to manage human, structural and real estate assets. IFMA has more than 14,200 members in 120 chapters located primarily in the United States and Canada. IFMA has seven membership categories — professional, associate, affiliate, allied, academic, lifetime and corporate sustaining.

IFMA holds educational seminars, programs, and conferences for members. Every year, IFMA produces World Workplace, an exposition with various educational opportunities for attendees. World Workplace is held in a different location throughout North America each year. Through the Certified Facility Manager program, IFMA offers facility managers an opportunity for professional certification, as well as a gauge of their competency level. Additional support is available through the IFMA Library, which contains extensive data and reports on facility management and provides access to authoritative sources for research. IFMA also sponsors research projects, producing several research reports yearly. All IFMA programs and services are available to members and nonmembers, and are offered to members at a discount.

IFMA has two main publications: *Facility Management Journal* and *IFMA News. Facility Management Journal*, published six times per year, is a unique publication with articles written by active facility managers. It is free to members and $75 per year for nonmembers. *IFMA News*, published monthly, focuses on IFMA happenings. Regular features include code and regulation information, book reviews, research reports, job openings and other subjects of interest. It is free to members and $75 per year for nonmembers.

National Association of College and University Business Officers

One Dupont Circle, Suite 500, Washington, DC
Phone: (202) 861-2500, Fax: (202) 861-2583

NACUBO is a nonprofit organization representing the interests of chief administrative officers at more than 1,200 institutions of higher education. The association promotes sound management and financial administration of colleges and universities through publications, information programs, and research, and provides information to members concerning government activities affecting higher education.

NACUBO also provides opportunities for basic and advanced training, career development, skill improvement, and other forms of professional growth in management and financial administration in higher education.

Bibliography and Recommended Resources

Chapter 1: The Building Proforma

From ASHRAE (The American Society of Heating, Refrigeration and Air-Conditioning Engineers, Atlanta, GA):

- *Preparation of Operating and Maintenance Documentation for Building Systems: Guideline 4-1993.*
- *Operating and Maintaining Buildings for Health, Comfort and Productivity*, 1993.
- *Engineering Indoor Environments*, 1994.

From Building Operators and Managers Association (BOMA):

- *BOMA Handbook*
- *The Refrigerant Manual: Managing the Phase-out of CFCs* (See the Appendix for more information on BOMA.)

Building Codes:

- UBC (Uniform Building Code), International Conference of Building Officials, 5360 Workman Mill Road, Whittier, CA 90601. *Note: Supplements include VAC Standards, Uniform Mechanical Code, Uniform Administrative Code, Uniform Code for Building Conservation, Uniform Fire Code, and the UFC Standards.*
- BOCA (Building Officials and Code Administrators International, Inc.), 405 W. Flossmoor Road, Country Club Hills, IL 60478
- SBCCI (Southern Building Code Congress International, Inc.), 40 Montclair Road, Birmingham, AL 35213

From R.S. Means Company, Inc.:

- Adaptive Environments and R.S. Means Co., Inc. *ADA Compliance Pricing Guide*, 1994.
- Applied Management Engineering, P.C. and Harvey H. Kaiser, Ph.D. *Maintenance Management Audit*, 1991.
- Colen, Harold R., P.E. *HVAC Systems Evaluation: Maintenance, Operations*, 1990.
- Farren, Carol E. *Planning and Managing Interior Projects*, 1988.
- Kaiser, Harvey H., Ph.D. *Facilities Manager's Reference*, 1989.

- Kearney, Deborah S., Ph.D. *The ADA in Practice*, 2nd Edition, 1995.
- Liska, Roger W., P.E., AIC. *Facilities Maintenance Management*, 1988.
- Magee, Gregory H., P.E. *Facilities Maintenance Standards*, 1988.
- Owen, David D. *Facilities Planning and Relocation*, 1993.
- Rowe, William H., III, AIA, P.E. *HVAC: Design Criteria, Options, Selection*, 2nd Edition, 1994.

Construction Cost Data Annuals:
- *Means Electrical Cost Data*
- *Means Facilities Construction Cost Data*
- *Means Facilities Maintenance and Repair Cost Data*
- *Means Mechanical Cost Data*
- *Means Plumbing Cost Data*

Chapter 2: Capital Planning

Association of Physical Plant Administrators of Colleges and Universities, National Association of Colleges and University Business Officers, and Coopers & Lybrand. "The Decaying American Campus: A Ticking Time Bomb." APPA, Alexandria, VA, 1989.

Brooks, Kenneth W., et. al. *From Program to Educational Facilities*, University of Kentucky Center for Professional Development, 1980.

Dunn, John, A. "Financial Planning Guidelines for Facility Renewal and Adoption." Society for College and University Planning, Ann Arbor, MI, 1989.

Kaiser, Harvey H., ed. "Planning and Managing Higher Education Facilities." Jossey-Bass, Inc., San Francisco, No. 61, Spring 1989.

Meyerson, Joel W. and Peter M. Mitchell, eds. "Financing Capital Maintenance." National Association of College and University Business Officers, Washington, DC, 1990.

Probasco, Jack. "Crumbling Campuses." *Business Officers,* Washington, DC, 1991.

"Issues Related to Higher Education Facilities." *CEFP Journal,* Columbus, OH, January/February, 1988.

Silverman, Robert A. "Bricks, Mortar, and Assets." Keystone Advisors, Cambridge, MA, 1990.

Chapter 3: Maintenance Outsourcing

Applied Management Engineering, P.C. and Harvey H. Kaiser. *Maintenance Management Audit: A Step by Step Workbook to Better Your Facility's Bottom Line*. R.S. Means Company, Inc., Kingston, MA, 1991.

Benchmarks II, Research Report #13. International Facilities Management Association, Houston, TX, 1994.

Cotts, David G. and Michael Lee. *The Facility Management Handbook*. AMACOM, American Management Association, New York, NY, 1992.

Leake, Ernie and John L. Stanley. *Benchmarking for Facility Management Workbook*. International Facility Management Association, Houston, TX, 1994.

Means Facility Maintenance and Repair Cost Data 1996. R.S. Means Company, Inc., Kingston, MA.

Westerkamp, Thomas A. *Maintenance Manager's Standard Manual*. Prentice Hall, Englewood Cliffs, NJ, 1993.

Chapter 4: Benchmarking Facility Maintenance

Applied Management Engineering, P.C. and Harvey H. Kaiser. *Maintenance Management Audit: A Step by Step Workbook to Better Your Facility's Bottom Line*. R.S. Means Company, Inc., 1991.

Benchmarks II, Research Report #13. International Facility Management Association (IFMA), Houston, TX, 1994.

Camp, Robert C. *Benchmarking: The Search for Industry Best Practices That Lead to Superior Performance*. ASQC Press, Milwaukee, WI, 1989.

Kinnsman, Margaret P. "Benchmarking with APPA's Strategic Assessment Model." *Facilities Manager*. APPA: The Association of Higher Education Facilities Officers. Alexandria, VA, January 1996.

"Prospectus on NACUBO's Benchmarking Survey." National Association of College and University Business Officers, Washington, DC, 1996.

Smith, Bertrand, Jr. "Benchmarking: Old Technique, New Frontier." *Facilities Manager*. APPA: The Association of Higher Education Facilities Officers. Alexandria, VA, Spring 1995.

"Strategic Assessment Model." APPA: The Association of Higher Education Facilities Officers and American Management Systems, Inc., Alexandria, VA, 1995.

Vega, Michael F., PE, CPE and Charles L. Van Tine, PE, CPE. "Improving Productivity Through Benchmarking." AFE: The Association for Facilities Engineering (formerly AIPE), Cincinnati, OH.

Watson, Gregory H. *The Benchmarking Workbook: Adapting Best Practices for Performance Improvement*. Productivity Press, Cambridge, MA, 1992.

Chapter 5: Computerized Maintenance Management Systems

Hounsell, Dan. "Making Promises Pay." *Maintenance Solutions*, May 1994.

Keller, Chris. "Analyzing CMMS Needs." *Maintenance Solutions*, January/February 1995.

Keller, Chris. "Smart CMMS Specifications." *Maintenance Solutions*, November/December 1995.

Keller, Chris. "Software Provides a Platform for Developing a Maintenance Information Network." *Maintenance Solutions*, May 1994.

Kinsley, Joseph R. "Implementing a Computerized Maintenance Management System." *Electrical Construction & Maintenance*, Vol. 92, No. 10, October 1992.

Chapter 6: Codes and Regulations

From ASHRAE (American Society of Heating, Refrigeration and Air-Conditioning Engineering, Atlanta, GA):

- *Preparation of Operating and Maintenance Documentation for Building Systems: Guideline 4-1993.*
- *Operating and Maintaining Buildings for Health, Comfort and Productivity*, 1993.
- *Engineering Indoor Environments*, 1994.

From Building Operators and Managers Association (BOMA):
- *BOMA Handbook*
- *The Refrigerant Manual: Managing the Phase-out of CFCs* (See Appendix for more information on BOMA.)

Building Codes
- UBC (Uniform Building Code), International Conference of Building Officials, 5360 Workman Mill Road, Whittier, CA 90601. Note: Supplements include VAC Standards, Uniform Mechanical Code, Uniform Administrative Code, Uniform Code for Building Conservation, Uniform Fire Code, and the UFC Standards.
- BOCA (Building Officials and Code Administrators International, Inc.), 405 W. Flossmoor Road, Country Club Hills, IL 60478
- SBCCI (Southern Building Code Congress International, Inc., 40 Montclair Road, Birmingham, AL 35213

From R.S. Means Company, Inc.:
- Adaptive Environments and R.S. Means Co., Inc. *ADA Compliance Pricing Guide*, 1994.
- Applied Management Engineering, P.C. and Harvey H. Kaiser. *Maintenance Management Audit*, 1991.
- Colen, Harold R., P.E. *HVAC Systems Evaluation: Comparing Systems, Solving Problems, Efficiency and Maintenance*, 1990.
- Farren, Carol E. *Planning and Managing Interior Projects*, 1988.
- Kaiser, Harvey H., Ph.D. *Facilities Manager's Reference*, 1989.
- Kearney, Deborah S., Ph.D. *The ADA in Practice*, 2nd Edition, 1995.
- Liska, Roger W., P.E., AIC. *Facilities Maintenance Standards*, 1988.
- Magee, Gregory H., P.E. *Facilities Maintenance Management*, 1988.
- Owen, David D. *Facilities Planning and Relocation*, 1993.
- Rowe, William H., III, AIA, P.E. *HVAC: Design Criteria, Options, Selection*, 2nd Edition, 1994.
- *Means Electrical Cost Data*
- *Means Facilities Cost Data*
- *Means Facilities Maintenance and Repair Cost Data*
- *Means Mechanical Cost Data*
- *Means Plumbing Cost Data*

Chapter 7: Foundations, Substructures and Superstructures

Emmons, Peter H. *Concrete Repair and Maintenance Illustrated*. R.S. Means Company, Inc., 1992.

Kaiser, Harvey H., Ph.D. *Facilities Manager's Reference*. R.S. Means Company, Inc., 1995.

Liska, Roger W., P.E., AIC. *Means Facilities Maintenance Standards*. R.S. Means Company, Inc., 1988.

Matulionis, Raymond C. and Joan C. Freitag, eds. *Preventive Maintenance of Buildings*. Van Nostrand Reinhold, NY, 1991.

Means Facilities Maintenance & Repair Seminar Workbook. R.S. Means Company, Inc., 1995.

Chapter 8: Roofing

Brotherson D.E., ed. *Roofing Systems, ASTM Special Technical Publication 603*, ASTM (American Society for Testing and Materials), Conshohocken, PA, 1976.

Brzozowski, Kenneth J. "Roof Maintenance Products and Techniques." *The Construction Specifier*, November, 1990.

Buckley, Tim. "Successful Roofing in Cold Weather." *Professional Roofing* National Roofing Contractors Association, Rosemont, IL, November 1994.

Burger, Bill. "Roofing Technology Update: BUR and Mod Bit." *Building Operating Management*. Trade Press Publishing Corp., Milwaukee, February 1996.

Carlson, Jim. "Repair Tips for Foil-Surfaced Mod Bit Membranes." *Professional Roofing*. National Roofing Contractors Association, Rosemont, IL, December 1994.

Decision Guide for Roof Slope Selection. Air Force Engineering & Services Center, Tyndall Air Force Base, FL. Prepared by Oak Ridge National Laboratory, Oak Ridge, TN, 1988.

Dupuis, Rene M., Ph.D., PE. "SPR Roof Systems: Field Survey and Performance Review." *Professional Roofing*. National Roofing Contractors Association, Rosemont, IL, March 1996.

Eastman, Martin and Tom Smith. "Single-Ply Repairs: Old Techniques Won't Help Ailing High-Tech Systems." *Professional Roofing*, May, 1990.

"Guide to Roof Repair & Maintenance." *RSI Magazine*, November, 1983.

Gumpertz, W.H., ed., *Single-Ply Roofing Technology, ASTM Special Technical Publication 790*. ASTM (American Society of Testing and Materials), 1982.

Herbert, R.D., III. *Roofing: Design Criteria, Options, Selection*. R.S. Means Company, Inc., 1989.

Kashiwagi, Dean T., Ph.D., PE. "Performance Issues of Sprayed Polyurethane Foam Roof Systems." *Professional Roofing*. National Roofing Contractors Association, Rosemont, IL, January 1996.

Laaly, H.O. "Preventive Maintenance and Roof Repair." *The Science and Technology of Traditional and Modern Roofing Systems*. Laaly Scientific Publishing, Los Angeles, CA, 1991.

The NRCA Roofing and Waterproofing Manual, Third Edition. National Roofing Contractors Association, Rosemont, IL, 1990.

"Roof Maintenance." *The Roofing Industry Education Institute*, Englewood, CO, 1991.

"Roof Survey, Inspections, Maintenance & Repair." *The Roofing Industry Educational Institute*. Englewood, CO, 1991.

SPRI and Building Operating Management. "Flexible Membrane Roofing." *Building Operating Management*. Trade Press Publishing Corp., Milwaukee, WI, August 1995.

U.S. Army Corps of Engineers, Construction Engineering Research Laboratory, Carter Doyle, Wayne Diller and Myer J. Rosenfield. *Handbook for Repairing Nonconventional Roofing Systems, USA-CERL Technical Report M-89/04*. December, 1988.

Warshaw, Ruth. "Understanding Single-Ply Roofing Options." *Building Operating Management*. Trade Press Publishing Corp., Milwaukee, WI, September 1993.

Chapter 9: Interior Finishes

Brezinski, Darlene. "Picking Paints That Perform: Sizing Up the Latest Generation of Interior Coatings." *Maintenance Solutions*. Trade Press Publishing Corp., Milwaukee, WI, January/February 1995.

Liska, Roger W., P.E., AIC. *Means Facilities Maintenance Standards*. R.S. Means Company, Inc., 1988.

Means Repair and Remodeling Estimating. R.S. Means Company, Inc., 1989.

Owen, David D. *Facilities Planning and Relocation*. R.S. Means Company, Inc., 1993.

Chapter 10: Conveying Systems

"The Americans with Disabilities Act, Its Impact on Existing Elevator Installations." Schindler Elevator Corporation, 1992. Copies are available at no charge by writing to:

ADA Handbook, ML-275
Schindler Elevator Corporation
P.O. Box 1935
Morristown, NJ 07962-1935

For general ADA guidance, the federal Architectural and Transportation Barriers Compliance Board has several publications, including an "ADA Handbook" and a "Title III Technical Assistance Manual" which offer official guidance. These can be ordered by calling 1-800-USE-ABLE.

Kozlowski, David. "Improving Elevator Service." *Building Operating Management*. Trade Press Publishing Corp., Milwaukee, March 1995.

Chapter 11: Mechanical

ASHRAE Journal (a monthly publication) The American Society of Heating, Refrigerating and Air-Conditioning Engineers Inc., Atlanta, GA.

ASHRAE Handbook & Product Directory: Equipment, HVAC Applications, HVAC Systems & Equipment. Each is updated every four years. Contact ASHRAE for the latest editions.

Burgess, William A., Michael J. Ellenbecker, Robert D. Treitman, *Ventilation for Control of the Work Environment*. John Wiley & Sons, Inc., 1989.

Colen, Harold R. P.E. *HVAC Systems Evaluation*. R.S. Means Company, Inc., 1990.

Coon, J. Walter, PE. *Fire Protection: Design Criteria, Options, Selection*. R.S. Means Company, Inc., (See Chapter 10 for definition of NFPA Guide, Recommended Practice, and Standard.) 1991.

Galeno, Joseph J. and Sheldon T. Greene. *Plumbing Estimating*. R.S. Means Company, Inc., 1991.

Moylan, John, ed. *Mechanical Estimating*, 2nd Edition. R.S. Means Company, Inc., 1992.

National Board of Boiler and Pressure Vessel Inspectors. "Boiler Preventive Maintenance." *Maintenance Solutions*. Trade Press Publishing Corp., Milwaukee, September 1995.

Rowe, William H., III AIA, PE. *HVAC: Design Criteria, Options, Selection*, 2nd Edition. R.S. Means Company, Inc., 1994.

Spring, Harry. "Fire Protection." *Building Operating Management*. Trade Press Publishing Corp., Milwaukee, April 1993.

Chapter 12: Electrical

Bachner, John Philip and Frederick Nicholson. "Office Lighting: A Bottom-Line Perspective." *Building Operating Management*. Trade Press Publishing Corp., Milwaukee, January 1994.

Carlson, Reinhold A., P.E. and Robert A. DiGiandomenico. *Understanding Building Automation Systems*. R.S. Means Company, Inc., 1991.

Electrical Estimating Methods. R.S. Means Company, Inc., 1995.

Laurie, Robert J. "Nine Great Ideas to Improve Your Electrical Maintenance." *Electrical Construction Maintenance*, Vol. 93, No. 6, June 1994.

Liska, Roger, W., P.E., AIC. *Facilities Maintenance Standards*, R.S. Means Company, Inc., 1988.

Westercamp, Thomas A. *Maintenance Manager's Standard Manual*. Prentice Hall, 1993.

Chapter 13: Landscaping

Fee, Sylvia Hollman. *Landscape Estimating*, 2nd Edition. R.S. Means Company, Inc., 1991.

Scott, Charles. "Maintenance and Landscape Design: It Looks Great, But Can You Mow It?" *Maintenance Solutions*. Trade Press Publishing Corp., Milwaukee, May 1994.

Walker, Cathy. "A Mower Maintenance Game Plan." *Maintenance Solutions*. Trade Press Publishing Corp., Milwaukee, January 1996.

Chapter 15: Maintenance & Repair Estimating

Means Facilities Cost Data. R.S. Means Company, Inc.

Means Facilities Maintenance and Repair Cost Data. R.S. Means Company, Inc.

Chapter 16: Preventive & Predictive Maintenance Estimating

Kaiser, Harvey H., Ph.D. *Facilities Manager's Reference*. R.S. Means Company, Inc., Kingston, MA, 1989.

Liska, Roger W., P.E., AIC. *Facilities Maintenance Standards*. R.S. Means Company, Inc., Kingston, MA, 1988.

Magee, Gregory H., P.E. *Facilities Maintenance Management*. R.S. Means Company, Inc., Kingston, MA, 1988.

Matulionis, Raymond C. and Joan C. Freitag, eds. *Preventive Maintenance of Buildings*. Van Nostrand Reinhold, New York, NY, 1991.

"Preventive Maintenance Programs That Work." *AIPE Facilities*. March/April 1991.

Chapter 17: General Maintenance Estimating

Bidding & Estimating. Building Service Contractors Association International, 10201 Lee Highway, Fairfax, VA 22030, 1991.

ISSA's 292 Cleaning Times. International Sanitary Supply Association, 7373 N. Lincoln Avenue, Lincolnwood, IL 60646, 1992.

Means Facilities Maintenance & Repair Cost Data. R.S. Means Company, Inc.

Chapter 18: Value Engineering

Conway, Don. "Justifying the Replacement of Obsolete Equipment." *AIPE Facilities*. 1984.

Chapter 19: Deferred Maintenance

Mendelbaum, Gerald B. "The Growing Cost of Deferred Maintenance." *AIPE Facilities*, September/October 1994.

Scholtes, Peter R. *The Team Handbook*. Joiner Associates, Madison, WI, 1988.

Chapter 20: Reserve Funds

CIRA Guidelines. The American Institute of CPAs.

HUD *Federal Preservation Law, Titles IV and VI*. Department of Housing and Urban Development, 820 1st St., N.E., Washington, DC 20002.

Chapter 21: Apartment Buildings & Condominiums

Dept. of Housing & Urban Development (HUD), 820 1st St., N.E., Washington, DC 20002.

Chapter 22: Educational Facilities

APPA, AFE (formerly AIPE), and NACUBO offer resources in facilities management in higher education. (See the Appendix for more information on these organizations.)

Chapter 23: Health Care Facilities

Becker, Franklin. *The Total Workplace: Facilities Management and the Elastic Organization*. Van Nostrand Reinhold, New York, NY, 1990.

Binder, Stephen. *Corporate Facility Planning*. McGraw-Hill, Inc., New York, NY, 1989.

Binder, Stephen. *Strategic Corporate Facilities Management*. McGraw-Hill, Inc., New York, NY, 1992.

Brauer, Roger L. *Facilities Planning: The User Requirements Method*, 2nd Edition. AMACOM, New York, NY, 1992.

Camp, Robert C. *Benchmarking: The Search for Industry Best Practices That Lead to Superior Performance*. ASQC Press, Milwaukee, WI, 1989.

Competencies for Facility Management Professionals. International Facility Management Association. Compiled for IFMA by Hale Associates, Houston, TX, 1992.

Cotts, David G. and Michael Lee. *The Facility Management Handbook*. AMACOM, New York, NY, 1992.

Covey, Steven. *The Seven Habits of Highly Effective People*. The Covey Leadership Center, 1993.

Farren, Carol E. *Planning and Managing Interior Projects*. R.S. Means Company, Inc., Kingston, MA, 1988.

Gilbreath, Robert D. *Managing Construction Contracts*. John Wiley and Sons, New York, NY, 1993.

Hammer, Jeffrey M. *Facility Management Systems*. Van Nostrand Reinhold, New York, NY, 1988.

Hemmes, Michael. *Managing Health Care Construction Projects*. American Hospital Publishing, Inc., 1993.

Kaiser, Harvey H., Ph.D. *The Facilities Manager's Reference*. R.S. Means Company, Inc., Kingston, MA, 1989.

Knutson, Joan and Ira Bitz. *Project Management: How to Plan and Manage Successful Projects*. AMACOM, New York, NY, 1991.

Magee, Gregory H. *Facilities Maintenance Management*. R.S. Means Company, Inc., Kingston, MA, 1988.

McLarney, V. James and Linda F. Chaff. *Effective Health Care Facilities Management*. American Hospital Publishing, Inc., Chicago, IL, 1991.

Owen, David D. *Facilities Planning and Relocation*. R.S. Means Company, Inc., Kingston, MA, 1993.

Rohde, Deborah J., Lawrence D. Pruybil, William O. Hochkammer. *Planning and Managing Major Construction Projects*. Health Administration Press, Ann Arbor, MI, 1985.

Rondeau, Edmond P. and Kevin Brown, Paul Lopides. *Facilities Management*. John Wiley and Sons, New York, NY, 1995.

Chapter 24: Historic Buildings

Doore, Patricia, Clem Labine, eds. *The Old House Journal New Compendium*. The Old House Journal Corp., 1983.

Glaab, Charles and A. Theodore Brown. *A History of Urban America*. Macmillan, New York, NY, 1982.

Kraut, John A., and Arnold Rice. *United States Since 1865*. Barnes & Noble, New York, NY, 1977.

Lebovich, William L. *America's City Halls*. The Preservation Press, National Trust for Historic Preservation, 1984.

National Park Service. *Twentieth Century Building Materials*. McGraw-Hill, 1995.

Recording Historic Buildings compiled by Harley J. McKee. U.S. Dept. of the Interior, National Park Service, Washington, DC, 1970.

Roth, Leland. *A Concise History of American Architecture*. Harper & Row, New York, NY, 1976.

The Old House Journal Restoration Manual, 1986.

Chapter 25: Hotels and Convention Centers

Roffman, Harold and David M. Stipanuk. *Hospitality Facilities Management and Design*. The Educational Institute of the American Hotel & Motel Association, 1992.

Chapter 27: Manufacturing Facilities

Blache, Klaus M., Ph.D. and Reneem Landgraff. "North American Maintenance Benchmarks." Institute of Industrial Engineers 8th Annual Conference Proceedings, 1991. Available from AFE (formerly AIPE). (See the Appendix for more on AFE.)

"Preventive Maintenance Programs That Work." *AIPE Facilities*. March/April 1991.

Chaper 28: Office Buildings

Klein, Judy Graf. *The Office Book*. Quarto Marketing, Ltd., 1982.

Index

Index